AF368918

Sustainable Development of Electrical Energy Storage Technologies in Energy Production

Sustainable Development of Electrical Energy Storage Technologies in Energy Production

Editor

Grigorios L. Kyriakopoulos

MDPI • Basel • Beijing • Wuhan • Barcelona • Belgrade • Manchester • Tokyo • Cluj • Tianjin

Editor
Grigorios L. Kyriakopoulos
National Technical University of Athens
Greece

Editorial Office
MDPI
St. Alban-Anlage 66
4052 Basel, Switzerland

This is a reprint of articles from the Special Issue published online in the open access journal *Sustainability* (ISSN 2071-1050) (available at: https://www.mdpi.com/journal/sustainability/special_issues/Electrical_Energy_Storage_Technologies).

For citation purposes, cite each article independently as indicated on the article page online and as indicated below:

LastName, A.A.; LastName, B.B.; LastName, C.C. Article Title. *Journal Name* **Year**, *Volume Number*, Page Range.

ISBN 978-3-0365-0928-0 (Hbk)
ISBN 978-3-0365-0929-7 (PDF)

Contents

About the Editor

Grigorios L. Kyriakopoulos (PhD in Chemical Engineering) is Teaching and Research Associate at the School of Electrical and Computer Engineering, NTUA. He has accomplished a solid education background by receiving 20 academic qualifications in the fields of: Chemical Engineering (MEng, MSc, PhD, PostDoc); Business and Management (BA, MA); Energy (MSc); Environmental Studies (BSc, 2 MSc); Hellenic Culture (BA, MA); Education (PGCE); Human Resource Management (CertHE); Psychology (CertHE, HND, PGCert). He serves as editorial and reviewer member at 21 journals, including that of MDPI journals: *Sustainability, Laws, Challenges, Journal of Risk* and *Financial Management*, while he is reviewer at a plethora of international scientific journals and technical committees member. He is (co)author of scientific papers related to his specialization research fields of: Engineering, Environment, Energy, Renewable Energy Sources, Waste Management, Circular Economy, Business Administration, Education, Culture, Extroversion/Internationalization of SMEs, Development Economics, and Behavioral Ecology.

Preface to "Sustainable Development of Electrical Energy Storage Technologies in Energy Production"

Dear readers,

Nowadays energy production systems have attracted a globally contentious scientific debate as it counts variable strategic stakeholders of competitive interests. Such indicative competitive interests are land use for energy crops against agricultural production, as well as the ongoing trend of energy production from Renewable Energy Sources (RES), comparing to the traditional overexploitation of conventional fossil-based energy sources in mainland national energy grids.

In this research context the book conveys new and flexible responses to the electrical energy storage (EES) systems in a wide range of applications: Technological advancements, environmental impacts, economies of scale achievement, involvement of RES in EES technologies, as well as EES diffusion and its impacting on regional, globalized and socio-economic contexts of analysis. This book merits its attractiveness since it is one of the contemporary initiatives worldwide that analyzes the multifaceted impacts and complements novel research approaches of wide geographical coverage and pluralistic thematic corpus. In particular this book accommodates 15 rigorously peer-reviewed papers out of a total of 25 manuscripts submitted for consideration and publication. Consequently, the book is systematically approaching the following research fields:

- Municipal Waste Management Strategy Review and Waste-to-Energy Potentials in New Zealand, by Jean-François Perrot and Alison Subiantoro

- New Carbon Emissions Allowance Allocation Method Based on Equilibrium Strategy for Carbon Emission Mitigation in the Coal-Fired Power Industry, by Qing Feng et al.

- Sustainable Waste Tire Derived Carbon Material as a Potential Anode for Lithium-Ion Batteries, by Joseph S. Gnanaraj et al.

- Renewable Energy and Economic Growth: Evidence from European Countries, by Stamatios Ntanos et al.

- Multi-Port High Voltage Gain Modular Power Converter for Offshore Wind Farms, by Sen Song et al.

- Day-Ahead Probabilistic Model for Scheduling the Operation of a Wind Pumped-Storage Hybrid Power Station: Overcoming Forecasting Errors to Ensure Reliability of Supply to the Grid, by Jakub Jurasz and Alexander Kies

- Environmental Behavior of Secondary Education Students: A Case Study at Central Greece, by Stamatios Ntanos et al.

- A Social Assessment of the Usage of Renewable Energy Sources and Its Contribution to Life Quality: The Case of an Attica Urban Area in Greece, by Stamatios Ntanos et al.

- Agricultural Commodities and Crude Oil Prices: An Empirical Investigation of Their Relationship, by Eleni Zafeiriou et al.

- First Approach to a Holistic Tool for Assessing RES Investment Feasibility, by José María Flores-Arias et al.

- Transforming Data Centers in Active Thermal Energy Players in Nearby Neighborhoods, by Marcel Antal et al.

- Socio-Cultural Impact of Energy Saving: Studying the Behaviour of Elementary School Students in Greece, by Sideri Lefkeli et al.

- Public Perceptions and Willingness to Pay for Renewable Energy: A Case Study from Greece, by Stamatios Ntanos et al.

- Optimal Investment Planning of Bulk Energy Storage Systems, by Dina Khastieva et al.

- A Spatial Decision Support System Framework for the Evaluation of Biomass Energy Production Locations: Case Study in the Regional Unit of Drama, Greece, by Konstantinos Ioannou et al.

All aforementioned research contributions of this book have disclosed representative research strategies, operation challenges and technological aspects of exploitation, models and simulations, R&D, as well as new and clean energy-oriented infrastructure. Taken into consideration the decisive role of RES towards the full development of EES and the wider penetration of innovative energy production schemes in the following years, it is prospected that RES in EES could materialize the energy production methodologies of the future. Such methodologies and innovative energy production schemes, if properly enabled, will ensure the optimization of energy sufficiency, the intensification of environmental protection, and the compliance with a wider social acceptability in favor of a greener technology and economy. Besides, the main limitations and the key challenges derived from these scientific approaches, they are nurturing a fresher scientific viewpoint of insightful novel prospects among developed and developing economies worldwide.

I am grateful to all researchers and contributors of this book for their intellectual researches, insightful propositions, pluralistic methodologies and creative argumentation deployed. Besides, I want to warmly thank the professional, editorial and administrative, personnel at the "Sustainability" MDPI editions for their qualitative work and overall supportiveness that made this book possible. I hope you to find the contents of this timely book fascinating and attractive. If you would like any further information on this book, I am at your disposal.

Dr. Grigorios L. Kyriakopoulos
Editor

 *sustainability* 

Article

Municipal Waste Management Strategy Review and Waste-to-Energy Potentials in New Zealand

Jean-François Perrot and Alison Subiantoro *

Department of Mechanical Engineering, The University of Auckland, 20 Symonds Street, Auckland 1010, New Zealand; jper848@aucklanduni.ac.nz
* Correspondence: a.subiantoro@auckland.ac.nz

Received: 7 August 2018; Accepted: 27 August 2018; Published: 31 August 2018

Abstract: Municipal waste management and Waste-to-Energy (WtE) potentials in New Zealand are discussed. The existing main waste management strategy of New Zealand is to reduce, reuse and recycle waste. Most of the remaining waste is currently disposed of in landfills. WtE options were explored in this study as a more sustainable waste treatment alternative in the country, while making use of the annual 30.8 petajoule of available waste energy in New Zealand. Four WtE technology options were discussed and compared, namely incineration, anaerobic digestion, gasification and pyrolysis. The aspects in comparison were air pollution, cost, side products, capacity, commercial maturity, energy efficiency and type of waste treated. Special emphasis was given to environment-friendliness and cost. From the comparison, it was found that anaerobic digestion seems to be the most attractive solution for the country as it is environment-friendly, economical and the concept is consistent with New Zealand's existing waste management strategy. The major limitations of anaerobic digestion are its low energy production efficiency and its limited waste treatment capacity. Hence, an effective national waste reduction and recycling strategy is crucial for the success of this waste management option.

Keywords: municipal waste; energy; New Zealand; incineration; anaerobic digestion; pyrolysis; gasification

---

1. Introduction

The World Bank predicted the global municipal waste generation will increase from 1.3 billion tons per year in 2012 to 2.2 billion tons per year in 2025 [1]. If not managed properly, this will create various social and environmental problems, including air, soil and water pollutions, the spread of diseases and the release of greenhouse gases, particularly methane, to the atmosphere that contributes to global warming.

In New Zealand, waste management is an especially pressing issue. According to the World Bank, the country was the most wasteful nation in the developed world and ranked number 10 globally in terms of municipal waste generation per capita in 2012 and is projected to remain so for the foreseeable future [2]. This high amount of waste is due to the fact that the nation is a big consumer of resources and import high volumes of goods, which is typical among the OECD (The Organisation for Economic Co-operation and Development) members [3], with a relatively small population of less than 5 million people in 2018 [4]. The waste issue in the country is recently worsened by China's decision in 2017 to restrict imports of plastic, which directly impacts the 30,000 tons of plastic waste New Zealand used to export annually [5].

Currently, the main strategy adopted by New Zealand's government regarding its waste management strategy is to reduce, recycle and reuse waste [6]. Most of the non-recyclable and/or non-reusable waste ends up in landfills [7]. Although landfilling maybe the most economical solution in the short term, its long-term impacts to the environment and its sustainability are problematic.

Among the options explored around the world to face the waste issue, energy generation from municipal waste is one of the most attractive solutions [8]. Furthermore, there is a good symmetry between the ever-increasing energy demand and the amount of municipal waste generated, making the Waste-to-Energy (WtE) option even more attractive. It is noted, however, that the concept has its own issues that will be discussed further below.

Comprehensive reviews on various WtE technologies are available in the literature [9,10]. Reviews on waste management issues in Europe [11,12], Africa [13], Asia [14] and in developing countries [15] are also available. Country-specific studies, such as that in China [16], India [17], Turkey [18], USA [19], Colombia [20], Malaysia [21], Thailand [22] and Saudi Arabia [23], among others, can be found too. In general, it can be seen that waste management is a growing challenge among the countries. A number of government policies have been put in place to overcome this issue with varying degrees of effectiveness, and various options have been considered to reduce the amount of waste that ends up in landfills, including recycling and WtE technologies. It is noted that incineration is still the most common method due to its effectiveness. To the authors' knowledge, no such study is available for New Zealand and its unique contexts. The status of zero-waste management strategy in New Zealand up to 2002 was mentioned very briefly in Reference [24]. A somewhat related analysis of waste management in New Zealand is available [3], but the emphasis was on key policies and government programs. Moreover, there have been new developments in the country since the article was published in 2009, including the significant population increase and the shift towards better environment management in recent years. The goal of this study is to investigate the potentials of WtE technology to solve the waste issue of New Zealand. To reach this objective, a review of the current waste management condition in the country is carried out. Data from various sources are collected to draw a comprehensive picture of the national waste situation. Afterwards, the country's waste strategy, including some of the relevant laws and regulations, are presented and its effectiveness is discussed. The available WtE technologies in the world are then reviewed. The technologies include: waste incineration, anaerobic digestion, pyrolysis and gasification. Their strengths and weaknesses are compared based on the contexts of New Zealand. The comparisons will focus on their air pollution, cost, side products, capacity, commercial maturity, efficiency and type of waste treated associated with each technology.

2. Waste Management in New Zealand

According to an OECD report in 2017 [25], the amount of municipal solid waste generated in New Zealand has increased steadily alongside population, gross domestic product and private final consumption in recent years. Its waste generation per capita is currently among the highest in the OECD and the majority of the waste is landfilled. From 2010–15, it was reported that the amount of waste per capita of New Zealand grew by more than 20% [26], as can be seen in Table 1. Here, municipal waste is defined as the waste collected and treated by municipalities. It covers the waste from households, including bulky waste, similar waste from commerce and trade, office buildings, institutions and small businesses, as well as yard and garden waste, street sweepings, the contents of litter containers and market cleansing waste (if managed as household waste). Hence, the figures exclude waste materials that are recovered, reused or recycled. As can be observed, the amount of waste grew with years although there was a slight dip from 2010–11. Based on this trend, it was predicted that the total amount of municipal waste generation in New Zealand will be doubled in 30 years.

Table 1. Municipal waste of New Zealand from 2010–15 (data are extracted from OECD database [26]).

Year	Total Amount of Waste (Millions of tons)	Amount of Waste Per Capita (kg)
2010	2.532	582.0
2011	2.512	573.0
2012	2.514	570.3
2013	2.684	604.2
2014	2.931	649.9
2015	3.221	701.3

A report from the country's Ministry for the Environment [7] explained that the reduction in the amount of municipal waste in the early 2010s was because individual, community and local and central government initiatives had been successful in diverting large amounts of waste from landfills through recovery and recycling of items, among others. Nevertheless, the report also states that there was still approximately 75% of the waste disposed of to landfills that could be potentially diverted. Further reductions were also possible at other stages of a product's life cycle. As shall be discussed later, this strategy of waste minimization is currently the government's main waste management strategy.

It is difficult to get accurate data on the proportion of the municipal waste that ends up in landfills. However, based on the data from the Ministry for the Environment [7], there were 2.531 million tons and 2.461 million tons of waste disposed of to landfills in 2010 and 2011, respectively. Comparing these data with OECD's data on the amount of waste produced [26] in Table 1, it can be seen that more than 97% of the waste are disposed of into landfills.

The waste composition data in New Zealand's landfills can be gathered from reports that have been released by the various regional councils in the country. Perrot [27] has compiled the available data from 2011–17 and the summary is presented in Table 2. The types of waste have been categorized into those with and without energy generation potentials. In total, 78.2% of the waste can be used for power generation in WtE plants. Organic materials represent almost one third of the total waste in landfills, while timber and plastic represent one fourth of the waste.

Table 2. Waste composition of landfills in New Zealand between 2011 and 2017 (data are adapted from regional council reports as compiled by Perrot [27]).

Type of Waste	Percentage
Waste with energy potential	
Organic	30.4%
Timber	14.1%
Plastic	12.1%
Paper	9.0%
Nappies and sanitary	6.3%
Textiles	5.4%
Rubber	0.9%
Waste with no energy potential	
Rubble and concrete	10.0%
Potential hazardous	4.4%
Glass	3.3%
Metal	3.1%

At the moment, the waste management plans realized by regional councils in New Zealand are mainly governed by the Waste Minimization Act of 2008 [28]. The act encourages the reduction of waste generated and disposed of in the country. The aim is to protect the environment from harm and provide environmental, social, economic and cultural benefits. To achieve this goal, several measures have been taken over the years:

- Imposition of a levy on all waste disposed of in landfills to generate funding to help local government, communities and businesses to minimize waste.
- Establishment of product stewardship schemes. This strategy gives the responsibility to a product designer, seller or user to minimize its impact on the environment.
- Regulations made to control the disposal of products, materials or waste.
- Empowerment of the responsibilities of regional authorities that need to write a waste assessment every six years.
- Creation of the Waste Advisory Board to give independent advice to the Minister for the Environment on matters related to waste minimization.

The country's Ministry for the Environment has also formulized New Zealand's Waste Strategy in 2002 and revised it in 2010 [6]. The strategy has a vision towards zero waste and a sustainable New Zealand through three main goals. They are to lower the costs and risks of waste to society, to reduce the environmental damage from the generation and disposal of waste, and to increase economic benefit by using material more efficiently. The strategy focuses more on waste prevention rather than on waste disposal and adopts this waste treatment hierarchy: reduction (most desirable option), reuse, recycle, recovery, treatment, and lastly, disposal (least desirable option).

In response, various regional councils in the country have taken these steps to achieve the zero-waste objective by 2040 [27]:

- the development of additional recycling facilities,
- the reduction of green waste quantities,
- the education of the population about the waste hierarchy and the need to consume less,
- the need to gather more data about the waste situation in each region,
- the reduction of illegal dumping,
- the development of resource recovery networks.

With these regulations and instruments in place, there has been considerable progress in access to recycling services and environmental controls around disposal facilities over the years. However, as can be seen from the waste generation data in Table 1, there has been less success in the overall reduction of waste generation in New Zealand.

In terms of waste treatment, landfilling is still the most common method of solid waste disposal in the country, although there have been discussions about alternative solutions, including various WtE options. Among the alternatives, WtE incineration technology is the most mature [10]. Nevertheless, New Zealand has traditionally been reluctant to commit into WtE incineration plants. Some arguments that have been raised against the technology by the Waste Management Institute of New Zealand [29] are:

- Community perception: Waste incineration has the potential to cause harm to both the environment and human health. However, the institute also noted that with today's modern and efficient technologies these health and environmental concerns can be largely avoided through treatment and mitigation of emissions. Therefore, through engagement and consultation with affected communities, these perceptual issues can be overcome.
- Cannibalizing recycling programs: The development of WtE plants contradicts with the recycling policy of New Zealand because those plants always need a lot of feedstock to run at maximum capacity.
- Consistency of feedstock issue: New Zealand has a relatively large land area as compared to its population. Hence, getting consistent volumes of waste are difficult. Only in an area such as Auckland, there is certainly no shortage of waste. However, there are many and varied disposal options already in existence competing for the same waste, many of which are likely to be far cheaper solutions.
- Cost issue: Landfilling remains New Zealand's predominant mechanism for waste disposal as it is cost-effective. Only with significant waste disposal levies through regulatory intervention, a WtE plant can be considered economically attractive.

3. Waste-to-Energy Technologies and Their Status in New Zealand

As laid out in the previous section, New Zealand is highly reliant on landfilling to treat its waste. At the same time, waste generation in the country is growing steadily and China's recent decision not to import plastic waste anymore will force the country to reconsider its waste management strategy sooner or later. The authors believe that it is time to review the WtE options to solve New Zealand's waste situation. The existing concerns against the technology, particularly to WtE incineration plants, are noted and will be addressed.

Obviously, another added benefit for New Zealand from WtE technologies is the additional renewable energy source available to reduce its dependency on fossil fuels. It is noted that around 85% of the electricity and 40% of the total energy supply in New Zealand were already generated from renewable sources in 2016 [30]. Fossil fuels are still used to produce electricity during dry seasons. Waste can be a very attractive fuel alternative as its supply is relatively unaffected by the seasons.

3.1. Incineration

The most mature technology to extract energy from waste is incineration [10]. The process involves burning of waste in furnaces and using the heat produced to generate useful power (in the form of electricity or heat). By-products of the process are ash and exhaust gas. The ash residue can be further processed to remove metals for recycling and the remainder can be used for construction materials. Most modern plants currently have an energy efficiency of around 30% [10]. Integrated WtE-Gas Turbines power plants have been proposed recently to increase the energy efficiency to more than 40% and in 2012, there are three such plants in Spain, Netherlands and Japan [31]. Although the technology is not yet matured, the integrated gas turbine system seems to be the future of WtE incineration technologies.

The main advantage of incineration as compared to the other options is its effectiveness to remove waste in terms of capacity, type of waste treated and volume reduction [32]. The potential economic values of the residues are an added benefit of the technology. The weaknesses of WtE incinerators are related to their cost, social and environmental issues [29]. WtE incineration plants are relatively expensive to build and communities are usually reluctant to live near incinerators. The incineration of waste releases polluting metals, dioxins and toxic gases that are harmful for human health to the atmosphere. However, it is noted that the quantities released have decreased in the past few years because of stricter rules imposed by governments and the technological progress made in this field [33,34].

WtE incineration technology has been implemented successfully in Europe. The biggest investors include France, Germany, the United Kingdom, Italy and Sweden [35]. Incineration allows these investors to reduce the volume of waste disposed in landfills by 90%. In European Union (EU) countries, 26% of waste is used to generate energy in 2015 [36]. This energetic valorization represents 1.3% of the final electricity consumption and 8.9% of the heat consumption. 409 WtE plants were counted in the EU in 2012 that burned 74 million of tons of waste producing 30 TWh of electricity and 74 TWh of heat [36].

The waste management case of Sweden [37] is particularly relevant for New Zealand and the concerns raised by the authorities. Like New Zealand, Sweden puts a lot of emphasis on recycling and has been very successful, with half of the waste in Sweden is recycled every year. Contrary to New Zealand, however, the rest of the waste is converted to energy in the 33 WtE plants in the country, leaving only about 1% of waste in landfills [37]. The waste treatment systems have been so effective that Sweden also imports waste from other countries that are willing to pay the price to feed its plants.

In New Zealand, there are currently only three high-temperature hazardous waste incinerators in operation. They are at Auckland International Airport, in New Plymouth and near Christchurch International airport [27].

3.2. Anaerobic Digestion

Another technology that is available to create energy from waste is anaerobic digestion. It is a process involving the decomposition of putrescible materials by bacteria acting without the presence of air. Biogas, which is a renewable energy comprised of methane and carbon dioxide, is generated from the process. The main advantages of anaerobic digestion over incineration are its lower emissions of carbon dioxide, the possible valorization of organic waste for soil conditioners and the reduced odor emissions [32]. From life cycle perspective, anaerobic digestion is more attractive than incineration too [38]. The main limitation is that the process is slow. Hence, it may not be practical to rely solely on anaerobic digestion to remove all the waste from a country. Furthermore, it can only treat biodegradable waste or organic fraction of municipal solid waste (OFMSW) [39].

Around the world, there were approximately 12,000 biogas plants in 2016 and is expected to grow to 15,000 by 2025 [40]. 90% of those plants are in Europe and the strongest biogas markets are in Germany, France, Italy and Poland [40]. However, this technology is less mature and is less popular than the WtE incineration alternative. Furthermore, there are unique challenges to its implementation in urban areas [41].

In New Zealand, the biogas sector is currently growing. In 2015, the estimated total amount of methane collected and transformed into biogas was around 4.7 PJ (petajoule), which can power 40,000 households [42]. There are 31 main biogas plants in New Zealand treating waste from landfills, sewage, wastewater, rural and industrial waste [42].

3.3. Gasification and Pyrolysis

Gasification process involves the conversion of an organic compound into a gas mixture (syngas) and a solid by-product (char). The conversion is made at a temperature of more than 650 °C [32]. Syngas has a high heating power and can be used for power generation or biofuel production. Char is a mixture of organic carbon and ash. Several gasification technologies exist around the world and what differentiates them is the reactors used and the operations realized [43].

Pyrolysis is thermal degradation with a limited supply of, or in the complete absence of, an oxidizing agent at a temperature of between 400 °C and 1000 °C [44]. The conversion can produce three products: pyrolysis gas, pyrolysis liquid and solid coke. The WtE potential efficiency of pyrolysis is similar to that of gasification and both techniques can treat all kind of municipal waste. However, these technologies are less matured as compared to anaerobic digestion and incineration [44].

In New Zealand, there have been small-scale gasification and pyrolysis waste treatment projects, particularly for wood waste [45,46]. However, to the authors' knowledge, there is no large-scale gasification/pyrolysis plant in the country as of 2018.

4. Waste-to-Energy Generation Potentials in New Zealand

From the above discussions, there are at least four major WtE technologies available, i.e., incineration, anaerobic digestion, gasification and pyrolysis. These technologies will now be compared. The comparison includes the energy potential and suitability of each technology in the context of New Zealand.

4.1. Energy Potential from Municipal Waste in New Zealand

In 2015, New Zealand generated 3.221 million tons of municipal waste [26]. To get an estimation of the energy that could be generated from this waste, a calculation was carried out. Table 3 shows the amount of each waste type generated annually in the New Zealand (data are adapted from Reference [27]), the heating value of each kind of waste (values are approximated based on Reference [47]) and the corresponding available energy. In total, there is more than 30.8 PJ of energy available per year from waste in New Zealand. In comparison, the total energy demand of the country was 577.6 PJ in 2016, of which 25.5 PJ, 270.9 PJ and 81.3 PJ was supplied by coal, oil and natural gas, respectively [30]. Therefore, the available energy from municipal waste in New Zealand is comparable to the energy demand provided by coal.

Table 3. Annual municipal waste energy potential of New Zealand.

	Percentage (%) [27]	Amount of Waste (tons)	Heating Value (MJ/kg) [47]	Energy Contained (GJ)
Organic	30.4	979,184	3	2,937,552
Paper, nappies and sanitary	15.3	492,813	16	7,885,008
Timber	14.1	454,161	6	2,724,966
Plastic	12.1	389,741	35	13,640,600
Rubber	0.9	28,989	14	405,846
Textiles	5.4	173,934	19	3,304,746
Total				**30,898,718**

To estimate the potential energy that can be generated from WtE plants, the total amount of waste energy is multiplied by the various efficiency values of the different WtE technologies. The calculated values for incineration, anaerobic digestion, gasification and pyrolysis are tabulated in Table 4. It should be noted that for each technology, the amount of energy generated varies for two main reasons: (1) the different efficiencies, and (2) the different kind of waste that it can treat. For example, anaerobic digestion can only treat organic waste, which explains the low value of energy generated as compared to other technologies, since plastic waste has a high heating value and represents around 12% of waste generated in New Zealand. As a consequence, anaerobic digestion needs to be combined with a good recycling strategy, if it is to be adopted nationwide. As expected, the most sophisticated technologies (WtE-GT integrated and advanced gasification) offer the most energy generation, and further studies in this area will allow the energetic optimization of future WtE plants. However, these technologies are less mature at the moment.

Table 4. Energy generation potentials from municipal waste in New Zealand.

	Incineration (Conventional)	Incineration (WtE-GT Integrated)	Anaerobic Digestion	Advanced Gasification	Pyrolysis
Energy production efficiency (%)	21.0% [47]	42.0% [31]	10.4% [48]	35.0% [49]	20.5% [50]
Annual energy production potential (GJ)	5,806,909	11,613,819	1,015,671 [1]	9,665,899	5,661,455

[1] Only organic waste is treated.

4.2. Suitability Comparison of the Technologies for New Zealand

To analyze the suitability of the available WtE technologies in the context of New Zealand, a comparison is carried out in the following aspects: air pollution, cost (capital and maintenance), side product, capacity of production, commercial maturity, energy efficiency and type of waste to be treated.

Studies of gas emissions from each type of plants are available in the literature [32,51–53]. There are technologies to treat the various gas emissions to reduce air pollution. These, however, have financial implications. For example, although gasification plants emit less carbon dioxide than incineration plants [54], it is more expensive to treat the emission from gasification plants than that from incinerators [55]. Among all the technologies, anaerobic digestion is the least air polluting since all gases are captured to produce methane [56]. Pyrolysis is less polluting than incineration because of the absence of oxygen during the process and the lower temperature used [57]. It is also less polluting than gasification for the same reason [50].

To compare the cost of each technology, Table 5 tabulates the costs of WtE plants based on data from GIZ (German Corporation for International Cooperation) [58]. It can be seen that anaerobic digestion is the most economical option. Conventional incineration is relatively more economical than gasification/pyrolysis, while the advanced incineration plant is currently very costly.

Table 5. Cost estimations of various WtE technologies.

	Incineration (Conventional)	Incineration (WtE-GT Integrated)	Anaerobic Digestion	Gasification/Pyrolysis
Annual capacity (tons) [58]	150,000	150,000	100,000	250,000
Investment cost (million EUR) [58]	55	160	16	100
Total cost (EUR/ton)	65	280	28	75

Regarding side products, incineration units usually recycle the bottom ash to recover metals from it and the remaining ash can be used as construction materials (aggregates) [10]. Anaerobic digestion's main side product is digestate, which can be used for fertilizer [32]. For gasification, only ashes remain from the process apart from the syngas are generated [54]. The side products of pyrolysis depend on the parameters used. The range and proportions of products is then quite varied and is constituted by unconverted carbon, charcoal, ash, pyrolysis oil and syngas [59].

Regarding the waste treatment capacities, the common scales of capacity of each technology are as follows [47]: (1) incineration can treat 1500 tons of waste per day, (2) pyrolysis and gasification can treat 10 and 100 tons of waste per day, respectively, and (3) around 500 tons of waste per day can be treated by anaerobic digestion [58]. The capacity of production obviously depends on the size of the treatment facility, but those figures give an idea of the current performances of the existing plants in the world.

In terms of the level of maturity of the technologies, waste incineration is the most mature, followed by anaerobic digestion, while pyrolysis and gasification are not yet mature [10,50].

Another aspect to consider is the energy production efficiency of each technology. It is noted that the efficiency of each technology depends on many factors such as the thermodynamic cycle employed, the scale of the plant and all the techniques used for optimization that are different for each plant. For conventional incinerators, steam turbines are commonly used and the electrical energy production efficiency is typically around 15–30% [47]. When used to provide heating, the efficiency of incinerators can reach 90% and for combined heat and power is 40% [47]. For anaerobic digestion, the maximum efficiency of anaerobic digesters to produce biogas based on the heating values is 28% [48]. Assuming a gas turbine with an efficiency of 30–40%, the overall efficiency is between 8.4–11.2%. The gasification technology has an efficiency of between 10–27% or 30–40% (advanced gasification), depending on the type of turbine used [50]. Pyrolysis has an efficiency of between 16–25%, assuming a gas turbine for the electricity production process [50].

In terms of the type of waste that can be treated, incineration is the most comprehensive as it can treat almost any type of waste. Gasification and pyrolysis plants may be able to treat all kind of waste with less environmental impact and at a lower cost as compared to incineration, but the technologies are less mature. Anaerobic digestion is the most limited option as it cannot treat non-biodegradable materials, such as plastic waste.

To summarize the comparison of the various aspects above, to find the most suitable technology for New Zealand, scores ranging from 0 (the worst) to 3 (the best) were assigned by ranking each of the technologies in each parameter. For example, anaerobic digestion is the most preferred option in the aspect of air pollution, followed by pyrolysis, gasification and lastly, incineration. Therefore, in the aspect of air pollution, anaerobic digestion was given a score of 3, pyrolysis got a score of 2, a score of 1 for gasification and incineration was given a zero score. Table 6 summarizes the scores of the four technologies in all the aspects considered without any special weightage for any of the parameters. The table shows that, without weightage, incineration and anaerobic digestion are generally the most attractive WtE options, followed by pyrolysis and, lastly, gasification. Incineration is attractive mainly because of its capacity and maturity. Anaerobic digestion is advantageous in terms of its environment-friendliness and cost.

Table 6. Comparison of WtE technologies without weightages.

	Incineration	Anaerobic Digestion	Gasification	Pyrolysis
Air pollution	0	3	1	2
Cost	2	3	0.5	0.5
Side products	1.5	1.5	0	3
Capacity	3	2	1	0
Maturity	3	2	0.5	0.5
Energy efficiency	1.5	0	3	1.5
Waste type	2	0	2	2
Total	**13**	**11.5**	**8**	**9.5**

In practice, New Zealand has two main concerns regarding its waste treatment strategy, i.e., environment-friendliness and economy. Therefore, more weightages should be assigned to the relevant parameters in comparison. The main parameter that is relevant to environment-friendliness is air pollution, while that for the economy is cost. A double weightage was therefore given to these two parameters. The results are presented in Table 7. The table shows that anaerobic digestion seems to be the most attractive solution for New Zealand in general. It is both environment-friendly and economical. The technology is relatively mature. The inability of anaerobic digestion to treat non-biodegradable waste is actually consistent with New Zealand's national waste management strategy to reduce, reuse and recycle waste. It should be noted, however, that the energy production efficiency of anaerobic digestion is relatively low as compared to the alternatives. Furthermore, it is seriously lacking in its waste treatment capacity. Therefore, in order for this option to be successful in its implementation nationwide, it is crucial to ensure the effectiveness of the existing national waste management strategy to reduce, reuse and recycle.

Table 7. Comparison with special weightages on environmental and economic aspects.

	Weightage	Incineration	Anaerobic Digestion	Gasification	Pyrolysis
Air pollution	2	0	6	2	4
Cost	2	4	6	1	1
Side products	1	1.5	1.5	0	3
Capacity	1	3	2	1	0
Maturity	1	3	2	0.5	0.5
Energy efficiency	1	1.5	0	3	1.5
Waste type	1	2	0	2	2
Total		**15**	**17.5**	**9.5**	**12**

5. Conclusions

This article discusses the municipal waste management and the Waste-to-Energy (WtE) potentials in New Zealand. Globally, New Zealand generates more waste per capita as compared to other OECD countries. Moreover, the total amount of waste generated has been steadily increasing over the years and is expected to continue to rise following the country's economic and population growths. The main waste management strategy of New Zealand is currently to reduce, reuse and recycle waste, while most of the non-reusable and non-recyclable waste ends up in landfills.

Various alternative technologies are available to treat waste in a more sustainable manner. WtE options were explored in this study for possible future implementation in New Zealand. This option is attractive as it is environmentally friendly and can further reduce the country's dependency on fossil fuels. According to the calculation, the energy potential of waste in New Zealand was more than 30.8 PJ per year, which is more than the amount of energy from coal in the country.

Four technologies were discussed and compared: incineration, anaerobic digestion, gasification and pyrolysis. The aspects in focus were air pollution, cost, side products, waste treatment capacity, commercial maturity, energy efficiency and type of waste treated. Without any special weightage to any of the parameters, incineration is the most attractive option. It is a very mature technology and is able to treat any type of waste effectively. However, environment-friendliness and cost are especially important for New Zealand. Therefore, more weightages were given to the air pollution and cost aspects. From the comparison, it was found that anaerobic digestion seems to be the most attractive solution for the country. Anaerobic digestion plants are environment-friendly and economical. It is also consistent with New Zealand's existing waste management strategy. The major limitations of anaerobic digestion are its low energy production efficiency and its limited waste treatment capacity. It is noted, therefore, that this option has to run in tandem with an effective national strategy to reduce, reuse and recycle waste in the country.

Author Contributions: Conceptualization, A.S.; Methodology, J.-F.P. and A.S.; Data Curation, J.-F.P.; Investigation, J.-F.P.; Writing-Original Draft Preparation, J.-F.P.; Writing-Review & Editing, A.S.; Supervision, A.S.; Project Administration, J.-F.P. and A.S.

Funding: This research received no external funding

Conflicts of Interest: The authors declare no conflicts of interest.

References

1. Solid Waste Management. Available online: http://www.worldbank.org/en/topic/urbandevelopment/brief/solid-waste-management (accessed on 19 July 2018).
2. Hoornweg, D.; Bhada-Tata, P. *What a Waste—A Global Review of Solid Waste Management*; Urban Development Series Knowledge Papers; The World Bank: Washington, DC, USA, 2012; pp. 40–44.
3. Davies, A.R. Clean and green? A governance analysis of waste management in New Zealand. *J. Environ. Plann. Manag.* **2009**, *52*, 157–176. [CrossRef]
4. Estimated Resident Population of New Zealand. Available online: https://www.stats.govt.nz/topics/population (accessed on 19 July 2018).
5. NZ Most Wasteful Country in Developed World. Available online: https://www.nzherald.co.nz/nz/news/article.cfm?c_id=1&objectid=11977131 (accessed on 19 July 2018).
6. The New Zealand Waste Strategy—Reducing Harm, Improving Efficiency. Available online: http://www.mfe.govt.nz/sites/default/files/wastestrategy.pdf (accessed on 19 July 2018).
7. Quantity of Solid Waste Sent to Landfill Indicator Update. Available online: http://www.mfe.govt.nz/more/environmental-reporting/reporting-act/waste/solid-waste-disposal-indicator/quantity-solid-waste (accessed on 19 July 2018).
8. Cucchiella, F.; D'Adamo, I.; Gastaldi, M. Sustainable waste management: Waste to energy plant as an alternative to landfill. *Energy Convers. Manag.* **2017**, *131*, 18–31. [CrossRef]
9. Astrup, T.F.; Tonini, D.; Turconi, R.; Boldrin, A. Life cycle assessment of thermal Waste-to-Energy technologies: Review and recommendations. *Waste Manag.* **2015**, *37*, 104–115. [CrossRef] [PubMed]
10. Lombardi, L.; Carnevale, E.; Corti, A. A review of technologies and performances of thermal treatment systems for energy recovery from waste. *Waste Manag.* **2015**, *37*, 26–44. [CrossRef] [PubMed]
11. Malinauskaite, J.; Jouhara, H.; Czajczyńskab, D.; Stanchev, P.; Katsou, E.; Rostkowski, P.; Thorne, R.J.; Colón, J.; Ponsá, S.; Al-Mansour, F.; et al. Municipal solid waste management and waste-to-energy in the context of a circular economy and energy recycling in Europe. *Energy* **2017**, *141*, 128. [CrossRef]
12. Taelman, S.E.; Tonini, D.; Wandl, A.; Dewulf, J. A holistic sustainability framework for waste management in European cities: Concept development. *Sustainability* **2018**, *10*, 2184. [CrossRef]
13. Couth, R.; Trois, C. Carbon emissions reduction strategies in Africa from improved waste management: A review. *Waste Manag.* **2010**, *30*, 2336–2346. [CrossRef] [PubMed]
14. Othman, S.N.; Noor, Z.Z.; Abba, A.H.; Yusuf, R.O.; Hassan, M.A.A. Review on life cycle assessment of integrated solid waste management in some Asian countries. *J. Clean. Prod.* **2013**, *41*, 251–262. [CrossRef]

15. Guerrero, L.A.; Maas, G.; Hogland, W. Solid waste management challenges for cities in developing countries. *Waste Manag.* **2013**, *33*, 220–232. [CrossRef] [PubMed]

16. Zhang, D.Q.; Tan, S.K.; Gersberg, R.M. Municipal solid waste management in China: Status, problems and challenges. *J. Environ. Manag.* **2010**, *91*, 1623–1633. [CrossRef] [PubMed]

17. Sharholy, M.; Ahmad, K.; Mahmood, G.; Trivedi, R.C. Municipal solid waste management in Indian cities—A review. *Waste Manag.* **2008**, *28*, 459–467. [CrossRef] [PubMed]

18. Metin, E.; Eröztürk, A.; Neyim, C. Solid waste management practices and review of recovery and recycling operations in Turkey. *Waste Manag.* **2003**, *23*, 425–432. [CrossRef]

19. Psomopoulos, C.S.; Bourka, A.; Themelis, N.J. Waste-to-energy: A review of the status and benefits in USA. *Waste Manag.* **2009**, *29*, 1718–1724. [CrossRef] [PubMed]

20. Alzate-Arias, S.; Jaramillo-Duque, A.; Villada, F.; Restrepo-Cuestas, B. Assessment of government incentives for Energy from Waste in Colombia. *Sustainability* **2018**, *10*, 1294. [CrossRef]

21. Fazeli, A.; Bakhtvar, F.; Jahanshaloo, L.; Sidik, N.A.C.; Bayat, A.E. Malaysia's stand on municipal solid waste conversion to energy: A review. *Renew. Sustain. Energy Rev.* **2016**, *58*, 1007–1016. [CrossRef]

22. Menikpura, S.N.M.; Sang-Arun, J.; Bengtsson, M. Assessment of environmental and economic performance of Waste-to-Energy facilities in Thai cities. *Renew. Energy* **2016**, *86*, 576–584. [CrossRef]

23. Ouda, O.K.M.; Raza, S.A.; Nizami, A.S.; Rehan, M.; Al-Waked, R.; Korres, N.E. Waste to energy potential: A case study of Saudi Arabia. *Renew. Sustain. Energy Rev.* **2016**, *61*, 328–340. [CrossRef]

24. Zaman, A.U. A comprehensive review of the development of zero waste management: Lessons learned and guidelines. *J. Clean. Prod.* **2015**, *91*, 12–25. [CrossRef]

25. Organisation for Economic Co-operation and Development (OECD). *OECD Environmental Performance Reviews: New Zealand 2017*; OECD Publishing: Paris, France, 2017; p. 23. ISBN 978-92-64-26829-6.

26. Waste: Municipal Waste, OECD Environment Statistics (Database). Available online: https://doi.org/10.1787/data-00601-en (accessed on 19 July 2018).

27. Perrot, J.-F. Municipal Waste Management Strategy Review and Waste-to-Energy Generation Potential in New Zealand. Master's Thesis, The University of Auckland, Auckland, New Zealand, 2018.

28. Waste Minimisation Act 2008. Available online: http://www.legislation.govt.nz/act/public/2008/0089/latest/DLM999802.html?src=qs (accessed on 20 August 2018).

29. What about Energy from Waste? Available online: http://www.infrastructurenews.co.nz/%EF%BB%BF-what-about-energy-from-waste/ (accessed on 19 July 2018).

30. Energy in New Zealand. Available online: http://www.mbie.govt.nz/info-services/sectors-industries/energy/energy-data-modelling/publications/energy-in-new-zealand/documents-images/energy-in-nz-2017.pdf (accessed on 23 July 2018).

31. Branchini, L. Advanced Waste-To-Energy Cycles. Ph.D. Thesis, Universita di Bologna, Bologna, Italy, 2012.

32. Bosmans, A.; Vanderreydt, I.; Geysen, D.; Helsen, L. The crucial role of Waste-to-Energy technologies in enhanced landfill mining: A technology review. *J. Clean. Prod.* **2013**, *55*, 10–23. [CrossRef]

33. Scungio, M.; Buonanno, G.; Stabile, L.; Ficco, G. Lung cancer risk assessment at receptor site of a waste-to-energy plant. *Waste Manag.* **2016**, *56*, 207–215. [CrossRef] [PubMed]

34. Damgaard, A.; Riber, C.; Fruergaard, T.; Hulgaard, T.; Christensen, T.H. Life-cycle-assessment of the historical development of air pollution control and energy recovery in waste incineration. *Waste Manag.* **2010**, *30*, 1244–1250. [CrossRef] [PubMed]

35. Waste-to-Energy Plants in Europe in 2015. Available online: http://www.cewep.eu/2017/09/07/waste-to-energy-plants-in-europe-in-2015/ (accessed on 20 July 2018).

36. Valorisation Énergétique des Déchets Opportunités et Défis (In French). Available online: http://www.europarl.europa.eu/RegData/etudes/BRIE/2015/554208/EPRS_BRI(2015)554208_FR.pdf (accessed on 20 July 2018).

37. The Swedish Recycling Revolution. Available online: https://sweden.se/nature/the-swedish-recycling-revolution/ (accessed on 20 July 2018).

38. Tong, H.; Shen, Y.; Zhang, J.; Wang, C.-H.; Ge, T.S.; Tong, Y.W. A comparative life cycle assessment on four waste-to-energy scenarios for food waste generated in eateries. *Appl. Energy* **2018**, *225*, 1143–1157. [CrossRef]

39. Ranieri, L.; Mossa, G.; Pellegrino, R.; Digiesi, S. Energy recovery from the organic fraction of municipal solid waste: A real options-based facility assessment. *Sustainability* **2018**, *10*, 368. [CrossRef]

40. Biogas to Energy: The World Market for Biogas Plants. Available online: https://www.ecoprog.com/fileadmin/user_upload/leseproben/extract__biogas_to_energy_ecoprog.pdf (accessed on 9 August 2016).
41. Di Matteo, U.; Nastasi, B.; Albo, A.; Garcia, D.A. Energy contribution of OFMSW (organic fraction of municipal solid waste) to energy-environmental sustainability in urban areas at small scale. *Energies* **2017**, *10*, 229. [CrossRef]
42. Overview of Biogas in NZ (WB06). Available online: https://www.biogas.org.nz/documents/resource/WB06-biogas-overview.pdf (accessed on 23 July 2018).
43. Kumar, A.; Jones, D.D.; Hanna, M.A. Thermochemical Biomass Gasification: A Review of the Current Status of the Technology. *Energies* **2009**, *2*, 556–581. [CrossRef]
44. Zaman, A.U. Life cycle assessment of pyrolysis–gasification as an emerging municipal solid waste treatment technology. *Int. J. Environ. Sci. Technol.* **2013**, *10*, 1029–1038. [CrossRef]
45. Das, O.; Sarmah, A.K.; Bhattacharyya, D. A novel approach in organic waste utilization through biochar addition in wood/polypropylene composites. *Waste Manag.* **2015**, *38*, 132–140. [CrossRef] [PubMed]
46. Arbestain, M.C.; Jones, J.R.; Condron, L.M.; Clough, T.J. Research and application of biochar in New Zealand. In *Agricultural and Environmental Applications of Biochar: Advances and Barriers*; Soil Science Society of America: Washington, DC, USA, 2017; pp. 423–444. ISBN 978-0-89118-967-1.
47. World Energy Resources, Waste to Energy. Available online: https://www.worldenergy.org/wp-content/uploads/2017/03/WEResources_Waste_to_Energy_2016.pdf (accessed on 20 July 2018).
48. Nordlander, E.; Holgersson, J.; Thorin, E.; Thomassen, M.; Jinyue, Y. Energy efficiency evaluation of two biogas plants. In Proceedings of the Third International Conference on Applied Energy, Perugia, Italy, 16–18 May 2011; pp. 1661–1674.
49. Waste Technologies: Waste to Energy Facilities. Available online: http://www.wasteauthority.wa.gov.au/media/files/documents/SWIP_Waste_to_Energy_Review.pdf (accessed on 20 August 2018).
50. Feasibility Study on Solid Waste to Energy Technological Aspects. Available online: http://funginstitute.berkeley.edu/wp-content/uploads/2014/01/SolidWasteToEnergy.pdf (accessed on 23 July 2018).
51. Kim, J.K.; Oh, B.R.; Chun, Y.N.; Kim, S.W. Effects of temperature and hydraulic retention time on anaerobic digestion of food waste. *J. Biosci. Bioeng.* **2006**, *102*, 328–332. [CrossRef] [PubMed]
52. Arena, U. Process and technological aspects of municipal solid waste gasification. *Waste Manag.* **2012**, *32*, 625–639. [CrossRef] [PubMed]
53. Beylot, A.; Hochar, A.; Michel, P.; Descat, M.; Ménard, Y.; Villeneuve, J. Municipal solid waste incineration in France: An overview of air pollution control techniques, emissions, and energy efficiency. *J. Ind. Ecol.* **2017**, 1–11. [CrossRef]
54. Kumar, A.; Samadder, S.R. A review on technological options of waste to energy for effective management of municipal solid waste. *Waste Manag.* **2017**, *69*, 407–422. [CrossRef] [PubMed]
55. Matsakasa, L.; Gao, Q.; Jansson, S.; Rova, U.; Christakopoulos, P. Green conversion of municipal solid wastes into fuels and chemicals. *Electron. J. Biotechnol.* **2017**, *26*, 69–83. [CrossRef]
56. Li, Y.; Park, S.Y.; Zhu, J. Solid-state anaerobic digestion for methane production from organic waste. *Renew. Sustain. Energy Rev.* **2011**, *15*, 821–826. [CrossRef]
57. Samoladab, M.C.; Zabaniotou, A.A. Comparative assessment of municipal sewage sludge incineration, gasification and pyrolysis for a sustainable sludge-to-energy management in Greece. *Waste Manag.* **2014**, *34*, 411–420. [CrossRef] [PubMed]
58. Waste-to-Energy Options in Municipal Solid Waste Management. Available online: https://www.giz.de/en/downloads/GIZ_WasteToEnergy_Guidelines_2017.pdf (accessed on 23 July 2018).
59. Czajczyńska, D.; Anguilanoc, L.; Ghazald, H.; Krzyżyńsk, R.; Reynold, A.J.; Spencer, N.; Jouhara, H. Potential of pyrolysis processes in the waste management sector. *Therm. Sci. Eng. Prog.* **2017**, *3*, 171–197. [CrossRef]

 *sustainability*

Article

New Carbon Emissions Allowance Allocation Method Based on Equilibrium Strategy for Carbon Emission Mitigation in the Coal-Fired Power Industry

Qing Feng [1], Qian Huang [1], Qingyan Zheng [2] and Li Lu [1,2,*]

[1] Business School, Sichuan University, Chengdu 610064, China; toryfrankfq@126.com (Q.F.); qianH95@163.com (Q.H.)

[2] Tourism School, Sichuan University, Chengdu 610064, China; cyan_zheng@foxmail.com

* Correspondence: lulirudy@scu.edu.cn; Tel.: +86-028-85996613

Received: 18 July 2018; Accepted: 17 August 2018; Published: 17 August 2018

Abstract: The carbon emissions from coal-fired power have become an increasing concern to governments around the world. In this paper, a carbon emissions allowances allocation based on the equilibrium strategy is proposed to mitigate coal-fired power generation carbon emissions, in which the authority is the lead decision maker and the coal-fired power plants are the follower decision makers, and an interactive solution approach is designed to achieve equilibrium. A real-world case study is then given to demonstrate the practicality and efficiency of this methodology. Sensitivity analyses under different constraint violation risk levels are also conducted to give authorities some insights into equilibrium strategies for different stakeholders and to identify the necessary tradeoffs between economic development and carbon emissions mitigation. It was found that the proposed method was able to mitigate coal-fired power generation carbon emissions significantly and encourage coal-fired power plants to improve their emissions performance.

Keywords: carbon emission allowance allocation; emission mitigation; coal-fired power generation; cap and tax mechanism

---

1. Introduction

Because of their major contribution to global climate change, there has been increased research to determine the best ways to reduce carbon emissions, which have been exponentially increasing due to the increased demand for energy [1–3]. Although renewable energy systems are more environmentally friendly than traditional energy systems, two thirds of the world's electricity is still generated using fossil fuel-based generation plants [4,5]; that is, coal, gas and oil-fired thermal power remain the main sources for electricity generation, especially in developing countries (e.g., China and India) [6,7]. Approximately 20% of the global electricity produced in 2016 was supplied by coal-fired power plants (CPP), with some developing countries having an even higher proportion [8]; for example, coal-fired power plants supply more that 65% of China's needs [9]. Therefore, for sustainable social development, it is necessary to mitigate or control CPPs' carbon emissions.

Research has shown that "hard-path" and "soft-path" approaches can be taken to mitigate carbon emissions [10,11]. "Hard-path" methods mainly focus on advanced clean coal technologies (CCT) such as integrated gasification combined cycles (IGCC), carbon capture and storage (CCS), ultra-supercritical technology (USC) and externally-fired combined cycle (EFCC) technologies [12–14]. For example, Hoya and Fushimi evaluated the performance of advanced IGCC power generation systems with low-temperature gasifiers and gas cleaning and found that the lowest net thermal efficiency rose to 57.2% and the minimum carbon emission factors fell to 39.7 kg-CO_2 MWh [15]. Kayal and Chakraborty designed and developed carbon-based metal organic framework (MOF) composite for

CO_2 capture and concluded that the MAX-MIL composite was able to adsorb a greater quantity of CO_2 compared with the original methods [16]. Even though these "hard-path" methods are highly efficient in reducing carbon emissions, commercial-scale applications are still extremely expensive [17,18], especially for developing countries, which tend to prefer "soft-path", less-expensive solutions [19,20]. The "soft-path" approach focuses on policy controls or operations management methods for carbon emissions mitigation. For example, Cao and Xu investigated the effects of cap-and-trade policy (CTP) and low carbon subsidy policy (LCSP) on carbon emissions reduction and concluded that carbon emissions reductions were positively correlated with the carbon trading price, but not with low carbon subsidies [21]. Shih and Frey developed a multi-objective chance-constrained optimization method under certainty to improve emission performance by adjusting coal blending ratios [22]. Wang et al. proposed a multi-objective unit commitment approach to simulate the impacts of manifold uncertainty on system operation with emission concern and suggested operational insights for mixed generation systems [23]. Xu et al. developed an equilibrium strategy based on a hydro-wind-thermal complementary system for carbon emission reduction and obtained some useful suggestions [24]. Although such studies have gone some way to alleviating the human activity caused global climate change effects, the reality is still not satisfactory due to the complexity and uncertainty of human activities; thus, further improvements are necessary.

There has been increased research interest in the carbon emissions allowance allocation (CEAA) method to mitigate carbon emissions [25,26]. Cap and trade and carbon taxes have been the two most popular emissions reduction mechanisms to curb CPP carbon emissions [27]. While cap and trade mechanisms are business friendly, as the trading price is determined by supply and demand, there is increased trading price uncertainty [28]. While carbon tax mechanisms are simpler and easier to implement and the tax increases financial revenue [27], which can be used to sponsoring of green projects such as renewable energy, there is no upper limit to the possible emissions reduction [28]. By combining the advantages of these two mechanisms, a new cap and tax mechanism was developed to mitigate carbon emissions. As a key determinant for the CEAA strategy, the allocation strategy is crucial to ensure carbon emissions mitigation. In this paper, a combination of free and taxable allocation strategies under a carbon emissions cap is adopted for the carbon quota allocations. The free emissions allowances are used to meet the CPP basic operations, and the taxable emissions allowances are employed to meet further CPP development.

Previous CEAA studies have tended to consider only a single CPP participant. However, in actual production activities, carbon emissions mitigation involves both the CPP and the authority, which usually have conflicting targets. For example, the CPP generally has a profit objective, while the authority, as a representative of public benefits, generally has environmental protection as the main starting point; therefore, traditional optimization methods are not effective. The equilibrium strategy, which has been proven to be powerful in addressing such conflicts, has been widely used in many fields. For example, Liu et al. developed a computable general equilibrium (CGE) model to explore the impacts of a carbon tax on the socio-economic system and had some useful results [29]. Tu et al. employed an equilibrium strategy to solve regional water resource allocation conflicts between different sub-areas under multiple uncertainties [30]. Kardakos et al. proposed an equilibrium optimization method to address an optimal bidding strategy problem that considered the mutual interactions between the various stakeholders in the electricity market [31]. The successful application of the equilibrium strategy in these areas motivated the use of this method in this paper to address the conflicts between the authority and CPPs to achieve regional sustainable development. However, as the equilibrium strategy is an abstract concept, a specific, quantitative method is needed to describe the situation. As bi-level programming has been proven to be the most efficient method for expressing equilibrium strategies and describing the interactions between multiple stakeholders, bi-level programming is integrated into the CEAA problem to determine the equilibrium between the authority and the CPPs.

Compared with previous studies, the equilibrium strategy established in this study, which integrates a bi-level multi-objective programming model, a carbon emissions allowance allocation method and uncertainty theory, has the ability to address the equilibrium between the authority and the CPPs, the conflict between economic development and environmental protection and the uncertainties simultaneously. The remainder of this paper is organized as follows. Section 2 discusses the features of the CEAA problem in preparation for establishing the mathematical model. In Section 3, a bi-level multi-objective mathematical model is built based on a real situation, after which in Section 4, a case study is given to demonstrate the practicality and effectiveness of the proposed methodology. Section 5 gives a detailed results analysis and in-depth discussion, and conclusions and future research are given in Section 6.

2. Key Problem Statement

With a carbon emissions allowance allocation and a cap and tax mechanism, the CEAA problem is complex for both the authority and the CPPs.

As a public representative, the authority must ensure stable local economic development and mitigate the associated carbon emissions. However, the authority also has the power to develop the policies that must be implemented by the CPP if they wish to keep their power generation rights. However, the authority has an obligation to consider the actual CPP situation when making decisions to avoid non-sustainable CPP development or a cessation of operations, which could be harmful to stable economic development. Therefore, the authority divides the total carbon emissions into free emissions, which allow the CPP to meet its production and operation commitments and ensures fairness, and taxable emissions, which supplement the free emissions and can be used to regulate the market. Therefore, the authority pursues a balance between financial benefits and carbon emissions reduction by satisfying the CPPs' basic rights and meeting the regional electricity needs.

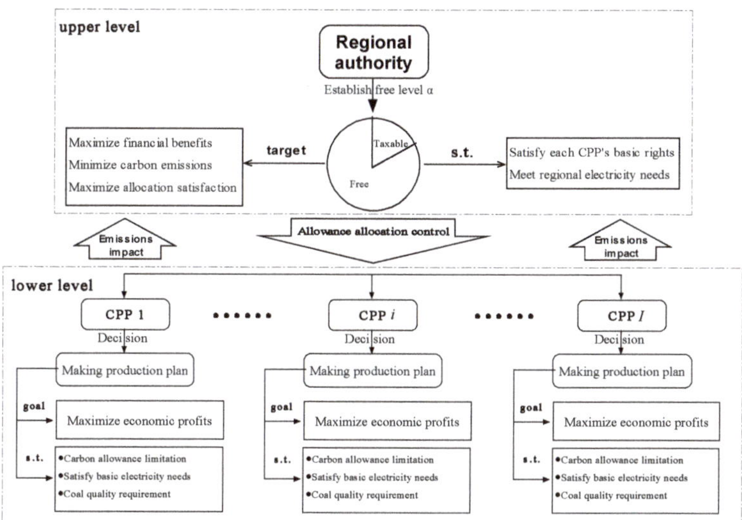

Figure 1. Concept model of the bi-level structure for the carbon emissions allowance allocation (CEAA) problem. CPP, coal-fired power plant.

According to the rational person hypothesis, the primary goal of the CPPs is to maximize profits while also considering emissions performance, boiler conditions, social responsibility and the component coal that can be purchased on the market. Therefore, as higher carbon emissions allowances mean higher production and higher profits, each CPP seeks to obtain as high a carbon emissions allowance as possible. However, as the authority seeks to mitigate carbon emissions, the CPPs that have better emissions performance are more competitive. Therefore, CPPs are allocated higher emissions quotas if they put some effort into improving the CPP emissions performance. This relationship between the authority and the CPPs for the CEAA problem is shown graphically in Figure 1.

3. Model Development

In this section, a bi-level optimization model for the CEAA problem in the coal-fired power industry is built.

3.1. Assumptions

The various assumptions involved in this paper are as follows:

1. The CEAA problem is a single production period decision; at the beginning of the next production period, the decision process is reset.
2. All decision makers are rational and seek to maximize returns under limited resources.

3.2. Authoritative Carbon Emissions Allowance Allocations

The complete carbon emissions allowances allocation problem involves authority allocation decisions and CPP coal purchase decisions. Therefore, in this study, the CEAA based on the cap and tax system has two levels (i.e., the authority and the CPPs), the details of which are discussed in the following (the required symbol descriptions are given in Table 1).

Table 1. Model variables and parameters.

Indices	Description
i	Index for coal power plants (CPPs), $i \in \Psi = \{1, 2, ..., I\}$.
j	Index for component coal, $j \in \Phi = \{1, 2, ..., J\}$.
k	Index for coal quality, $k \in \Omega = \{1, 2, ..., K\}$.
Crisp parameters	Description
μ	Taxes that the corporation should pay for each unit of power generation.
γ	Carbon tax price of the exceeding part when exceeding the allocated free carbon emissions allowances.
CE_c	The cap of carbon emissions allowances.
CE_f	The allocated free carbon emissions allowances.
CE_t	The allocated taxable carbon emissions allowances.
$CEA_i^{\min}$	The minimum carbon emissions allowances demand of CPP i.
$CEA_i^{\max}$	The maximum carbon emissions allowances demand of CPP i.
D	Amount of power needed to maintain regional development.
p	Price of a unit of electric power.
CT_{jk}	Operation cost of pollutant-control measures for reducing pollutant k in CPP i.
η_{ik}	Removal rate of air pollutant k in the CPP i by taking pollutant-control measures.
ED_i	Amount of power that CPP i has the responsibility to produce to meet the basic demand in the region.
Q_{ij}^u	Amount of component coal j that can be procured by CPP i.
Uncertain parameters	Description
$\widetilde{T}_i$	Carbon-power conversion coefficient of a unit of carbon emissions allowances for CPP i.
$\widetilde{T}_{ij}$	Coal-power conversion coefficient of component coal j at CPP i.
$\widetilde{C}_j$	Procurement cost of component coal j.
$\widetilde{CEF}_{ij}$	Carbon emission factor of component coal j to carbon at CPP i.
$\widetilde{EF}_{jk}$	Emission factor of component coal j to air pollutant k.
$\widetilde{LCQ}_{ik}$	Lower bounds of coal quality k for meeting the operation allowance of the i-th CPP.
$\widetilde{CQ}_{jk}$	Coal quality k of component coal j.
$\widetilde{UCQ}_{ik}$	Upper bounds of coal quality k for meeting the operation allowance of the i-th CPP.
Decision variables	Description
X_i	Free carbon emissions allowances allocated to CPP i, which are determined by the authority.
Y_i	Taxable carbon emissions allowance for CPP i allocated by the authority.
Z_{ij}	Amount of component coal j burned by CPP i.
Policy control parameters	Description
θ	Minimal allowance satisfactory degree chosen by the authority.
α	The free carbon emission level, which is determined by the authority.
β	Attitude of the authority towards the historical data.
λ	Attitude of the authority towards carbon emissions reduction.

3.2.1. Pursuing Possible Financial Benefits

While protecting the environment, the authority is also responsible for using taxpayer receipts to ensure steady local economic and social development. The goal is to develop an optimal carbon emissions allowance allocation scheme that returns the maximum revenue from both the value added tax (VAT) and the carbon tax.

Let $\widetilde{T}_{ij}$ denote the coal-power conversion coefficient of component coal j at CPP i. Even though the uncertain parameters $\widetilde{T}_{ij}$ are very difficult to determine exactly due to many objective factors such as coal quality, based on historical data, the value can be estimated within a certain range; therefore, this type of uncertain situation is considered fuzzy and is described using trapezoidal fuzzy numbers, which are written as $T_{ij} = (r_{ij}^1, r_{ij}^2, r_{ij}^3, r_{ij}^4)$, where $r_{ij}^1 \leq r_{ij}^2 \leq r_{ij}^3 \leq r_{ij}^4$ [32–34]. As the fuzzy parameters cannot be directly calculated, the expected value operator method is employed to transform the trapezoidal fuzzy number into its corresponding expected value [35]; therefore, the corresponding expected value for conversion parameters T_{ij} is calculated, where $\widetilde{T}_{ij} \rightarrow E\left[\widetilde{T}_{ij}\right] = \frac{1-\varphi}{2}(r_{ij}^1 + r_{ij}^2) + \frac{\varphi}{2}(r_{ij}^3 + r_{ij}^4)$ and $0 \leq \varphi \leq 1$. The revenue function for the authority is formulated as follows:

$$\max FB = \mu \sum_{i=1}^{I} \sum_{j=1}^{J} E[\widetilde{T}_{ij}] Z_{ij} + \gamma \sum_{i=1}^{I} Y_i \tag{1}$$

where FB is the potential financial benefits function from the CPPs of the authority.

3.2.2. Minimizing Total Carbon Emissions Allowance

As electricity demand is growing because of global economic development, more coal-fired power will be required to meet demand. However, as the carbon emissions produced from the CPP can cause irreversible climate change effects [36], the authority has an important responsibility to protect the environment while at the same time ensuring steady local development. As the total carbon emissions in the region are equal to the carbon emissions allowances cap, the total carbon emissions function can be written as follows:

$$\min TC = CE_c \tag{2}$$

where TC is the total carbon emissions function from the CPPs.

3.2.3. Maximizing Allocation Free Carbon Allowance Satisfaction

Previous research has found that fairness is a critical factor for sustainable development [37]. Therefore, in this allowance allocation problem, a satisfactory degree method is proposed to measure the fairness [25,38]. The higher the free carbon emissions allowances granted to a CPP, the higher the satisfactory degree; therefore, the authority defines the allocation satisfaction function for each CPP as follows:

$$SD_i = \begin{cases} 0, & X_i \leq CEA_i^{\min} \\ \frac{X_i - CEA_i^{\min}}{CEA_i^{\max} - CEA_i^{\min}}, & CEA_i^{\min} \leq X_i \leq CEA_i^{\max} \\ 1, & X_i \geq CEA_i^{\max} \end{cases} \tag{3}$$

where SD_i is the allocation degree of satisfaction for CPP i in this study.

To ensure the sustainable development of the region and allocation fairness for each CPP, an objective function is used to maximize the minimal allocation satisfaction, as shown in Equation (4).

$$\max SD = \min \{SD_i\} \tag{4}$$

where SD is the minimal carbon emissions allowance allocation satisfaction for each CPP.

3.2.4. Allocation Constraints

A carbon emissions allowances cap (i.e., CE_c) is determined as part of the allocated free allowances (i.e., CE_f) and taxable allowances (i.e., CE_t), as shown in Equation (5).

$$CE_c = CE_f + CE_t \tag{5}$$

The free carbon emissions allowances depend on the level of free carbon emissions determined by the authority; therefore, the corresponding free and taxable emissions allowances are allocated to each CPP, as shown in Equations (6)–(8).

$$CE_f = \alpha CE_c \tag{6}$$

$$CE_f = \sum_{i=1}^{I} X_i \tag{7}$$

$$\sum_{i=1}^{I} Y_i \leq CE_t \tag{8}$$

3.2.5. Demand Constraints

To protect taxpayer rights, while the authority cannot allocate carbon emissions allowances the CPP is unable to carry, they have an obligation to ensure the CPP's basic rights (i.e., a minimum carbon emissions allowance that maintains the basic CPP operations), as can be seen in Equation (9).

$$CEA_i^{\min} \leq X_i + Y_i \leq CEA_i^{\max} \quad \forall i \in \Psi \tag{9}$$

3.2.6. Power Supply Constraints

The authority has an obligation to guarantee an adequate supply of electricity. Based on historical data, total electricity production is $\sum_{i=1}^{I} E[\widetilde{T}_i]X_i \geq D$. However, there may be some errors if only historical data are used. As electricity supply is essential for economic and social development, insufficient power supplies could cause economic decline and social panic. In addition, due to the inherent complexity and external uncertainty of power generation and fluctuating demand, electricity production needs to be estimated as accurately as possible. Therefore, the actual predicted electricity production is $\sum_{i=1}^{I} \sum_{j=1}^{J} E[\widetilde{T_{ij}}]Z_{ij}$, and the two parts are then combined using a harmonic parameter. Therefore, this constraint is as shown in Equation (10).

$$\beta \sum_{i=1}^{I} E[\widetilde{T}_i]X_i + (1 - \beta) \sum_{i=1}^{I} \sum_{j=1}^{J} E[\widetilde{T_{ij}}]Z_{ij} \geq D \tag{10}$$

where β is the attitude of the authority towards the historical data, which allows the authority to comprehensively consider both the historical data and the forecast values.

3.3. CPP Coal Purchase Scheme

3.3.1. Economic Profits

Each CPP seeks the largest possible profit under the restrictions of the carbon emissions allowances allocation. CPP profits are income from electricity sales revenue minus the sales tax and component coal procurement, pollution treatment and the carbon emissions allowance tax. Therefore, the profits function for each CPP is formulated as follows:

$$\max EP_i = (p - \mu) \sum_{j=1}^{J} E[\widetilde{T_{ij}}] Z_{ij} - \sum_{j=1}^{J} E[\widetilde{C_j}] Z_{ij} - \sum_{j=1}^{J} \sum_{k=1}^{K} CT_{jk} E[\widetilde{EF_{jk}}] \eta_{ik} Z_{ij} - \gamma Y_i \tag{11}$$

where EP_i is the potential economic profits function of each CPP; $E[\widetilde{T_{ij}}]$ is the expected value of T_{ij}; $E[\widetilde{C_j}]$ is the expected value of C_j; $E[\widetilde{EF_{jk}}]$ is the expected value of EF_{jk}.

3.3.2. Carbon Emissions Allowance Constraints

The total carbon emissions at each CPP must not exceed the free (i.e., X_i) and taxed (i.e., Y_i) carbon emissions quotas allocated by the authority, otherwise the CPP is severely punished or even deprived of its power generation rights. Therefore, the constraints are expressed as in Equation (12).

$$\sum_{j=1}^{J} E[\widetilde{CEF_{ij}}] Z_{ij} \leq X_i + Y_i \ \ \forall i \in \Psi \tag{12}$$

where $E[\widetilde{CEF_{ij}}]$ is the expected value of CEF_{ij}.

3.3.3. Coal Quality Requirement

For the CPPs, the five properties (i.e., volatile matter content, heat rate, ash content, moisture content and sulfur content) of burning coal need to meet the boiler requirements [39,40]. Therefore, these properties are limited within a special range using a coal blending method to ensure normal boiler operations, as shown in Equation (13):

$$E[\widetilde{LCQ_{ik}}] \sum_{j=1}^{J} Z_{ij} \leq \sum_{j=1}^{J} E[\widetilde{CQ_{jk}}] Z_{ij} \leq E[\widetilde{UCQ_{ik}}] \sum_{j=1}^{J} Z_{ij}, \ \ \forall i \in \Psi, \forall k \in \Omega \tag{13}$$

where $k = 1$ denotes volatile matter content, $w = 2$ denotes heat rate, $w = 3$ denotes ash content, $w = 4$ denotes moisture content and $w = 5$ denotes sulfur content.

3.3.4. Social Responsibility Limitation

Modern enterprises not only consider profits, but also have necessary social responsibilities. As electricity is essential to social and economic development, the supply of basic electricity is the most basic social responsibility for each CPP. Therefore, this constraint ensures an electricity supply-demand balance, as seen in Equation (14).

$$\sum_{j=1}^{J} E[\widetilde{T_{ij}}] Z_{ij} \geq ED_i \ \ \forall i \in \Psi \tag{14}$$

3.3.5. Component Coal Purchase Quantity Limitations

There is a limit to each component coal that can be purchased by the CPP, which is affected by the coal production and coal consumption in other industries. There are also nonnegative constraints on the decision variable Z_{ij} as the component coal burned by the CPP cannot be nonnegative. By combining these two parts, this constraint is expressed as follows:

$$0 \leq Z_{ij} \leq Q_{ij}^u \ \ \forall i \in \Psi, \forall j \in \Phi \tag{15}$$

3.4. Global Model

By integrating Equations (1)–(15), the global optimization model for the carbon emissions allowance allocation based on the cap and tax mechanism is built, as shown in Equation (16). There is interaction between the authority and the CPPs as the authority's decisions affect the CPPs' decisions. The authority seeks to mitigate carbon emissions, and the CPPs seek profit maximization. However, as the decisions by each CPP also affect the authority's decisions and the other CPPs, conflicts arise when all stakeholders attempt to achieve an optimal solution based on their own respective optimization targets; therefore, a compromise is necessary to achieve equilibrium between the authority and the CPPs. Initially, based on historical information and its own objectives, the authority decides on an initial carbon emissions allowances allocation scheme, which is sent to the CPPs. Each CPP than formulates its own production plan in line with the allocated carbon emissions allowance, the coal quality requirements, its social responsibility and the market conditions. These CPP plans are then fed back to the authority, which adjusts its initial decisions in consideration of the emissions performance of each CPP, after which an improved allocation plan is sent to the CPPs again. The above process is repeated until all stakeholders reach equilibrium. Therefore, this problem is expressed mathematically as a bi-level programming model, as follows:

$$
\begin{aligned}
&\max FB = \mu \sum_{i=1}^{I} \sum_{j=1}^{J} E[\widetilde{T_{ij}}]Z_{ij} + \gamma \sum_{i=1}^{I} Y_i \\
&\min TC = CE_c \\
&\max SD = \min\{SD_i\} \\
&s.t. \begin{cases}
CE_c = CE_f + CE_t \\
CE_f = \alpha CE_c, \ \alpha \in [0,1] \\
CE_f = \sum_{i=1}^{I} X_i \\
\sum_{i=1}^{I} Y_i \leq CE_t \\
CEA_i^{\min} \leq X_i + Y_i \leq CEA_i^{\max}, \ \forall i \in \Psi \\
\beta \sum_{i=1}^{I} E[\widetilde{T_i}]X_i + (1-\beta) \sum_{i=1}^{I} \sum_{j=1}^{J} E[\widetilde{T_{ij}}]Z_{ij} \geq D \\
\max EP_i = (p - \mu) \sum_{j=1}^{J} E[\widetilde{T_{ij}}]Z_{ij} - \sum_{j=1}^{J} E[\widetilde{C_j}]Z_{ij} - \sum_{j=1}^{J} \sum_{k=1}^{K} CT_{jk}E[\widetilde{EF_{jk}}]\eta_{ik}Z_{ij} - \gamma Y_i \\
\quad s.t. \begin{cases}
\sum_{j=1}^{J} E[\widetilde{CEF_{ij}}]Z_{ij} \leq X_i + Y_i, \ \forall i \in \Psi \\
E[\widetilde{LCQ_{ik}}] \sum_{j=1}^{J} Z_{ij} \leq \sum_{j=1}^{J} E[\widetilde{CQ_{jk}}]Z_{ij} \leq E[\widetilde{UCQ_{ik}}] \sum_{j=1}^{J} Z_{ij}, \ \forall i \in \Psi, \forall k \in \Omega \\
\sum_{j=1}^{J} E[\widetilde{T_{ij}}]Z_{ij} \geq ED_i, \ \forall i \in \Psi \\
0 \leq Z_{ij} \leq Q_{ij}^u \ \forall i \in \Psi, \forall j \in \Phi
\end{cases}
\end{cases}
\end{aligned}
\tag{16}
$$

4. Case Study

4.1. Case Description

Jiangsu Province, one of the most economically-prosperous areas in China, is located in southeast China. Because of the rapid industrialization and urbanization, the demand for electricity primarily supplied by coal-fired power plants has increased dramatically, leading to a commensurate increase in carbon emissions. Due to the pressure from international bodies and local public opinion, the authority in Jiangsu Province has planned to reduce carbon emissions in the next five-year plan. To reduce the computation burden, only three major CPPs (i.e., the Xiaguan CPP, the Huarun CPP and the Yancheng CPP; the locations for which are shown in Figure 2) in Jiangsu Province were chosen in this case study to demonstrate the practicability and efficiency of the proposed method.

Figure 2. Location of the case region.

4.2. Model Transformation

Based on the actual background and characteristics of the CEAA problem, this paper transforms the multi-objective optimization problem into a single-objective optimization problem by determining a primary objective and treating the secondary objectives as corresponding constraints with appropriate threshold values according to the research of Zeng et al. [41]. As Jiangsu Province is still developing, continued economic development remains the primary objective of the authority. However, under pressure to ensure sustainable development, the environment must also be protected; therefore, the authority transforms the minimization of the total carbon emissions allowance objective into a corresponding environmental constraint. Analogously, to ensure a fair market environment, the authority transforms the objective to maximize free carbon allowance allocation satisfaction into its corresponding constraint. Therefore, Model (16) is transformed into a corresponding single-objective form, as shown in Equation (17):

$$
\max FB = \mu \sum_{i=1}^{I} \sum_{j=1}^{J} E[\widetilde{T_{ij}}] Z_{ij} + \gamma \sum_{i=1}^{I} Y_i
$$

$$
s.t. \begin{cases}
CE_c \leq \lambda CE \\[4pt]
SD_i \geq \theta \\[4pt]
CE_c = CE_f + CE_t \\[4pt]
CE_f = \alpha CE_c, \ \alpha \in [0,1] \\[4pt]
CE_f = \sum_{i=1}^{I} X_i \\[4pt]
\sum_{i=1}^{I} Y_i \leq CE_t \\[4pt]
CEA_i^{\min} \leq X_i + Y_i \leq CEA_i^{\max}, \ \forall i \in \Psi \\[4pt]
\beta \sum_{i=1}^{I} E[\widetilde{T_i}] X_i + (1-\beta) \sum_{i=1}^{I} \sum_{j=1}^{J} E[\widetilde{T_{ij}}] Z_{ij} \geq D \\[4pt]
\max EP_i = (p-\mu) \sum_{j=1}^{J} E[\widetilde{T_{ij}}] Z_{ij} - \sum_{j=1}^{J} E[\widetilde{C_j}] Z_{ij} - \sum_{j=1}^{J} \sum_{w=1}^{W} CT_{jw} E[\widetilde{EF_{jw}}] \eta_{iw} Z_{ij} - \gamma Y_i \\[4pt]
\quad s.t. \begin{cases}
\sum_{j=1}^{J} E[\widetilde{CEF_{ij}}] Z_{ij} \leq X_i + Y_i, \ \forall i \in \Psi \\[4pt]
E[\widetilde{LCQ_{ik}}] \sum_{j=1}^{J} Z_{ij} \leq \sum_{j=1}^{J} E[\widetilde{CQ_{jk}}] Z_{ij} \leq E[\widetilde{UCQ_{ik}}] \sum_{j=1}^{J} Z_{ij}, \ \forall i \in \Psi, \forall k \in \Omega \\[4pt]
\sum_{j=1}^{J} E[\widetilde{T_{ij}}] Z_{ij} \geq ED_i, \ \forall i \in \Psi \\[4pt]
0 \leq Z_{ij} \leq Q_{ij}^u \ \forall i \in \Psi, \forall j \in \Phi
\end{cases}
\end{cases}
\tag{17}
$$

where λ is the attitude of the authority towards carbon emissions reduction; θ is the minimal allowance satisfactory degree chosen by the authority; and CE is the actual carbon emission amount in the last production cycle. However, the bi-level optimization model is still a non-deterministic polynomial hard problem even in its most simple form. The different levels of the decision makers control or influence the decisions of the others through their own decisions; in other words, equilibrium between the upper and lower levels needs to be achieved through constant stakeholder interaction. Therefore, an interactive algorithm was designed to resolve this complexity. First, a feasible region for the upper level model was built based on the constraints and a feasible solution randomly produced as the initial solution. Then, this initial feasible solution was sent to the lower level and a corresponding solution obtained, which was fed back to the upper level. After receiving the feedback from the lower level, the upper level made corresponding adjustments to obtain an improved solution, which was then sent to the lower level again. This procedure continued to iterate until the termination condition was met. In this paper, the termination condition was set as $\sum_{i=1}^{I} \left(\left| X_i^n - X_i^{n-1} \right| + \left| Y_i^n - Y_i^{n-1} \right| \right) \Big/ \sum_{i=1}^{I} (X_i^n + Y_i^n) \leq 1\%$. This process was accomplished using the following procedure.

Step 1: Randomly generate a set of initial feasible solutions (X_i^1, Y_i^1) in the feasible zone of the upper level.

Step 2: Solve the lower level problem using the simplex method by inputting X_i and Y_i into the lower level optimization model.

Step 3: Obtain the optimal solution from Z_{ij}, which is fed back to the upper-level optimization model.

Step 4: Solve the upper level problem using the simplex method, and obtain the improved solution (X_i, Y_i).

Step 5: The improved solution is sent to the lower level model again.

Step 6: Repeat Step 3 and Step 4 until the termination condition is reached.

4.3. Data Collection

The basic data shown in Table 2 were obtained from the annual reports of the three power plants. Macro data, such as the actual carbon emissions CE from the last production period were taken from the Statistical Yearbook of Jiangsu Province, China, as shown in Table 3. The uncertain data in the model are described using trapezoidal fuzzy numbers based on fuzzy set theory. These fuzzy data were determined from interviews with experts and engineers, a well as historical data [37]. Therefore, the uncertain parameters in this paper were collected in fuzzy form, as shown in Table 4.

Table 2. Crisp parameters of each CPP.

	Xiaguan CPP	Huarun CPP	Yancheng CPP
Emission reduction measure			
For SO_2 ($w = 1$)	LDS	LDS	LDS
For NO_x ($w = 2$)	SCR	SCR	SCR
For PM_{10} ($w = 3$)	CDE and EP	EF and EP	EF and EP
Emission reduction efficiency, η_{iw}			
For SO_2 ($w = 1$) (%)	96.2	96.1	95.9
For NO_x ($w = 2$) (%)	85.9	85.7	85.4
For PM_{10} ($w = 3$) (%)	98.8	98.5	98.4
Emission reduction cost, CT_{jw}			
For SO_2 ($w = 1$) (RMB/kg)	2.4	2.2	1.7
For NO_x ($w = 2$) (RMB/kg)	16.7	15.8	14.2
For PM_{10} ($w = 3$) (RMB/kg)	3.5	2.8	2.1
Minimum power supply, ED_i (10^9 kWh)	2.7	4.2	1.9
Minimum allowance demand, CEA_i^{min} (10^6 tonne)	3	4	2
Maximum allowance demand, CEA_i^{max} (10^6 tonne)	6	8	6
Coal quality requirement, $[LCQ_{ik}, LCQ_{ik}]$			
Volatile matter (% weight)	>6 and <27	>7 and <29	>9 and <32
Heat rate (GJ/tonne)	>22.3	>22.1	>21.9
Ash content (% weight)	<20	<22	<23
Moisture content (% weight)	<5	<6	<7
Sulfur content (% weight)	<0.8	<0.9	<1.0

Notes: LDS (i.e., lime spray dryer system), SCR (i.e., selective catalytic reduction), EP (i.e., electrostatic precipitator), CDE (i.e., cyclone dust extractor) and FF (i.e., fabric filter).

Table 3. Other parameters employed in the proposed model.

Taxable carbon allowance price, γ (RMB/tonne)	30
Added-value tax, μ (RMB/kWh)	0.01
Price of unit electric, p (RMB/kWh)	0.45
Total basic electric supply, D (10^9 kWh)	8.8×10^9
Actual carbon emission amount in the last production cycle, CE (tonne)	1.98×10^7

Table 4. Parameters of component coals in fuzzy form.

	Tongmei	Shenhua	Yitai	Zhongmei
Coal characteristics, $\widetilde{CQ}_{jk}$				
Volatile matter (% weight)	(6.9, 7.5, 8.3, 9.3)	(35.2, 37.4, 38.9, 40.5)	(23.6, 25.4, 27.1, 27.9)	(24.9, 27.4, 29.8, 29.9)
Heat rate (GJ/tonne)	(22.6, 22.9, 23.5, 23.8)	(21.2, 21.7, 22.1, 22.2)	(20.4, 20.9, 21.3, 21.8)	(18.2, 18.6, 20.1, 21.1)
Ash content (% weight)	(19.9, 20.6, 21.4, 22.1)	(14.7, 15.8, 16.6, 16.9)	(10.4, 11.5, 12.1, 14)	(16.3, 17.4, 18.9, 19.4)
Moisture content (% weight)	(4.3, 4.4, 4.8, 4.9)	(5.1, 5.4, 5.7, 6.2)	(1.9, 2.3, 2.7, 3.1)	(2.4, 2.9, 3.2, 3.9)
Sulfur content (% weight)	(0.4, 0.5, 0.7, 0.8)	(0.1, 0.2, 0.3, 0.6)	(0.6, 0.7, 1.1, 1.2)	(0.1, 0.2, 0.2, 0.3)
Emission factor, $\widetilde{EF}_{jw}$				
For SO_2 ($w = 1$) (kg/tonne)	(4.3, 4.9, 5.3, 7.9)	(5.6, 6.3, 7.4, 7.9)	(6.4, 6.9, 7.6, 8.7)	(8.6,8.8,9.2,10.2)
For NO_x ($w = 2$) (kg/tonne)	(1.8, 2.3, 2.8, 3.1)	(2.3, 2.9, 3.5, 3.7)	(6.1, 6.4, 6.7, 6.8)	(8.8, 9.1, 9.4, 9.9)
For PM_{10} ($w = 3$) (kg/tonne)	(0.1, 0.2, 0.2, 0.3)	(0.2, 0.4, 0.6, 0.8)	(0.4, 0.7, 1.2, 1.3)	(1.1, 1.3, 1.5, 1.7)
Coal-power conversion coefficient, $\widetilde{T}_{ij}$				
Xiaguan CPP (kWh/tonne)	(2490, 2535, 2595, 2660)	(2415, 2425, 2435, 2445)	(2320, 2330, 2340, 2370)	(2120, 2145, 2160, 2175)
Huarun CPP (kWh/tonne)	(2485, 2520, 2550, 2565)	(2375, 2390, 2415, 2420)	(2265, 2295, 2315, 2325)	(2090, 2105, 2140, 2145)
Yancheng CPP (kWh/tonne)	(2420, 2455, 2480, 2525)	(2325, 2335, 2345, 2355)	(2225, 2240, 2255, 2280)	(2045, 2055, 2085, 2095)
Carbon emission factor, $\widetilde{CEF}_{ij}$				
Xiaguan CPP (kg/tonne)	(2045, 2075, 2095, 2145)	(1955, 1965, 1975, 1985)	(1875, 1890, 1905, 1930)	(1730, 1745, 1760, 1765)
Huarun CPP (kg/tonne)	(2090, 2105, 2115, 2130)	(1980, 1995, 2010, 2015)	(1905, 1910, 1925, 1940)	(1750, 1760, 1775, 1795)
Yancheng CPP (kg/tonne)	(2120, 2130, 2145, 2165)	(1995, 2015, 2030, 2040)	(1940, 1945, 1955, 1960)	(1780, 1785, 1790, 1805)
Procurement cost, $\widetilde{C}_j$ (RMB/tonne)	(665, 675, 685, 695)	(630, 635, 645, 650)	(570, 590, 615, 665)	(525, 540, 555, 580)

5. Results and Discussion

5.1. Results and Sensitivity Analysis

The collected data were input into the proposed model (i.e., Equation (17)) and the solution approach run on MATLAB software, from which the optimal carbon emissions allowance allocation for the authority was determined.

5.1.1. Different Free Carbon Emission Levels

As the authority needs to strengthen the control over the free carbon emissions allocations, a single result is unsatisfactory; therefore, several scenarios are considered. To illustrate the practicality and validity of this method, as a representative situation, Table 5 shows the results of the sensitivity analysis on the free carbon emission levels when $\theta = 0.5$ and $\lambda = 1$. In this situation, for fairness, the authority's attitude towards allocation satisfaction, was set at 0.5, and the carbon emissions reduction level was set at one, the most relaxed carbon emissions reduction attitude. It can be seen that in this situation, the authority earns 3.436×10^8 RMB when $\alpha = 0.82$, which is the lowest free emissions level. Figure 3 illustrates the changes to the financial benefits for the authority (i.e., FB), the total carbon emissions (i.e., TC) in the region, each CPP's profits (i.e., EP_i), the allocation satisfaction level (i.e., SD_i), the free emissions allocation allowance (X_i), the taxable emissions allocation allowance (Y_i) and the carbon emissions allowance ($X_i + Y_i$) against changing the free carbon emissions level. The financial benefits for the authority gradually increase, and the profits of each CPP decrease as the free carbon emissions level decreases. Further, to maximize profits, all CPPs strive to obtain a higher market share and finally reach equilibrium, at which point, all CPPs have used all their free carbon emission allowances and are seeking to obtain as high an allowance as possible from the authority. In addition, as the free emissions level decreases, the CPPs have lower free carbon emissions allowances and lower satisfactory degrees. From Figure 3, it can be seen that the Huarun CPP remained the most profitable CPP under the different free emissions levels and the highest carbon emissions allowance, which included both the free and taxable emissions allowances. In addition, the Yancheng CPP profits were lower than the Xiaguan CPP profits, although they have roughly equivalent carbon emissions quotas. It was concluded that the Yancheng CPP may be detrimental to the sustainable development of the coal-fired power industry in this region. Therefore, suitable free carbon emissions level can achieve both stable economic development and a fair market environment.

Table 5. Sensitivity analysis on free carbon emissions level α when $\theta = 0.5$ and $\lambda = 1$.

α	FB (10^8 RMB)	CPP	X_i (10^6 tonnes)	Y_i (10^5 tonnes)	EP_i (10^8 RMB)	Z_{i1} (10^6 tonnes)	Z_{i2} (10^6 tonnes)	Z_{i3} (10^6 tonnes)	Z_{i4} (10^6 tonnes)	SD_i
1	2.367	Xiaguan	5.94	0.00	11.30	2.00	0.89	0.00	0.00	0.9789
		Huarun	7.90	0.00	13.91	2.00	1.00	0.88	0.00	0.9757
		Yancheng	5.96	0.00	9.72	1.00	1.00	0.92	0.00	0.9901
					34.94					
0.94	2.723	Xiaguan	5.70	2.25	11.22	2.00	0.89	0.00	0.00	0.9006
		Huarun	7.34	5.84	13.78	2.00	1.00	0.89	0.00	0.8358
		Yancheng	5.57	3.79	9.59	1.00	1.00	0.92	0.00	0.8917
					34.58					
0.88	3.080	Xiaguan	5.42	5.44	11.19	2.00	0.91	0.00	0.00	0.8073
		Huarun	7.03	9.28	13.71	2.00	1.00	0.90	0.00	0.7563
		Yancheng	4.98	9.04	9.33	1.00	1.00	0.88	0.00	0.7442
					34.24					
0.82	3.436	Xiaguan	5.31	5.92	11.07	2.00	0.88	0.00	0.00	0.7711
		Huarun	6.74	11.64	13.56	2.00	1.00	0.87	0.00	0.6838
		Yancheng	4.19	18.08	9.23	1.00	1.00	0.94	0.00	0.5469

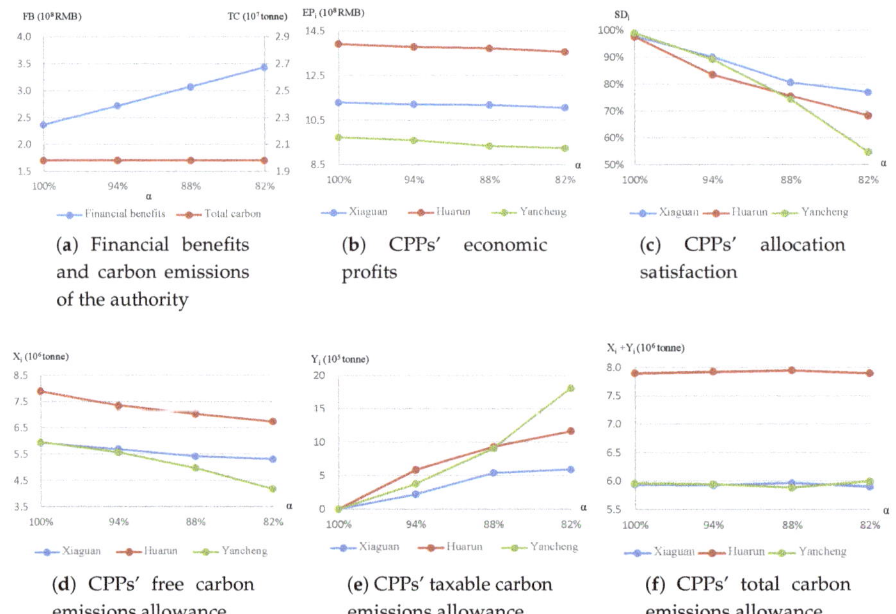

(**a**) Financial benefits and carbon emissions of the authority

(**b**) CPPs' economic profits

(**c**) CPPs' allocation satisfaction

(**d**) CPPs' free carbon emissions allowance

(**e**) CPPs' taxable carbon emissions allowance

(**f**) CPPs' total carbon emissions allowance

Figure 3. Comparative analysis under different free emission levels. (**a**) Financial benefit and carbon emissions of the authority; (**b**) CPPs' economic profits; (**c**) CPPs' allocation satisfaction; (**d**) CPPs' free carbon emissions allowance; (**e**) CPPs' taxable carbon emissions allowance; (**f**) CPPs' total carbon emissions allowance.

5.1.2. Different Carbon Emission Reduction Levels

As the carbon emissions reduction level is an another important factor for carbon emissions mitigation by the authority, several scenarios were again considered under different λ. As a representative example, Table 6 shows the sensitivity analysis results for the carbon emissions reduction levels when $\theta = 0.5$ and $\alpha = 0.9$ to verify the validity of the model. The carbon emissions allowance cap was divided into free and taxable emissions quotas and the free carbon emissions level set at 0.9. The results in Table 6 show that the authority achieves a minimum of 2.43×10^8 RMB when $\lambda = 0.82$, which was the lowest carbon emissions reduction level. Figure 4 shows the results for the comparative analysis under different carbon emissions reduction levels. From Figure 4a, it can be seen that both the financial benefits and total carbon emissions decrease as the environmental protection constraints are tightened (i.e., changing λ from one to its lowest level); however, the decrease in the carbon emissions ratio is larger than the decrease in the financial benefits. For example, when $\lambda = 0.94$ is compared with $\lambda = 1$, the financial benefit ratio decreases by 5.9% and the carbon emissions ratio decreases by 6.4%. Further, the ratios decrease by 6.4% and 6.8% when $\lambda = 0.88$ and by 6.8% and 7.3% when $\lambda = 0.82$ for the two factors. From this analysis, it was concluded that tightening the environmental protection constraints is more beneficial to sustainable development and that more relaxed environmental protection constraints cause more damage to sustainable development. Similar to the different free carbon emissions level scenarios, to maximize economic profits, each CPP is eager to gain a higher market share and eventually reaches equilibrium. In addition, it is clear that the CPP profits decrease with a decrease in the carbon emissions reduction level. The profits and free emissions allowance at the Huarun CPP are still the highest, and the Yancheng CPP is the lowest under the different carbon emissions reduction levels.

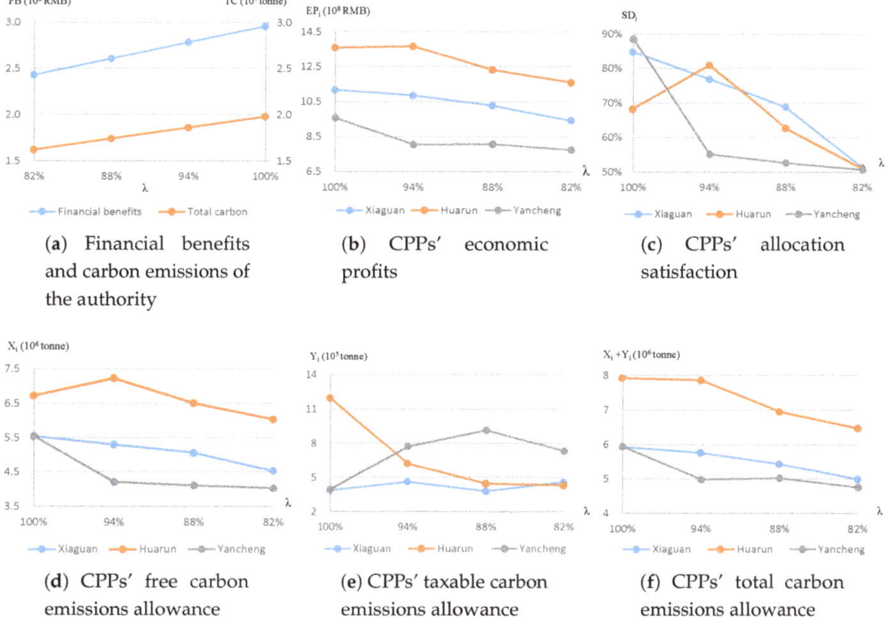

Figure 4. Comparative analysis under different carbon emission reduction levels. (**a**) Financial benefits and carbon emissions of the authority; (**b**) CPPs' economic profits; (**c**) CPPs' allocation satisfaction; (**d**) CPPs' free carbon emissions allowance; (**e**) CPPs' taxable carbon emissions allowance; (**f**) CPPs' total carbon emissions allowance.

Table 6. Sensitivity analysis on carbon emission reduction level λ when $\theta = 0.5$ and $\alpha = 0.9$.

λ	FB (10^8 RMB)	CPP	X_i (10^6 tonnes)	Y_i (10^5 tonnes)	EP_i (10^8 RMB)	Z_{i1} (10^6 tonnes)	Z_{i2} (10^6 tonnes)	Z_{i3} (10^6 tonnes)	Z_{i4} (10^6 tonnes)	SD_i
1	2.961	Xiaguan	5.55	3.88	11.18	2.00	0.89	0.00	0.00	0.8488
		Huarun	6.73	11.97	13.59	2.00	1.00	0.89	0.00	0.6826
		Yancheng	5.54	3.95	9.57	1.00	1.00	0.91	0.00	0.8858
0.94	2.787	Xiaguan	5.31	4.64	10.86	2.00	0.81	0.00	0.00	0.7695
		Huarun	7.24	6.24	13.66	2.00	1.00	0.85	0.00	0.8092
		Yancheng	4.21	7.73	8.04	1.00	1.00	0.42	0.00	0.5514
0.88	2.608	Xiaguan	5.07	3.82	10.28	2.00	0.64	0.00	0.00	0.6888
		Huarun	6.51	4.46	12.32	2.00	1.00	0.38	0.00	0.6275
		Yancheng	4.11	9.14	8.06	1.00	1.00	0.44	0.00	0.5263
0.82	2.429	Xiaguan	4.54	4.60	9.42	1.94	0.48	0.00	0.00	0.5142
		Huarun	6.04	4.31	11.59	2.00	1.00	0.13	0.00	0.5105
		Yancheng	4.03	7.32	7.73	1.00	1.00	0.31	0.00	0.5069

5.1.3. Different Allocation Satisfaction Levels

A fair market environment is conducive to regional sustainable development. In this paper, a satisfactory degree method is proposed to measure fairness. Similarly, several scenarios were conducted under different allocation satisfaction levels. As an example, Table 7 shows the results of the sensitivity analysis for the allocation satisfaction levels when $\alpha = 0.9$ and $\lambda = 1$, from which it can be seen that the authority achieves a minimum of 2.96066×10^8 RMB when $\theta = 0.8$, which is the lowest allocation satisfaction level. Figure 5 illustrates the changes in the financial benefits (i.e., FB) for the authority and each CPP's profits (i.e., EP_i) when the allocation satisfaction level changes. However, as can be seen, the allocation satisfaction does not significantly impact the financial benefits of the authority or the CPP profits. For example, when $\theta = 0.5$, the financial benefits are 2.96112×10^8 RMB, and as θ changes to 0.6, the financial benefits are 2.96081×10^8 RMB, a decrease of only 0.011%. When θ is set at 0.7 and 0.8, the ratio decreases only slightly by 0.004% and 0.001%. From Figure 5, similar situations can be seen for each of the CPP profits.

Table 7. Sensitivity analysis on allocation satisfaction level θ when $\alpha = 0.9$ and $\lambda = 1$.

θ	FB (10^8 RMB)	CPP	X_i (10^6 tonnes)	Y_i (10^5 tonnes)	EP_i (10^8 RMB)	Z_{i1} (10^6 tonnes)	Z_{i2} (10^6 tonnes)	Z_{i3} (10^6 tonnes)	Z_{i4} (10^6 tonnes)	SD_i
0.5	2.96112	Xiaguan	5.55	3.88	11.18	2.00	0.89	0.00	0.00	0.8488
		Huarun	6.73	11.97	13.59	2.00	1.00	0.89	0.00	0.6826
		Yancheng	5.54	3.95	9.57	1.00	1.00	0.91	0.00	0.8858
0.6	2.96081	Xiaguan	5.56	3.48	11.14	2.00	0.88	0.00	0.00	0.8519
		Huarun	7.40	5.14	13.78	2.00	1.00	0.88	0.00	0.8496
		Yancheng	4.87	11.18	9.42	1.00	1.00	0.94	0.00	0.7165
0.7	2.96068	Xiaguan	5.52	3.67	11.10	2.00	0.87	0.00	0.00	0.8404
		Huarun	7.35	5.65	13.76	2.00	1.00	0.88	0.00	0.8369
		Yancheng	4.95	10.48	9.46	1.00	1.00	0.94	0.00	0.7378
0.8	2.96066	Xiaguan	5.40	4.76	11.05	2.00	0.86	0.00	0.00	0.8007
		Huarun	7.21	7.19	13.73	2.00	1.00	0.89	0.00	0.8017
		Yancheng	5.21	7.85	9.54	1.00	1.00	0.94	0.00	0.8027

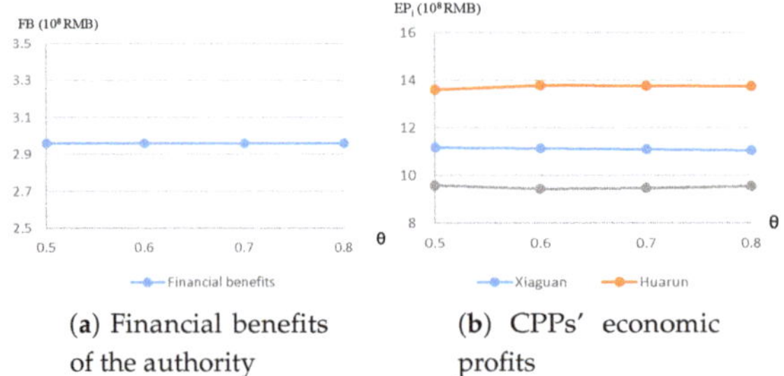

(**a**) Financial benefits of the authority

(**b**) CPPs' economic profits

Figure 5. Comparative analysis under different allocation satisfaction levels. (**a**) Financial benefits of the authority; (**b**) CPPs' economic profits.

5.2. Discussion

Based on the above results and analysis, the proposed method contributes to research on carbon emissions mitigation in the coal-fired power field and can assist authorities in establishing reasonable carbon emissions allowance allocation policies as the uncertain factors (coal characteristics, emissions factor, coal-power conversion coefficient, carbon emissions factor and procurement costs) are considered. The coal characteristics are uncertain due to the impact of the natural condition and the mining processes, and the uncertain emissions factor depends on the uncertain coal characteristics. The coal-power conversion coefficient and the carbon emissions factor are uncertain because of the uncertain combustion efficiencies. The procurement costs are uncertain because of the impact of price coordination and market fluctuations. At the same time, there are deviations in the collected data; that is, these uncertain parameters are influenced by both subjective and objective factors. There has been significant research conducted in dealing with such uncertainties. For example, Cheng et al. proposed an interval recourse liner programming (IRLP) to mitigate constraint violation problems in resources and environmental systems management (REM) under uncertainties [42]. Huang et al. developed an inexact fuzzy stochastic chance constrained programming (IFSCCP) method to address various uncertainties in evacuation management problems [43]. However, based on the actual background and characteristics of the CEAA problem in this paper, fuzzy theory was employed to fit reality and an expected value operator used to transform the fuzzy variables into corresponding expected values. Through this process, the results of the proposed method are more convincing.

In addition, the proposed method was shown to describe the interactive relationship between the authority and the CPPs effectively and to resolve the conflicts between economic development and environmental protection. Such situations are also found in other carbon emissions mitigation fields. For instance, there are similar interactive relationships between the authority and biomass power plants in the biomass power industry, in which there is also economic development and environmental protection conflicts. To mitigate these carbon emissions, authorities need to apply the appropriate CEAA strategy based on cap and tax mechanisms for biomass power plants, and the biomass power plants should have suitable biomass blending plans to achieve their required profits under the carbon quotas imposed by the authority.

5.3. Management Recommendations

Based on the above analysis and discussion, some management recommendations are given.

First, for regions that largely depend on CPP-generated electricity, a new cap and tax mechanism should be established to ensure the required environmental protection. Without such a mechanism, CPPs would arbitrarily emit carbon dioxide as they would lack the motivation to improve their emissions performances. Using the proposed methodology, the cap and tax mechanism is able to motivate CPPs to develop low carbon power generation. Further, under carbon emissions allowance allocation constraints, CPPs may be encouraged to improve their clean-energy technologies to decrease operating costs, which could further mitigate carbon emissions and gain higher profits.

Second, the authority can design suitable carbon emissions allowance allocation plans using the proposed method; that is, the authority can select the desired free carbon emissions levels and carbon emissions reductions levels based on the actual situation. Therefore, when using the proposed model, it is recommended that the authorities in developed regions set the lowest free carbon emissions level and the strictest carbon emissions reduction levels to encourage environmentally-friendly power generation. On the other hand, for developing regions, the authority can set relatively loose emissions reduction goals at the start to ensure steady local economic development. They can then continue to tighten the environmental protection parameter to aim for sustainable development.

6. Conclusions

This paper studied a coal-fired power generation carbon emission allowance allocation problem and proposed a bi-level multi-objective model that considered the mutual coordination and conflicts between an authority and CPPs. Using the proposed method, a carbon emissions allowance allocation with a cap and tax mechanism was established to ensure steady economic development and carbon emissions mitigation. This model has the ability to describe the interactions of all stakeholders whose decisions may affect the sustainable development of a region and therefore can assist them to develop corresponding strategies to adapt to the changes made by the other stakeholders. A case study on a coal-fired power generation system with three main CPPs was employed to illustrate the practicality and efficiency of the proposed method. Sensitivity analyses on free emission levels and carbon emissions reduction levels were also conducted, which could assist authorities to select the most appropriate local development strategy. The analysis and discussion demonstrated that considering both free emissions levels and carbon emissions reduction levels is able to assist in balancing economic development and environmental protection and that a cap and tax mechanism could play a significant role in the environmentally-friendly development of coal-fired power generation systems.

The following further research directions could be taken: (1) integrate technical innovation and management optimization to improve coal-fired power generation systems; (2) examine how the carbon emissions trading mechanism could be integrated with the proposed method.

Author Contributions: All authors contributed substantially to the research presented in this paper. Q.F., Q.H., Q.Z. and L.L. conceived of and designed the research and methodology. Q.F. contributed the research idea. L.L. proposed the framework of the paper. Q.H. and Q.Z. analyzed the data and wrote the paper. All authors read and approved the final manuscript.

Funding: This word is supported by the National Natural Science Foundation of China (Grant No. 71704124) and the Humanity and Social Science Youth Foundation of the Ministry of Education of China (Grant No. 17YJC630096 and No. 17YJC890021).

Acknowledgments: The authors would like to thank the reviewers for their constructive comments and suggestions for improving this paper.

Conflicts of Interest: The authors declare no conflict of interest.

References

1. Peters, G.P.; Andrew, R.M. The challenge to keep global warming below 2 °C. *Nat. Clim. Chang.* **2013**, *3*, 4–6. [CrossRef]
2. Hu, X.; Moura, S.J. Integrated optimization of battery sizing, charging, and power management in plug-in hybrid electric vehicles. *IEEE Trans. Control Syst. Technol.* **2015**, *24*, 1036–1043. [CrossRef]
3. Hu, X.; Murgovski, N. Optimal dimensioning and power management of a fuel cell/battery hybrid bus via convex programming. *IEEE/ASME Trans. Mechatron.* **2015**, *20*, 457–468. [CrossRef]
4. Service, R.F. Cleaning up coal-cost-effectively. *Science* **2017**, *356*, 798. [CrossRef] [PubMed]
5. Service, R.F. Fossil power, guilt free. *Science* **2017**, *356*, 796. [CrossRef] [PubMed]
6. Choudhary, D.; Shankar, R. An steep-fuzzy ahp-topsis framework for evaluation and selection of thermal power plant location: A case study from india. *Energy* **2012**, *42*, 510–521. [CrossRef]
7. Wu, Y.; Geng, S. Decision framework of solar thermal power plant site selection based on linguistic choquet operator. *Appl. Energy* **2014**, *136*, 303–311. [CrossRef]
8. BP Global. BP Technology Outlook. Available online: https://www.bp.com/content/dam/bp/pdf/technology/bp-technology-outlook.pdf (accessed on 17 August 2018).
9. CEC. *National Electric Power Industry Statistical Bulletin*; China Electricity Council: Beijing, China, 2017.
10. Mao, X.Q.; Zeng A. Co-control of local air pollutants and co 2 from the chinese coal-fired power industry. *J. Clean. Prod.* **2014**, *67*, 220–227. [CrossRef]
11. Lonsdale, C.R.; Stevens R.G. The effect of coal-fired power-plant SO_2 and nox control technologies on aerosol nucleation in the source plumes. *Atmos. Chem. Phys.* **2012**, *12*, 11519–11531. [CrossRef]
12. Beer, J.M. Combustion technology developments in power generation in response to environmental challenges. *Prog. Energy Combust. Sci.* **2000**, *26*, 301–327. [CrossRef]

13. Longwell, J.P.; Rubin, E.S.; Wilson, J. Coal: Energy for the future. *Prog. Energy Combust. Sci.* **1995**, *21*, 269–360. [CrossRef]

14. Beer, J.M. High efficiency electric power generation: The environmental role. *Prog. Energy Combust. Sci.* **2007**, *33*, 107–134. [CrossRef]

15. Hoya, R.; Fushimi, C. Thermal efficiency of advanced integrated coal gasification combined cycle power generation systems with low-temperature gasifier, gas cleaning and CO_2 capturing units. *Fuel Process. Technol.* **2017**, *164*, 80–91. [CrossRef]

16. Kayal, S.; Chakraborty, A. Activated carbon (type maxsorb-iii) and mil-101(cr) metal organic framework based composite adsorbent for higher CH_4 storage and CO_2 capture. *Chem. Eng. J.* **2018**, *334*, 780–788. [CrossRef]

17. Porter, R.; Fairweather, M. The range and level of impurities in CO_2 streams from different carbon capture sources. *Int. J. Greenh. Gas Control* **2015**, *36*, 161–174. [CrossRef]

18. Wang, J.; Ryan, D.; Anthorny, E.J.; Wigston, A. *Effects of Impurities on Geological Storage of CO_2*; Techical Report; IEA GHG: Cheltenham, UK, 2011.

19. Goto, K.; Yogo, K.; Higashii, T. A review of efficiency penalty in a coal-fired power plant with post-combustion CO_2 capture. *Appl. Energy* **2013**, *111*, 710–720. [CrossRef]

20. Li, Y. Dynamics of clean coal-fired power generation development in China. *Energy Policy* **2012**, *51*, 138–142.

21. Cao, K.; Xu, X. Optimal production and carbon emission reduction level under cap-and-trade and low carbon subsidy policies. *J. Clean. Prod.* **2017**, *167*, 505–513. [CrossRef]

22. Shih, J.; Frey, H. Coal blending optimization under uncertainty. *Eur. J. Oper. Res.* **1995**, *83*, 452–465. [CrossRef]

23. Wang, B.; Wang, S. Multi-objective unit commitment with wind penetration and emission concerns under stochastic and fuzzy uncertainties. *Energy* **2016**, *111*, 18–31. [CrossRef]

24. Xu, J.; Wang, F. Carbon emission reduction and reliable power supply equilibrium based daily scheduling towards hydro-thermal-wind generation system: A perspective from china. *Energy Convers. Manag.* **2018**, *164*, 1–14. [CrossRef]

25. Xu, J.; Qiu, R.; Lv, C. Carbon emission allowance allocation with cap and trade mechanism in air passenger transport. *J. Clean. Prod.* **2016**, *131*, 308–320. [CrossRef]

26. Zhao, S.; Shi, Y.; Xu, J. Carbon emissions quota allocation based equilibrium strategy toward carbon reduction and economic benefits in china's building materials industry. *J. Clean. Prod.* **2018**, *189*, 307–325. [CrossRef]

27. He, P.; Zhang, W. Production lot-sizing and carbon emissions under cap-and-trade and carbon tax regulations. *J. Clean. Prod.* **2015**, *103*, 241–248. [CrossRef]

28. Wittneben, B. Exxon is right: Let us re-examine our choice for a cap-and-trade system over a carbon tax. *Energy Policy* **2009**, *37*, 2462–2464. [CrossRef]

29. Liu, L.; Huang, C.; Huang, G.; Baetz, B.; Pittendrigh, S. How a carbon tax will affect an emission-intensive economy: A case study of the province of Saskatchewan, Canada. *Energy* **2018**, *159*, 817–826. [CrossRef]

30. Tu, Y.; Zhou, X. Administrative and market-based allocation mechanism for regional water resources planning. *Resour. Conserv. Recycl.* **2015**, *95*, 156–173. [CrossRef]

31. Kardakos, E.G.; Simoglou, C.K.; Bakirtzis, A.G. Optimal offering strategy of a virtual power plant: A stochastic bi-level approach. *IEEE Trans. Smart Grid* **2016**, *7*, 794–806. [CrossRef]

32. Puri, M.L. Fuzzy random variables. *J. Math. Anal. Appl.* **1986**, *114*, 409–422. [CrossRef]

33. Liu, B.; Liu, Y.K. Expected value of fuzzy variable and fuzzy expected value models. *IEEE Trans. Fuzzy Syst.* **2002**, *10*, 445–450.

34. Ferrero, A.; Salicone, S. The random-fuzzy variables: A new approach to the expression of uncertainty in measurement. *IEEE Trans. Instrum. Meas.* **2004**, *53*, 1370–1377. [CrossRef]

35. Xu, J.; Zhou, X. *Fuzzy-Like Multiple Objective Decision Making*; Springer: Berlin/Heidelberg, Germany, 2011.

36. Solomon, S.; Plattner, G.K. Irreversible climate change due to carbon dioxide emissions. *Proc. Natl. Acad. Sci. USA* **2009**, *106*, 1704–1709. [CrossRef] [PubMed]

37. Lv, C.; Xu, J. Equilibrium strategy based coal blending method for combined carbon and pm10 emissions reductions. *Appl. Energy* **2016**, *183*, 1035–1052. [CrossRef]

38. Xu, J.; Yang, X.; Tao, Z. A tripartite equilibrium for carbon emission allowance allocation in the power-supply industry. *Energy Policy* **2015**, *82*, 62–80. [CrossRef]

39. Dai, C.; Cai, X.H. A simulation-based fuzzy possibilistic programming model for coal blending management with consideration of human health risk under uncertainty. *Appl. Energy* **2014**, *133*, 1–13. [CrossRef]

40. Samimi, A.; Zarinabadi, S. Reduction of greenhouse gases emission and effect on environment. *Aust. J. Basic Appl. Sci.* **2012**, *8*, 1011–1015.
41. Zeng, Z.; Xu, J.; Wu, S.; Shen, M. Antithetic Method-Based Particle Swarm Optimization for a Queuing Network Problem with Fuzzy Data in Concrete Transportation Systems. *Comput. Aided Civ. Infrastruct. Eng.* **2014**, *29*, 771–800. [CrossRef]
42. Cheng, G.; Huang, G.H.; Dong, C. Interval Recourse Linear Programming for Resources and Environmental Systems Management under Uncertainty. *J. Environ. Inform.* **2017**, *30*, 119–136. [CrossRef]
43. Huang, C.; Nie, S.; Guo, L.; Fan, Y.R. Inexact Fuzzy Stochastic Chance Constraint Programming for Emergency Evacuation in Qinshan Nuclear Power Plant under Uncertainty. *J. Environ. Inform.* **2017**, *30*, 63–78. [CrossRef]

 *sustainability*

Article

Sustainable Waste Tire Derived Carbon Material as a Potential Anode for Lithium-Ion Batteries

Joseph S. Gnanaraj [1,*], Richard J. Lee [1], Alan M. Levine [1], Jonathan L. Wistrom [2], Skyler L. Wistrom [2], Yunchao Li [3], Jianlin Li [4], Kokouvi Akato [5], Amit K. Naskar [5] and M. Parans Paranthaman [3,*]

[1]   Energy Division, RJ Lee Group, 350 Hochberg Road, Monroeville, PA 15146, USA; RLee@RJLeeGroup.com (R.J.L.); ALevine@rjleegroup.com (A.M.L.)
[2]   Practical Sustainability, 1402 N College Drive, Maryville, MO 64468, USA; jonathanwistrom@yahoo.com (J.L.W.); sky_wistrom@hotmail.com (S.L.W.)
[3]   Chemical Sciences Division, Oak Ridge National Laboratory, Oak Ridge, TN 37831, USA; yli107@utk.edu
[4]   Energy & Transportation Science Division, Oak Ridge National Laboratory, Oak Ridge, TN 37831, USA; lij4@ornl.gov
[5]   Materials Science & Technology Division, Oak Ridge National Laboratory, Oak Ridge, TN 37831, USA; kakato@utk.edu (K.A.); naskarak@ornl.gov (A.K.N.)
*   Correspondence: JSGnanaraj@gmail.com (J.S.G.); paranthamanm@ornl.gov (M.P.P.); Tel.: +1-860-405-4914 (J.S.G.); +1-865-574-5045 (M.P.P.)

Received: 17 July 2018; Accepted: 7 August 2018; Published: 10 August 2018

Abstract: The rapidly growing automobile industry increases the accumulation of end-of-life tires each year throughout the world. Waste tires lead to increased environmental issues and lasting resource problems. Recycling hazardous wastes to produce value-added products is becoming essential for the sustainable progress of society. A patented sulfonation process followed by pyrolysis at 1100 °C in a nitrogen atmosphere was used to produce carbon material from these tires and utilized as an anode in lithium-ion batteries. The combustion of the volatiles released in waste tire pyrolysis produces lower fossil CO_2 emissions per unit of energy (136.51 gCO_2/kW·h) compared to other conventional fossil fuels such as coal or fuel–oil, usually used in power generation. The strategy used in this research may be applied to other rechargeable batteries, supercapacitors, catalysts, and other electrochemical devices. The Raman vibrational spectra observed on these carbons show a graphitic carbon with significant disorder structure. Further, structural studies reveal a unique disordered carbon nanostructure with a higher interlayer distance of 4.5 Å compared to 3.43 Å in the commercial graphite. The carbon material derived from tires was used as an anode in lithium-ion batteries exhibited a reversible capacity of 360 mAh/g at C/3. However, the reversible capacity increased to 432 mAh/g at C/10 when this carbon particle was coated with a thin layer of carbon. A novel strategy of prelithiation applied for improving the first cycle efficiency to 94% is also presented.

Keywords: battery grade carbon; waste tires; lithium-ion batteries; pouch cells; disordered carbon microstructure; surface coating

---

1. Introduction

Rechargeable lithium-ion batteries (LIBs) are being used as the most promising power source for small-scale applications such as consumer portable electronics, power tools and large-scale applications such as advanced power load leveling for smart grids to meet the energy demands of modern mobile technology, electric vehicles (EVs), and hybrid electric vehicles (HEVs) [1]. Graphite is the most widely used anode material due to its high thermal and chemical stabilities, and the practical reversible specific capacity reaching closer to the theoretical capacity of 372 mAh/g corresponding to a fully lithiated stoichiometric LiC_6 compound [2]. Currently, graphite is being used as an anode in LIBs that powers electric vehicles such as Tesla Model S [3,4]. However, fast charging LIBs pose safety concerns due to lithium platting on the surface of the graphite anode leading to the formation of lithium dendrites, causing inner short-circuiting and capacity fading [5].

Other types of carbonaceous materials such as hard carbons (HCs) and soft carbons (SC) have been investigated as alternate anode materials to enhance the performance characteristics of LIBs [4]. The unique amorphous structure in SC enables fast charging in LIBs even when the micron-sized particles are used, but it suffers from very low specific capacity of 250 mAh/g. The disordered structure in HC is a short-range order which cannot be graphitized even upon high temperature treatments [6]. Hard carbon is less susceptible to exfoliation due to the random orientation of the small graphitic grains. The nanovoids present between the grains reduce isotropic volume expansion. Thus, nanovoids and defects provide additional gravimetric capacity, allowing the capacity to exceed the theoretical capacity of graphite. HCs have demonstrated the ability to store more lithium than graphite and do not exfoliate during repeated cycling in LIBs [7]. Together, these properties make HCs a high capacity high cycle life material. Nevertheless, it suffers from large irreversible capacity loss, which is generally attributed to the high surface area, exposed edge planes in high fraction that increase the absolute quantity of solid electrolyte interphase (SEI) formed, reducing the coulombic efficiency in the first few cycles, and voltage hysteresis [8]. The first cycle irreversible capacity loss in LIBs has been studied extensively and is attributed to the formation of a passivating SEI during the first lithiation process, due to the electrolyte reduction at the negatively polarized graphite surface and the deposition of a number of organic and inorganic compounds, trapping lithium irretrievably in the inner pores of carbon, through chemical bonding with surface functional groups or by reaction with adsorbed oxygen/water molecules [9–14]. Various surface pretreatment methods, such as carbon coating, chemical fluorination, oxidation, doping, etching, and acid treatments, are reported to improve the electrochemical characteristics of the active carbon material [13]. The surface treatment allows a more defined control in terms of surface chemistry, composition, and reactivity, eliminate surface functional groups and reduce the surface area and thereby reducing the irreversible capacity loss due to SEI formation [15,16].

Graphitic carbon material production requires expensive synthesis conditions and very high temperature treatments close to 3000 °C, to provide the mobility for carbon atoms to rearrange and crystallize carbon into a graphite structure [17]. However, carbon produced from waste tires requires a simple sulfonation process followed by pyrolysis in a nitrogen atmosphere at 1100 °C [18–22]. Sulfonated tire-derived carbons have been tested as anodes in a half-coin cell lithium-ion battery configuration [18]. This process needs to be scaled up and demonstrated in a pouch full-cell configuration. The waste tires as a raw material also add a high value to the end-of-life tire rubber and provide a sustainable solution to the huge amount of waste tires generated worldwide on an annual basis. The carbon production cost and energy savings can be less than half of the graphite production cost, due to low temperature pyrolysis and the availability of low-cost raw waste tire rubber material. The demand to produce new tires to meet the needs of the rapidly growing auto industry increased to 4.3 percent every year to 2.9 billion tires in 2017 [22]. This paper demonstrates the successful scale up production of carbon material in kg quantities with 20–40 μm size particles from waste tires and reported its electrochemical performances as a promising anode in both coin and pouch LIB cell configurations. The micronized carbon materials coated with a thin layer of carbon using a chemical vapor deposition (CVD) approach and investigated the effects of carbon coating towards the improvements in electrochemical performances in LIB cells is also reported. In addition, a simple prelithiation strategy to improve the first cycle efficiency to 94% is also presented.

2. Experimental

2.1. Material Synthesis

Tire rubber crumb of size 1.5 was soaked at 70 °C for 12 h in a concentrated sulfuric acid bath which yielded sulfonated rubber slurry [18]. The slurry was then filtered, washed, and pyrolyzed at 1100 °C. The furnace was ramped up from room temperature to 400 °C at a rate of 1 °C/min and further increased to 1100 °C at a rate of 2 °C /min and held for 2 h and then allowed to cool down to room temperature under flowing nitrogen gas. Similarly, the sulfonated slurry was also heated to 1400 °C. The carbon sample was removed and ground to 20–40 μm size particles and used for further studies [21]. The carbon sample was also coated with a thin layer of carbon using Toluene as an aromatic hydrocarbon feedstock in a fluidized bed chemical vapor deposition (CVD) reactor (see Figure 1) at 800 °C under flowing nitrogen. The Toluene source was kept at 70 °C and bubbled with nitrogen gas at a flow rate of 2 L/min for 30 min. The thickness of the carbon coating was 10-nm. The commercial graphite of ~20 μm size particles was obtained from MTI Corp., Richmond, CA, USA and used as a standard electrode material for comparison.

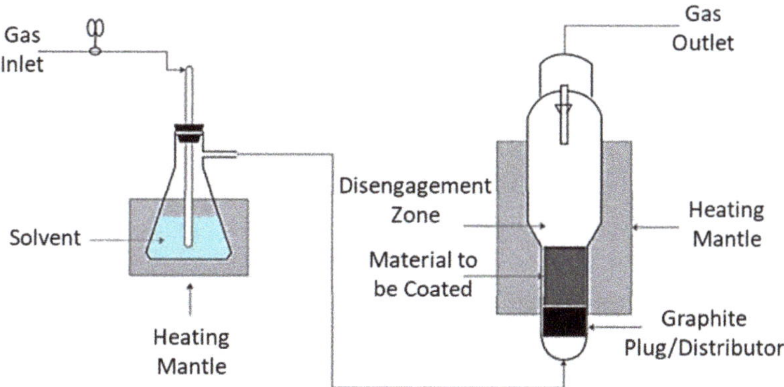

Figure 1. Schematic fluidized bed reactor for carbon coating.

2.2. Material Characterization

The carbon materials recovered from tires were characterized for their particle size, surface area, porosity, pore volume, morphology and crystallography. Nitrogen adsorption desorption isotherms were obtained with a Quantachrome NovaWin1000 Surface Area and Porosity Analyzer at 77.4 K. The specific surface area was determined by the Brunauer-Emmett-Teller (BET) method. The pore size distribution was obtained by the Barret-Joyner-Halenda (BJH) method. Raman spectra were collected with a Horiba LabRam HR using an excitation wavelength of 473 nm, a 600 g/mm grating and high spatial and spectral resolutions (800-mm monochromator). The X-ray powder diffraction (XRD) analysis was performed on a Rigaku Miniflex600 diffractometer with Cu Kα radiation. A Zeiss Merlin VP scanning electron microscope (SEM), Thornwood, NY, USA operated at 3 kV was used to characterize the surface morphologies of the samples. Interlayer distances of the carbons were determined by a Hitachi HD-2300 A scanning transmission electron microscopy (STEM), Clarksburg, MD, USA with a field emission source operated at 200 kV in bright-field imaging mode at a 2.1 Å resolution.

2.3. Electrochemical Characterization of Half Cells

The electrode was prepared using a doctor blade coating technique by casting a slurry containing 80 wt % tire derived carbon material or graphite, 10 wt % conductive carbon (C-NERGY Super C45,

Imerys Graphite & Carbon), and 10 wt % polyvinylidene difluoride (PVDF) binder in n-methyl-2-pyrrolidone (NMP) solvent onto a copper foil. The typical loading amount of active material was 2–2.5 mg/cm^2. The electrochemical performance and cycling tests were carried out in a CR2032 coin cell configuration. The coin cells were fabricated in an argon-filled glove box, using tire derived carbon with and without carbon coating or graphite electrode as the working electrode (10-mm diameter), metallic lithium foil as the counter electrode (13-mm diameter), Celgard 2325 as the separator and the solution of 1.0 M LiPF$_6$ in ethylene carbonate (EC)–dimethyl carbonate (DMC)–diethyl carbonate (DEC) (1:1:1 by volume) as the electrolyte. The coin cells were charge and discharge cycled using an Arbin BT2000 potentiostat/galvanostat multichannel system at room temperature under various constant current rates between C/5 (0.20 mA/cm^2) and C/50 (0.02 mA/cm^2) with the voltage cut off at 5 mV and 3.0 V.

2.4. Electrochemical Testing of Full Cells

The electrodes for pouch cells were made using a slot-die coater (Frontier Industrial Technology, Towanda, PA, USA) at Oak Ridge National Laboratory's (ORNL) Battery Manufacturing Facility (BMF) by following the coating procedures reported in detail elsewhere [23]. The cathode slurry contains 90% the active LiNi$_{0.5}$Mn$_{0.3}$Co$_{0.2}$O$_2$ (NMC532, Toda America, Battle Creek, MI, USA), 5% carbon black (Vulcan XR72C, Billerica, MA, USA), and 5% PVDF (PowerFlex, Kynar, Arkema Inc., Prussia, PA, USA) in NMP solvent. The anode slurry consists of 88% tire derived carbon material, 4% conductive carbon C45 and 8% PVDF binder (9300, Kureha America, New York, NY) in NMP solvent. The slurries were mixed separately in a high speed Ross planetary mixer (Charles Ross & Son Company, Hauppauge, NY, USA). The cathode and anode slurries were coated onto aluminum foil and copper foil respectively on a single side using the slot-die coater and transferred to a vacuum oven for drying at 120 °C overnight. A Celgard 2325 separator was used with an electrolyte composed of 1.2 M LiPF$_6$ in EC: DEC (3:7 wt ratio). The electrochemical performance characteristics and cycling tests were carried out in a pouch cell configuration. Pouch cells of dimensions: 84.4-mm long × 56.0-mm wide were fabricated in the dry room facility at BMF maintained at <0.5% (−53 °C dew point) humidity. The cycling performance tests were conducted at various charging and discharging current rates between C/5 and C/50 and the voltage between 2.0, 2.5 and 4.2 V using a potentiostat (VSP, BioLogic, Seyssinet-Pariset, France).

3. Results and Discussion

3.1. Material Characterization

Raman spectroscopy is being used as a nondestructive standard characterization tool to study crystalline, nanocrystalline, and amorphous carbons [24]. The position and width of the G and D bands and the intensity ratio of I$_D$/I$_G$ are widely used for characterizing the degree of graphitization and the quantity of defects present in the graphitic materials [24–26]. The Raman spectra of commercial graphite exhibit a huge sharp G-band frequency centering at 1585 cm^{-1} and a small weak D-band at 1368 cm^{-1} (Figure 2a). The G peak is due to the bond stretching modes of all pairs of sp^2 atoms in both rings and chains. The D peak is due to the breathing modes of sp^2 atoms in the rings. The intensity ratio of I$_D$/I$_G$ is calculated to be 0.123, with a small amount of disorder structure present in the commercial graphite compared to the perfect graphite where D band intensity is almost zero and therefore the ratio I$_D$/I$_G$ = 0. The measured full-width-at-half-maximum (FWHM) of G band is 19.2 cm^{-1} compared to the perfect graphite which has 14 cm^{-1} on a basal plane and 18 cm^{-1} in the prismatic edges [26].

The Raman spectrum of tire derived carbon sample pyrolyzed at 1100 °C and 1400 °C shown in Figure 1b,c illustrate well-resolved two broad G and D bands. The carbon pyrolyzed at 1100 °C exhibits a broad G band centering at 1599 cm^{-1} and D band at 1363 cm^{-1} (Figure 2b) while pyrolyzed at 1400 °C exhibits also a broad G band centering at 1604 cm^{-1} and D band at 1363 cm^{-1} (Figure 2c); however the peaks in the sample pyrolyzed at 1400 °C are relatively narrower than those that in 1100 °C. The average

I_D/I_G ratio obtained are 0.86 and 0.99 and FWHM(G) are 127 cm^{-1} and 89 cm^{-1} respectively for carbon pyrolyzed at both temperatures indicating a disorder crystalline structure [24,27]. The ratio of I_D/I_G is proportional to the number of aromatic rings. The increased relationship in the I_D/I_G ratio and G band position for the tire-derived carbons shows that more sp^2 amorphous carbon turns into nanocrystalline graphite at higher temperatures. The D band has a high sensitivity to disorder in the carbon material is usually attributed to the graphitization process and the formation of crystal growth [28]. This intensity of the D band increase through heat treatment could be related to the nano-crystallite graphite growth in the tire carbon caused by the defects and amorphous carbon.

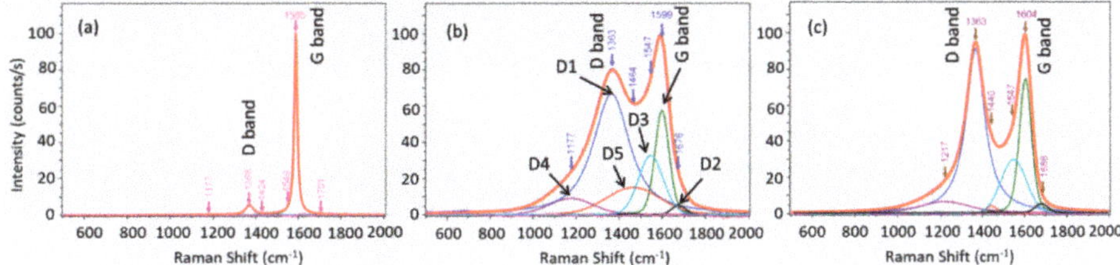

Figure 2. Raman spectra with peak deconvolution of (**a**) commercial graphite, and carbon produced from used tire with the pyrolysis temperatures of (**b**) 1100 °C, and (**c**) 1400 °C.

Deconvolution of the first order Raman spectrum of tire derived carbon pyrolyzed at 1100 °C was fitted with 6 curves attributed to the vibration modes of hexagonal carbon ring active at G-band (1599 cm^{-1}) represents an ideal graphitic structures, while D1 (1363 cm^{-1}), D2 (1676 cm^{-1}), D3 (1547 cm^{-1}), D4 (1177 cm^{-1}) and D5 (1464 cm^{-1}) bands relate to the defects in the disordered carbon structure (Figure 2b). Similarly, the tire derived carbon pyrolyzed at 1400 °C was also fitted with 6 curves: G-band (1603 cm^{-1}), D1 (1363 cm^{-1}), D2 (1688 cm^{-1}), D3 (1547 cm^{-1}), D4 (1217 cm^{-1}), and D5 (1440 cm^{-1}) (Figure 2c). The D1 band appears the most intense is normally assigned to the A$_{1g}$ symmetry in the graphitic lattice vibration mode. The origin of the D1 band in disordered carbon is due to the vibration mode of large number of small graphitic crystallites in the polycrystalline tire derived carbon atoms, at the edge of the graphene layers [29,30]. Another feature observed in the spectra is related to the line broadening of the G and D1 bands, indicating that the distribution of phonons activated by disorder corresponds to crystallites of different sizes. The intensity of the D1 curve increased by 36% and G band by 29% and the FWHM is reduced for D1 and G bands by 45% for the sample pyrolyzed at 1400 °C compared to 1100 °C illustrating higher temperatures pyrolysis graphitizes the nanocrystals in a short range order and making it as disordered graphitic. Other defect bands D3 and D4 bands in carbon pyrolyzed at 1100 °C are assigned to the amorphous carbon and hydrocarbons connected to the basic structural units of graphitic carbon respectively [31,32]. The intensity of D3 and D4 curves are decreased by 10% and 25% respectively when the pyrolysis temperature increased from 1100 °C to 1400 °C. The D5 band is detected at 1464 cm^{-1} for carbon pyrolyzed at 1100 °C and 1440 cm^{-1} for carbon pyrolyzed at 1400 °C assigned to hydrocarbons trapped in the carbon pores [33].

Figure 3a illustrates a low magnification scanning electron microscopy (SEM) image of tire derived carbon particles of various micron sizes and shapes with macro- and meso-pores visible on the sample surface. Figure 3b illustrates the high-resolution scanning transmission electron microscopy (HR-STEM) image of uncoated carbon pyrolyzed at 1100 °C, of both graphite layers and amorphous carbon phase pattern. Figure 3c illustrates a cross-sectional layered structure of carbon with the contrast profile of ~4.5 Å interlayer spacing. Figure 3d illustrates the selected area electron diffraction (SAED) patterns of uncoated carbon showing the coexistence of short range ordered graphite phase and amorphous regions. The fringes represent the stacks of graphene layers of polyaromatic structures [34].

These fringes are very short and completely disoriented. Figure 3e illustrates a HR-STEM image of carbon coated tire derived carbon, showing a uniform carbon coating of ~10 nm thickness on the surface of the carbon. The carbon coating enables the reduction of the surface area from 110 to 20 m^2/g.

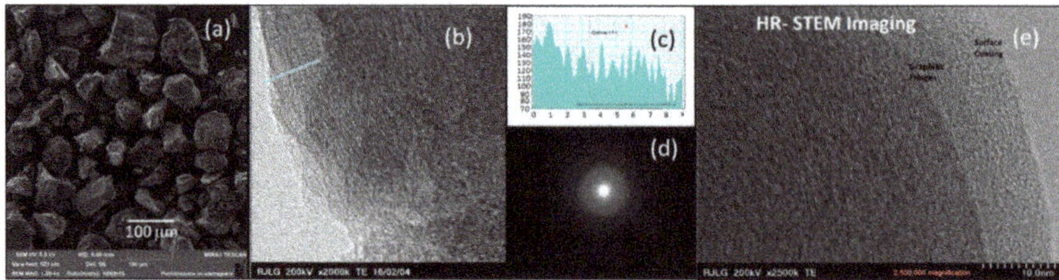

Figure 3. Tire derived carbon (**a**) scanning electron microscope (SEM) image, (**b**) high-resolution scanning transmission electron microscopy (HR-STEM) images showing both planar and cross-sectional layered structures, (**c**) the contrast profile of ~4.5 Å spacings, (**d**) Selected Area Electron Diffraction pattern (SAED), and (**e**) HR-STEM image of carbon coated tire derived carbon.

X-ray photoemission spectroscopy (XPS) spectra of the tire-derived carbons are shown in Figure 4. The C1s spectra in Figure 4a for both carbon samples of carbon coated and uncoated show a sharp peck at 284.8 eV, which is due to the sp^2 configuration. The fitting results for C1s spectra show that the Carbon peak is narrow consistent with C–C bond and there are no C–O or C=O groups on the surface. Figure 4 b shows the S2p scans of the samples. The peak at about 164 eV is related to the thiol group. A small doublet spike at around 165.2 eV could be due to the trace amount of sulfate group present. It was shown that the sulfate groups are completely removed from the sample when the pyrolysis temperature is increased further to 1600 °C [22]. The XPS elemental analysis for the samples is shown in Table 1. The carbon coating increases the carbon percentage by 4.7% while the O2s and S2p are reduced by 16.78% and 0.84%, respectively.

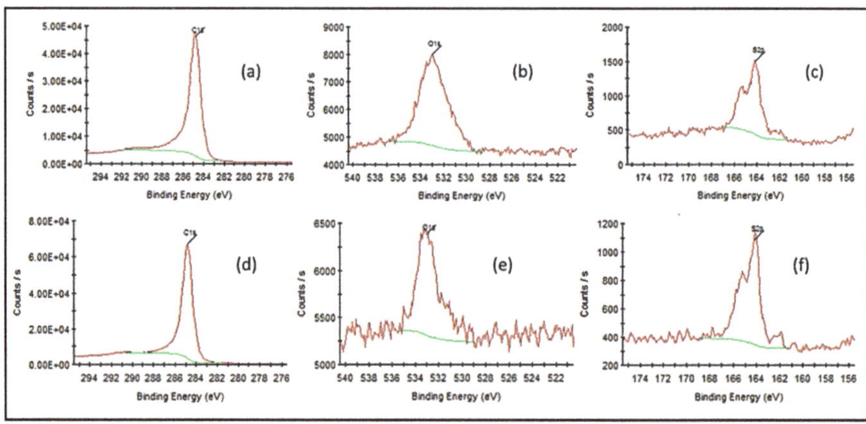

Figure 4. X-ray photoelectron spectroscopy (XPS) data of tire derived carbon, uncoated (top), and carbon coated (bottom) of scans of (**a,d**) C1s, (**b,e**) O2s, and (**c,f**) S2p.

Material characterization studies clearly demonstrate that the carbon material synthesized from waste tires has a unique disordered nanoporous structure with a large interlayer distance than that of graphite. This novel carbon material synthesis technology from tires demonstrates more sustainable and efficient use of resources.

Table 1. XPS surface concentration (in atomic percentages) of coated and uncoated tire derived carbon samples.

XPS Scans	Atomic %		Peak BE		FWHM eV		Area (P) CPS eV		SF	
	Coated	Uncoated	Coated	Uncoated	Coated	Uncoated	Coated	Uncoated	Coated	Uncoated
C1s	97.82	93.3	284.79	284.79	1.10	1.15	87,813.38	69,670.55	1	1
O2s	1.23	5.2	533.02	532.9	1.77	2.67	2701.93	9541.84	2.93	2.93
S2p	0.96	1.5	164.19	164.19	1.03	0.91	1546.58	2012.37	1.67	1.67

3.2. Electrochemical Measurements of Half Cells

Uncoated carbon materials of 38 μm and 20 μm particle sizes and carbon coated carbon materials of 20 μm particle size were utilized in half coin cells having lithium metal as the counter electrode in standard electrolyte solutions were cycled at various current rates are presented in Figure 5. The cell with larger particle size (38 μm) exhibited an initial first cycle coulombic efficiency of 51% at C/3 charge/discharge current rate. The first cycle coulombic efficiency was increased to 60% when the starting particle size was reduced to 20 μm and to 66% for carbon coated tire derived carbon samples. Further cycling exhibited the coulombic efficiency close to 100% for all the samples. The reversible capacity of uncoated carbon material is 247 mAh/g at C/5 charge/discharge current rate and it increased to 364 mAh/g for smaller size particles (20 μm) and it increased further to 437 mAh/g with carbon coated sample. Lowering the charge discharge cycling rate to C/7.5 and C/50 also increased the reversible capacity to 300 mAh/g and 430 mAh/g respectively corresponding to 65% and 92% respectively to the first cycle capacity. These results are interesting because the lithium utilization at C/50 current rate cycling is 92% which means the actual irreversible capacity loss is only 8%. This indicates that the apparent first cycle irreversible capacity loss was not due to the SEI formation rather it is related to the polarization caused by the poor conductivity of the carbon material. When the carbon was coated with a thin layer of carbon the capacity increased to 432 mAh/g at C/10 current rate. The increased capacity for the carbon coated sample can be related to the increased conductivity due to high carbon concentration of 98% and lower oxygen percentage of 1.23% as found from XPS studies (Table 1).

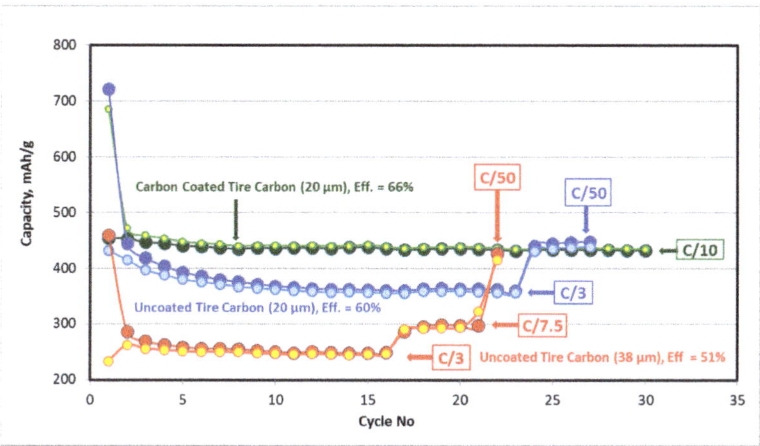

Figure 5. Cycling performance of lithium ion coin cells having lithium metal foil as the counter electrode and uncoated tire derived carbon of 38 μm and 20 μm size particle size and carbon coated tire derived carbon of 20 μm size particles cycled at various current rates.

Figure 6 compares charge capacities of uncoated and carbon coated tire derived carbon electrode material of 20 μm particle size pyrolyzed at 1100 °C and commercial graphite (20 μm) working electrodes in half coin cells. The uncoated carbon electrode exhibited 364 mAh/g capacity which is 3% higher capacity than that of the commercial graphite (353 mAh/g) while carbon coated exhibited 20% higher capacity of 437 mAh/g. The carbon coating also improved the first cycle columbic efficiency by 10%. The irreversible capacity and the surface area of the active material are directly proportional (higher capacity for low surface area of 20 m^2/g with carbon coated samples compared to the lower capacity for high surface area of 110 m^2/g with uncoated carbon samples).

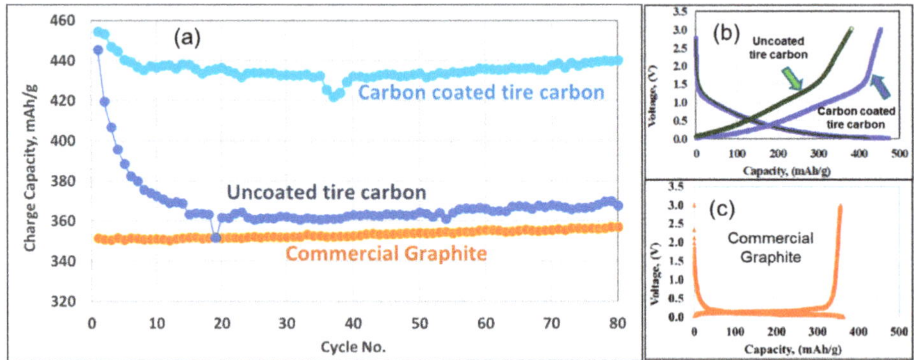

Figure 6. Comparison of cycling performances of uncoated and carbon coated tire derived carbon with commercial graphite as working electrodes vs. lithium metal counter electrode: (**a**) capacity plot, and (**b,c**) voltage profiles.

The carbon coating also reduced the interfacial resistance between the particles and improved the surface conductivity and thus enhanced the reversible capacity [11]. Furthermore, the carbon coating has reduced the voltage hysteresis in carbon coated samples (Figure 6b) compared to the uncoated carbon sample. The difference in the charge discharge voltage behavior of graphite (Figure 6c) vs. the carbons was found to correlate well with their unique structure [8]. The lithium insertion mechanism at higher voltage starting from ~1.2 V is more of an adsorption behavior while an intercalation process occurs at lower voltages [8,9]. The carbons vs. Li counter electrodes in coin-type cells reveal exceptionally good cycleability, with only moderate capacity fading.

3.3. Electrochemical Measurements of Pouch Full Cells

The use of tire derived carbon as a promising anode material in a lithium-ion battery pouch full cell configuration having NMC532 as cathode in standard electrolyte solutions has been demonstrated in Figure 7. The formation cycling tests were first performed at C/5, C/20, and C/50 charge and discharge current rates. After the formation cycling, the cells were cycled at C/5 charge discharge rate lost 0.12% capacity per cycles when the voltage cut off was limited to 2.0 V while the rest of the cycles lost only 0.08% capacity per cycle at 2.5 V limit. After 160 cycles at C/5, the cells were further cycled at lower current rate of C/50 showed an increase of 12% in reversible capacity. The cycling tests demonstrate carbon as a promising anode material in full cell configuration. Further electrochemical characterizations were carried out to design better cell chemistry.

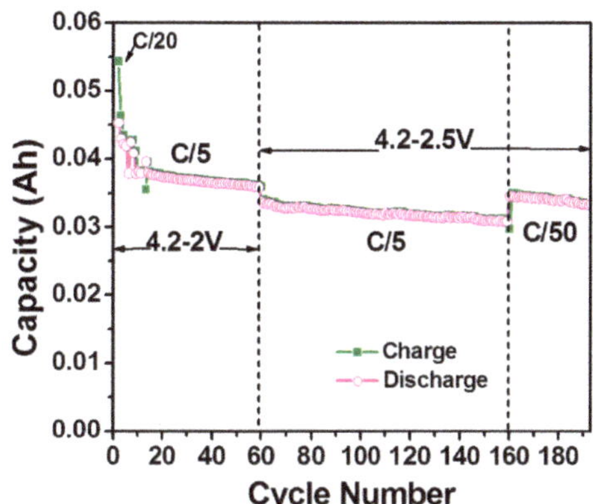

Figure 7. Cycling performance of the carbon anode material pyrolyzed at 1100 °C vs. NMC532 cathode at various current rates.

3.4. Prelithiation of Carbon Anodes

Prelithiation is an important strategy that attracts more and more interest to compensate for lithium loss in the formation cycle in lithium-ion batteries. Experimentally, it is more challenging to pre-lithiate anode materials than to pre-lithiate cathode materials since it requires a more reactive lithium source and the process is hard to control. There are some reports using stabilized lithium metal power (SLMP), however, this type of material is expensive and hard to disperse in NMP solution during the slurry preparation [35–37]. It will require a special complex procedure to apply pre-lithiated anodes.

The goal of using pre-lithiation treatment is to achieve first cycle efficiency of over 90% for the electrode materials. Here, the pre-lithiation is conducted by direct contacting of lithium metal with the electrolyte wetted with casted electrode, as shown in Figure 8a [38]. Different contact periods were studied to optimize the process conditions. Since the electrode was shorted after contact with Li, the assembled cell has almost 0 V OCV and at this time, first cycle efficiency would be that of the real second cycle since the first discharge is skipped. It is noticeable that all pre-lithiated cells have a very high efficiency (over 94%) from the beginning, compared to 60% efficiency of the cell without pre-lithiation. There are no major differences in the electrochemical performances during the first 22 cycles for different pre-lithiated electrodes shown in Figure 8. However, the long-term cycling test is needed to fully evaluate the influence of electrode properties.

In the meantime, new methods could be developed to pre-lithiate electrode materials, and they might be conducted in the future. Half cells with electrodes could be assembled and discharged to certain voltage, followed by disassembling of the cell and washing of the electrodes, then full cells can be made with pre-lithiated electrodes. This method can precisely control the amount of lithium pre-intercalated into the electrodes. Another approach is to use Li_3N as a lithium source additive when preparing the slurry and make a pouch cell [39]. Li_3N will decompose at 0.44 V during the formation cycle and the N_2 gas generated could be released before the final sealing of the pouch cell. Successful implementation of this technology would result in a significant reduction in carbon emissions and possibly result in a lower-cost higher performance lithium-ion battery [40–45].

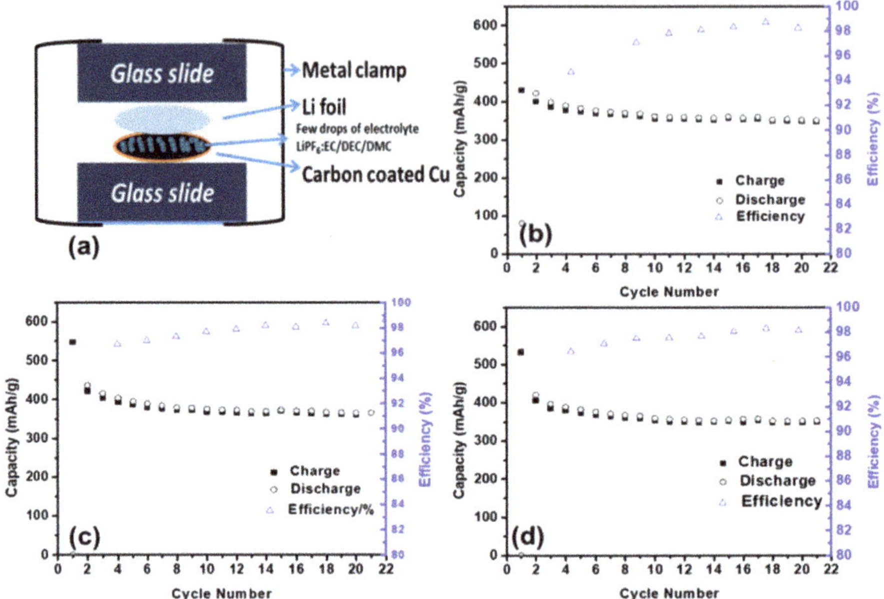

Figure 8. (**a**) Schematic of direct contact pre-lithiation process, electrochemical test data with (**b**) 20 min, (**c**) 19 h, (**d**) 23 h contact.

4. Conclusions

In summary, battery grade tire-derived carbon material was successfully prepared in high purity in large quantities and demonstrated as a potential anode material for LIBs. The combustion of waste tire rubber produces lower fossil CO_2 emissions per unit of energy (136.51 gCO_2/kW·h) compared to other conventional fossil fuels such as coal or fuel–oil, usually used in power generation [40]. The surface coating of the carbon improved the reversible lithium capacity by 20% and yielded 437 mAh/g, which is 20% higher than that of the commercial graphite. A novel prelithiation approach also yielded the first cycle efficiency of 94%. This novel green process can promote environmental sustainability to add value to the waste tires and benefit the battery industry to solve future energy crises.

Author Contributions: J.S.G. planned the experiments and wrote the paper, R.J.L., and A.M.L. identified the research questions, J.L.W., and S.L.W. carried out scale up carbon material synthesis, Y.L., J.L., and K.A. carried out the experimental work, A.K.N. and M.P.P. came up with the concept and analyzed the data.

Acknowledgments: The research (Y.L., M.P.P.) was supported by the U.S. Department of Energy (DOE), Office of Science, Office of Basic Energy Sciences, Materials Sciences and Engineering Division. Part of research at RJ Lee Group was funded by internal funding.

Conflicts of Interest: The authors declare no competing financial interest.

References

1. Diouf, B.; Pode, R. Potential of lithium-ion batteries in renewable energy. *Renew. Energy* **2015**, *76*, 375–380. [CrossRef]
2. Novák, P.; Goers, D.; Spahr, M.E. Carbon Materials in Lithium-Ion Batteries. In *Carbons for Electrochemical Energy Storage Systems*; Beguin, F., Frackowiak, E., Eds.; CRC Press: Boca Raton, FL, USA, 2010; Chapter 7, pp. 263–328, ISBN 978-1-4200-5307-4.
3. Battery University, Safety of Lithium-ion Batteries. 2017. Available online: http://batteryuniversity.com/learn/article/safety_of_lithium_ion_batteries (accessed on 2 June 2017).
4. Hori, H.; Shikano, M.; Kobayashi, H.; Koike, S.; Sakaebe, H.; Saito, Y.; Tatsumi, K.; Yoshikawa, H.; Ikenaga, E. Analysis of hard carbon for lithium-ion batteries by hard X-ray photoelectron spectroscopy. *J. Power Sources* **2013**, *242*, 844–847. [CrossRef]
5. Shi, H. Coke vs. graphite as anodes for lithium-ion batteries. *J. Power Sources* **1998**, *75*, 64–72. [CrossRef]
6. Buiel, E.; Dahn, J.R. Li-insertion in hard carbon anode materials for Li-ion batteries. *Electrochim. Acta* **1999**, *45*, 121–130. [CrossRef]
7. Gnanaraj, J.S.; Levi, M.D.; Levi, E.; Salitra, G.; Aurbach, D.; Fischer, J.E.; Claye, A. Comparison Between the Electrochemical Behavior of Disordered Carbons and Graphite Electrodes in Connection with Their Structure. *J. Electrochem. Soc.* **2001**, *148*, A525–A536. [CrossRef]
8. Dahn, J.R.; Zheng, T.; Liu, Y.; Xue, J.S. Mechanisms for Lithium Insertion in Carbonaceous Materials. *Science* **1995**, *270*, 590–593. [CrossRef]
9. Peled, E. The Electrochemical Behavior of Alkali and Alkaline Earth Metals in Nonaqueous Battery Systems—The Solid Electrolyte Interphase Model. *J. Electrochem. Soc.* **1979**, *126*, 2047–2051. [CrossRef]
10. Gnanaraj, J.S.; Thompson, R.W.; Iaconatti, S.N.; DiCarlo, J.F.; Abraham, K.M. Formation and Growth of Surface Films on Graphitic Anode Materials for Li-Ion Batteries. *Electrochem. Solid-State Lett.* **2005**, *8*, A128–A132. [CrossRef]
11. Lee, J.H.; Lee, H.Y.; Oh, S.M.; Lee, S.J.; Lee, K.Y.; Lee, S.M. Effect of carbon coating on electrochemical performance of hard carbons as anode materials for lithium-ion batteries. *J. Power Sources* **2007**, *166*, 250–254. [CrossRef]
12. Guerin, K.; Fevrier-Bouvier, A.; Flandrois, S.; Simon, B.; Biensan, P. On the irreversible capacities of disordered carbons in lithium-ion rechargeable batteries. *Electrochim. Acta* **2000**, *45*, 1607–1615. [CrossRef]
13. Kumar, T.P.; Kumari, T.S.D.; Stephan, A.M. Carbonaceous anode materials for lithium-ion batteries-the road ahead. *J. Ind. Inst. Sci.* **2009**, *89*, 393–424.
14. Xing, W.; Dahn, J.R. Study of irreversible capacities for Li insertion in hard and graphitic carbons. *J. Electrochem. Soc.* **1997**, *144*, 1195–1201. [CrossRef]
15. Xu, K. Electrolytes and Interphases in Li-Ion Batteries and Beyond. *Chem. Rev.* **2014**, *114*, 11503–11618. [CrossRef] [PubMed]
16. Buiel, E.; Dahn, J.R. Reduction of the Irreversible Capacity in Hard-Carbon Anode Materials Prepared from Sucrose for Li-Ion Batteries. *J. Electrochem. Soc.* **1998**, *145*, 1977–1981. [CrossRef]
17. Tamashausky, A.V. *An Introduction to Synthetic Graphite*; Asbury Graphite Mills Inc.: Pittsburgh, PA, USA, 2006; Available online: https://asbury.com/pdf/SyntheticGraphitePartI.pdf (accessed on 12 April 2018).
18. Naskar, A.K.; Bi, Z.; Li, Y.; Akato, S.K.; Saha, D.; Chi, M.; Bridges, C.A.; Paranthaman, M.P. Tailored recovery of carbons from waste tires for enhanced performance as anodes in lithium-ion batteries. *RSC Adv.* **2014**, *4*, 38213–38221. [CrossRef]
19. Naskar, A.K.; Paranthaman, M.P.; Bi, Z. Pyrolytic Carbon Black Composite and Method of Making the Same. U.S. Patent 9,441,113 B2, 13 September 2016.
20. Boota, M.; Paranthaman, M.P.; Naskar, A.K.; Li, Y.; Akato, K.; Gogotsi, Y. Waste Tire Derived Carbon—Polymer Composite Paper as Pseudocapacitive Electrode with Long Cycle Life. *ChemSusChem* **2015**, *8*, 3576–3581. [CrossRef] [PubMed]
21. Gnanaraj, J.S.; Lee, R.J.; Levine, A.M.; Wistrom, J.L.; Wistrom, S.L.; Li, Y.; Li, J.; Naskar, A.K.; Paranthaman, M.P. A Low-Cost Carbon Composite Anode Material from Recycled Waste Tires for Lithium-Ion Batteries. In Proceedings of the MRS Spring Meeting & Exhibit about Symposium EE4—Electrode Materials and Electrolytes for Lithium and Sodium Ion Batteries, Phoenix, AZ, USA, 28 March–1 April 2016. Paper EE4.5.06.

22. Li, Y.; Paranthaman, M.P.; Akato, K.; Naskar, A.K.; Levine, A.M.; Lee, R.J.; Kim, S.; Zhang, J.; Dai, S.; Manthiram, A. Tire-derived Carbon Composite Anodes for Sodium-ion Batteries. *J. Power Sources* **2016**, *316*, 232–238. [CrossRef]

23. Li, J.; Daniel, C.; An, S.J.; Wood, D. Evaluation residual moisture in lithium-ion battery electrodes and its effect on electrode performance. *MRS Adv.* **2016**, *1*, 1029–1035. [CrossRef]

24. Ferrari, A.C.; Robertson, J. Interpretation of Raman spectra of disordered and amorphous carbon. *Phys. Rev. B* **2000**, *61*, 14095. [CrossRef]

25. Manoj, B.; Kunjomana, A.G. Chemical leaching of an Indian bituminous coal and characterization of the products by vibrational spectroscopic techniques. *Int. J. Miner. Metall. Mater.* **2012**, *19*, 279–283. [CrossRef]

26. Maslova, O.A.; Ammar, M.R.; Guimbretière, G.; Rouzaud, J.-N.; Simon, P. Determination of crystallite size in polished graphitized carbon by Raman spectroscopy. *Phys. Rev. B* **2012**, *86*, 134205. [CrossRef]

27. Ferrari, A.C.; Robertson, J. Raman spectroscopy of amorphous, nanostructured, diamond-like carbon, and nanodiamond. *Phil. Trans. R. Soc. Lond. A* **2004**, *362*, 2477–2512. [CrossRef] [PubMed]

28. Deldicque, D.; Rouzaud, J.-N.; Velde, B.; Raman, A. A Raman—HRTEM study of the carbonization of wood: A new Raman-based paleothermometer dedicated to archaeometry. *Carbon* **2016**, *102*, 319–329. [CrossRef]

29. Manoj, B.; Kunjomana, A.G. Study of Stacking Structure of Amorphous Carbon by X-Ray Diffraction Technique. *Int. J. Electrochem. Sci.* **2012**, *7*, 3127–3134.

30. Zhou, Q.; Zhao, Z.; Zhang, Y.; Meng, B.; Zhou, A.; Qiu, J. Graphene Sheets from Graphitized Anthracite Coal: Preparation, Decoration, and Application. *Energy Fuels* **2012**, *26*, 5186–5192. [CrossRef]

31. Sadezky, A.; Muckenhuber, H.; Grothe, H.; Niessner, R.; Pöschl, U. Raman spectroscopy of soot and related carbonaceous materials: Spectral analysis and structural information. *Carbon* **2005**, *43*, 1731–1742. [CrossRef]

32. Pawlyta, M.; Rouzaud, J.N.; Duber, S. Raman microspectroscopy characterization of carbon black: Spectral analysis and structural information. *Carbon* **2015**, *84*, 479–490. [CrossRef]

33. Ramya, A.V.; Manoj, B.; Mohan, A.N. Extraction and characterization of wrinkled graphene nanolayers from commercial graphite. *Asian J. Chem.* **2016**, *28*, 1031–1034. [CrossRef]

34. Rosalin, F. Crystallite Growth in Graphitizing and Non-Graphitizing Carbons. *Proc. R. Soc. Lond.* **1951**, *209*, 196–218. [CrossRef]

35. Forney, M.W.; Ganter, M.J.; Staub, J.W.; Ridgley, R.D.; Land, B.J. Prelithiation of Silicon–Carbon Nanotube Anodes for Lithium Ion Batteries by Stabilized Lithium Metal Powder (SLMP). *Nano Lett.* **2013**, *13*, 4158–4163. [CrossRef] [PubMed]

36. Wang, Z.; Fu, Y.; Zhang, Z.; Yuan, S.; Amine, K.; Battaglia, V.; Liu, G. Application of Stabilized Lithium Metal Powder (SLMP) in graphite anode—A high efficient prelithiation method for lithium-ion batteries. *J. Power Sources* **2014**, *260*, 57–61. [CrossRef]

37. Jarvis, C.R.; Lain, M.J.; Yakovleva, M.V.; Gao, Y. A prelithiated carbon anode for lithium-ion battery applications. *J. Power Sources* **2006**, *162*, 800–802. [CrossRef]

38. Liu, N.; Hu, L.; McDowell, M.T.; Jackson, A.; Cui, Y. Prelithiated Silicon Nanowires as an Anode for Lithium Ion Batteries. *ACS Nano* **2011**, *5*, 6487–6493. [CrossRef] [PubMed]

39. Park, K.; Yu, B.-C.; Goodenough, J.B. Cathode Additive for High-Energy-Density Lithium-Ion Batteries. *Adv. Energy Mater.* **2016**, *6*, 2534. [CrossRef]

40. Martinez, J.D.; Puy, N.; Murillo, R.; Garcia, T.; Navarro, M.V.; Mastral, A.M. Waste tyre pyrolysis—A review. *Renew. Sustain. Energy Rev.* **2013**, *23*, 179–213. [CrossRef]

41. Li, J.; Qian, Y.; Wang, L.; He, X. Nitrogen-Doped Carbon for Red Phosphorous Based Anode Materials for Lithium Ion Batteries. *Materials* **2018**, *11*, 134. [CrossRef] [PubMed]

42. Zhou, Y.; Wang, Q.; Zhu, X.; Jiang, F. Three-Dimensional SnS Decorated Carbon Nano-Networks as Anode Materials for Lithium and Sodium Ion Batteries. *Nanomaterials* **2018**, *8*, 135. [CrossRef] [PubMed]

43. Yuan, G.; Xiang, J.; Jin, H.; Wu, L.; Jin, Y.; Zhao, Y. Anchoring ZnO Nanoparticles in Nitrogen-Doped Graphene Sheets as a High-Performance Anode Material for Lithium-Ion Batteries. *Materials* **2018**, *11*, 96. [CrossRef] [PubMed]

44. Li, H.; Liu, Z.; Yang, S.; Zhao, Y.; Feng, Y.; Bakenov, Z.; Zhang, C.; Yin, F. Facile Synthesis of ZnO Nanoparticles on Nitrogen-Doped Carbon Nanotubes as High-Performance Anode Material for Lithium-Ion Batteries. *Materials* **2017**, *10*, 1102. [CrossRef] [PubMed]
45. Hwang, J.; Ihm, J.; Lee, K.-R.; Kim, S. Computational Evaluation of Amorphous Carbon Coating for Durable Silicon Anodes for Lithium-Ion Batteries. *Nanomaterials* **2015**, *5*, 1654–1666. [CrossRef] [PubMed]

Article

Renewable Energy and Economic Growth: Evidence from European Countries

Stamatios Ntanos [1], Michalis Skordoulis [2,*], Grigorios L. Kyriakopoulos [3], Garyfallos Arabatzis [2], Miltiadis Chalikias [4], Spyros Galatsidas [2], Athanasios Batzios [5] and Apostolia Katsarou [6]

[1] Department of Business Administration, School of Business, Economics and Social Sciences, University of West Attica, 12244 Egaleo, Greece; sdanos@ath.forthnet.gr

[2] Department of Forestry and Management of the Environment and Natural Resources, School of Agricultural and Forestry Sciences, Democritus University of Thrace, 68200 Orestiada, Greece; garamp@fmenr.duth.gr (G.A.); sgalatsi@fmenr.duth.gr (S.G.)

[3] School of Electrical and Computer Engineering, National Technical University of Athens, 15780 Zografou, Greece; gregkyr@chemeng.ntua.gr

[4] Department of Tourism Management, School of Business, Economics and Social Sciences, University of West Attica, 12244 Egaleo, Greece; mchalikias@hotmail.com

[5] Department of Agricultural Economics, School of Agriculture, Faculty of Agriculture, Forestry and Natural Environment, Aristotle University of Thessaloniki, 54124 Thessaloniki, Greece; thanos.batzios@gmail.com

[6] Department of Education Sciences in Early Childhood, Democritus University of Thrace, 68100 Alexandroupolis, Greece; aposkats2@psed.duth.gr

* Correspondence: mskordoulis@gmail.com

Received: 10 July 2018; Accepted: 24 July 2018; Published: 26 July 2018

Abstract: This paper aims at examining the relationship between energy consumption deriving from renewable energy sources, and countries' economic growth expressed as GDP per capita concerning 25 European countries. The used dataset involves European countries' data for the period from 2007 to 2016. The statistical analysis is based on descriptive statistics, cluster analysis, and autoregressive distributed lag (ARDL), and reveals that all variables are related; this suggests a correlation between the dependent variable of GDP and the independents of renewable energy sources (RES) and Non-RES energy consumption, gross fixed capital formation, and labor force in the long-run. Furthermore, the results show that there is a higher correlation between RES' consumption and the economic growth of countries of higher GDP than with those of lower GDP. The obtained results are consistent with other papers reviewed in this study.

Keywords: renewable energy; energy consumption; Gross Domestic Product; economic growth

1. Introduction

During the last decades, a continuing shift from conventional to non-conventional sources of energy has been taking place [1,2]. Due to negative environmental effects of conventional energy forms of production and usage, and the finite yields of conventional sources of energy, the need for renewable energy sources (RESs) usage is becoming urgent [3]. Thus, special policies are developed to further promote the expansion of RESs. A critical environmental target of the European Union is to cover 20% of its energy needs with RESs until 2020. In a global context of analysis, about 19% of total energy usage is produced from RESs, while the strategic planning is to increase their usage by 50% by 2050 [4].

Relative to the abovementioned shift to RESs, a key question that arises is how their usage is valued under economic constraints, a question that has motivated a plethora of research interest worldwide [4]. Thus, there is ongoing research developed to investigate the relationship between RES and economic growth mainly attributed to the gross domestic product (GDP) [4].

The relevant literature nexus basically contains studies grouped on geographical distribution [2,4–10]. In most of the publications, methodological analysis included statistical models and case controls. Almost all the reviewed publications unveil a relationship that is developed between the level of national economic growth and the quantities of RESs used, especially over the long-run [4–11]. Several publications also investigate the association between economic development and RES-produced energy, to determine those variables that stimulate the others [4–11]. In most of the cases GDP is the dependent variable, while RES consumption, gross fixed capital formation (GFCF), and labor force are the main commonly examined predictors. A positive correlation is also recorded between GDP, RES usage, GFCF, and labor force in most of the cases [4–11]. These results emphasize the significant role of RES to economic growth.

Our research examines the interaction between economic growth and the energy produced by RES. Data concerning 25 European countries for the period between 2007 and 2016 is statistically analyzed; our main results indicate that there are two main groups of countries classified on RES consumption. RES consumption seems to be correlated with the economic growth of the countries examined. However, this correlation is highly bounded on countries with higher GDP.

2. Literature Review

2.1. Non-European Countries

Salim et al. [12] perform data analysis on OECD countries for the period 1980–2011. Their analysis reveals a long-run relationship that links energy consumption (both green and conventional) with industrial production and economic growth. In addition, evidence of a two-way relationship between conventional energy consumption and GDP growth was found in the short run. According to these results, the expansion of RESs is a viable solution to tackle energy security and climate change, while the gradual replacement of fossil fuels with RESs promotes a sustainable energy economy [12]. Under the same research context, Apergis and Payne [13] use data from 80 countries. They develop a statistical model where GDP is the dependent variable, while the main explanatory variables are electricity consumption from RESs and from conventional sources, among several macroeconomic variables. The results reveal a two-way relationship between RES consumption, conventional energy, and GDP, both in the short and long run. There is also a two-way short-run relationship between renewable and non-renewable forms of energy, which means that it is feasible to swap between energy forms.

In a similar publication, the effect of RESs on economic growth, including biomass, geothermal, wind and solar energyd and hydroelectricity is examined in 38 countries, which are classified into three groups [14]. For the 24 countries of the first group, RESs are significantly correlated to economic growth. Concerning the countries belonging to the second group, energy consumption is negatively associated to economic development. For the other countries of the third group, there is no impact of RES consumption on economic growth.

In the same research field, Inglesi-Lotz [15] analyzes panel data. In this dataset, GDP is the dependent variable, while the following independent variables are namely applied: RES consumption, RES percentage on the examined countries' energy mix, labor force, research and development costs, and capital formation. The results of the statistical analysis show that a 1% increase in RES consumption can increase the total GDP by 0.105%, while a 1% increase in the share of RESs in a country's energy mix can increase total GDP by 0.089%.

In the search of RES consumption determinants, Omri and Nguyen [16] use data concerning 64 countries for the years between 1990 and 2011. In all of the countries' clusters, results point out that environmental degradation is a significant variable. Moreover, the price of oil impacts the consumption of RESs. Lastly, changes in GDP affect the consumption of RESs only for very high and very low-income countries, while changes in trade openings have a statistically significant effect on the RES consumption in all the examined countries except for those of high income.

In the same scientific field, Al-Mulali et al. [17] conclude that RES consumption plays a key role in promoting economic growth for most of the countries as RESs are a net source of energy, having minimal environmental impacts. In addition, RES energy is necessary because it contributes to energy security since those countries are loosely bounded on imported fossil fuels [17].

Another paper focuses on energy consumption from renewable and non-renewable sources of energy in relation to international trade [18]. This survey examines the correlation between energy production and consumption from both renewable and non-renewable sources and international trade, examining 69 countries for the period between 1980 and 2010. The results contain indicators of a two-way causality between production and trade. These results show that any changes in trade can influence the economic outcome and, accordingly, changes in production can affect trade. In the short run, no link is reported between GDP and green energy consumption. However, there is a long-run two-way relationship between the variables.

Sadorsky [19] develops economic models for studying the consumption of energy from RES in relation to income, especially among emerging economies. The researcher perceives a relationship between income and green energy consumption as a prerequisite for the design of an effective energy policy. The model uses data from 18 emerging economies and focuses on the per capita consumption of energy from RESs and the per capita income. According to the analysis, a rise in GDP leads to an increase in energy consumption from renewables. High income is identified as significant by the model over the long run, suggesting that slight changes to GDP highly impact the consumption of energy from renewable sources. In conclusion, the results show that a slight increase in per capita income is directly causing a significant increase in RES consumption in the case of emerging economies.

By looking at the correlation between renewable energy and GDP in the case of the United States, the consumption is directly related to biomass-made energy [20]. No other relationship exists between real GDP and all other renewable energy sources examined, namely total RES consumption, geothermal energy consumption, hydropower consumption, and biomass energy consumption. The results of this study conclude that the generation of energy from waste is of utmost importance [20].

Fang [21] examines the impact of RES consumption on China's GDP. By using regression techniques, the analysis concludes that a 1% increase in RES consumption leads to 0.162% GDP growth, 0.44% for rural households and 0.368% for urban households. However, it is suggested that the expansion of the RES share inevitably provides politico-economic hurdles. Obstacles are attributed to the environmental impacts of fossil fuels use, which contain costs for society, such as human health degradation, infrastructure, decline in forests and fisheries, and increased spending associated with tackling climate change [21]. Concerning the economic stance, the researcher states that when all the external economies are considered, some RESs, especially wind power, are proven cheaper than conventional sources. Regarding political barriers, lack of coordination, unsupportive policy encouragement, as well as government grants that are inadequately developed in terms of geographical coverage or sound research contexts of analysis are reported [21].

In another study for the Chinese context, the relationship between RES consumption and economic growth is examined for the period 1977–2011 [22]. The results show a two-way relationship in the long-run between the RES consumption and GDP. This finding suggests that China's growing economy is conducive to the development of the renewable energy sector, offering promising contribution to further economic growth [22].

Shahbaz et al. [11] explore the relationship between RES consumption and GDP in Pakistan. Their analysis uses an econometric log-linear distribution model where the per capita GDP is the independent variable, while the per capita energy consumption, labor force, and gross fixed capital formation are the independent ones. Long-run results show that the model of renewable energy consumption can stimulate economic growth stronger than the opposite model, which shows that economic growth motivates the use of RESs. According to the researchers, the government should propose an integrated energy policy towards the exploitation of new RESs. Furthermore, local governments should promote investments in RESs to overcome the national energy crisis [11].

On the contrary, a similar study using regression models for the case of Turkey did not report any impact from RES-produced electricity consumption on GDP [23]. However, the non-renewable electricity consumption variable was statistically significant in the long-run, as a 1% increase in energy consumption led to a minor increase in GDP. Researchers therefore confirm the country's dependence on conventional energy sources and propose an expanding investment in conventional energy sources, by reducing the contributing share of RESs at the national energy mix [23].

Under this non-European context, another significant interdisciplinary research shows that the existing regulatory framework for the electricity and renewables sectors is an extremely important parameter that especially affects the role of regulatory agencies in Northern Africa and Middle East countries, under the promotion by the EU [24]. In this research, authors stress that in countries where an independent regulator operates, increased regulatory credibility exists in comparison to countries in which such a body does not exist. Besides, in those countries where regulatory framework is limiting administrative expropriation it, consequently, creates a more suitable environment for attracting investments in electricity production using renewables [24].

In another research work, the strategic environmental assessment (SEA) has been introduced as a comprehensive framework for the assessment of policies and plans [25]. The authors denote that suitable energy models, ecological assessment models, and multi-criteria approaches exist with immense potential for interconnection, which could be advanced into powerful SEA tools for integrated policy assessment [25].

In a global context [26], it is reported that since developed economies are confronting a persistent decline in economic growth, it is of utmost importance for the technological spur of renewables to support the energy autonomy of national energy mixes without creating irreversible environmental pollution. Particularly, the use of the Hurlin-Granger causality test shows that there is a unidirectional causality linkage from agricultural production to electricity consumption for non-coastal regions, and furthermore, there is a bidirectional relationship between agricultural electricity consumption and output for numerous coastal regions.

Recently, Sutter [27] organized the main concerns to which Americans are mindful of energy efficiency as accordingly: first, for their cars, appliances, homes, and machinery, and second, as the mandating of the use of renewable fuels in electricity production. The main representative foundation upon renewable energy resources utilization in the US is the Department of Energy (DOE).

In this respect, Cherukupalli [28] reports that though a dynamic progress in all areas of electro-technology is achieved in developed countries, a significant part of the population has low or no access to electricity. The author [28] also stresses that life and economic activity in large areas of the developing world remains unchanged, as access to reliable electricity is an extremely important aspect of enjoyment, benefits, and modern conveniences.

Finally, another study was conducted in the US context by Haerer and Pratson [29] who investigate the employment trends in the US electricity sector for the period 2008–2012. The authors denote that even though electricity production that was generated from coal declined, the percentage of jobs lost were offset two times higher by the increased employment rates from the expansion of RES-operated industries, especially, natural gas, solar, and wind. Furthermore, US sustain a well-developed, life cycle analysis-based energy sector, along with advanced electricity supply chain. On the other hand, the adaptability of the aforementioned outcomes to developing economies-where switching fuel types is a complex policy, or there exists RES-deprivation-shows slow adaptability and ineffective economies-of-scale achievement.

2.2. European Countries

In a recent publication, wind energy (terrestrial and offshore), hydro (small and large), and nuclear energy are the most preferable technologies [30]. The results identify a failure to reach emission reduction targets. To achieve this, it is necessary to increase the share of all zero emission technologies by 1–2%. It is stressed that offshore wind energy is the largest renewable energy technology (20.28%). Relevant studies conclude that an efficient, safe, and environmentally friendly energy future in the EU is closely related to RESs [30,31].

Another study investigates the consumption of fossil fuels and RESs, greenhouse gases, and economic growth, using EU countries' data, unveiling a relationship between greenhouse emissions and GDP [32]. Moreover, economic growth cannot affect emissions reduction, since from one point onwards, there is an upsurge in the phenomenon.

A study about developing European countries identifies an overall long-run relationship between energy consumption and GDP, while a short-run two-way relationship is reported for Hungary, Poland, Turkey, and Romania [33]. Consequently, RES consumption plays a determining role in supporting economic growth for numerous countries [17].

Another important aspect is the role of commerce to RES consumption promotion [16,18]. In emerging economies, it is reported that a slow increase in the per capita GDP can lead to a significant increase in RES consumption [19].

In the European context, Borkowska and Klimczak [34] argue that like other network industries (telecommunication, gas, rail), electricity was transformed from a monopoly to an imperfectly competitive market. The authors examine the electricity reform in Poland, showing that the reform failed in key aspects, such as industry restructuring, the creation of a transparent wholesale spot energy market, and the foundation of other institutions necessary to create clear rules to achieve reasonably competitive performance results. The lack of a sound regulatory implementation was the cause of the predominance of state ownership in the electricity industry and strong politicization of the industry [34].

In the Greek context, Tellidou and Bakirtzis [35] employ a research study upon the energy market performance to exercise the monopoly power. Test results on a two-node power system with two competing generator-agents demonstrated the negative effects of a monopoly [35].

Furthermore, Philippou et al. [36] first explore the opening of the electricity market in Greece. A decade later, Andrianesis et al. [37] denote that Greece's electricity market can be divided in two zones, due to the generation-consumption system configuration that creates an important transmission barrier from north to south. The authors also used a simplified model of the Greek electricity system that includes only the thermal power plants. An analysis upon the commodities of energy and ancillary services under the marginal pricing approach is presented [37]. A zonal pricing approach for energy provides the right incentive for the installation of new generation near consumption, if the zonal configuration reflected the actual system and operational conditions. Authors extend the zonal approach to include time response-based ancillary services, which were commodities that can be traded in the day-ahead market and are co-optimized with energy. Focusing on the day-ahead scheduling (DAS) market problem, which is formulated as a security-constrained unit-commitment problem, its objective is to co-optimize energy and reserves by considering the generation units' commitment costs. A dual analysis of the problem and calculation of shadow prices offers useful insight into how prices for each commodity are set in the presence of binding resource, transmission, or zonal reserve constraints.

Finally, in a technological applicability of RESs in the European context, Chatzisideris et al. [38] focus on organic photovoltaics (OPV), which are an emerging thin-film PV technology that promises to greatly improve the environmental and economic performances of PV technologies. While generic results are applied to all PV technologies, it is shown that PV systems installed in Greek houses, perform economically better than those in Denmark. Focusing on organic PV systems developed in an industrial-scale cost setting (1.53 €/Wp), it is argued that they deliver significant electricity bill savings for residential houses in Greece (38%) under current conditions, while they may not be sufficiently attractive for residential houses in Denmark (6.5%) mainly due to the different PV regulatory schemes. It is also noteworthy that investors interested in renewable energy technologies need to pursue scaling up the manufacturing capacity of OPV technologies, as well as assess a large number of countries to identify and prioritize financially attractive settings for PV self-consumption [38].

3. Data and Methods

To examine RES' consumption correlation with countries' economic growth, Eurostat's latest dataset (updated in May 2018) concerning 25 European countries for the period between 2007 and 2016 is used [39]. Due to the risk of different measurement methodology that can lead to biased results, other data sources, such as OECD or IEA data, are not inserted into our dataset.

Using IBM Corp's SPSS V.20 software, this study employs a hierarchical cluster analysis by using Ward's method in order to classify the examined countries based on their GDP and RES consumption. Cluster analysis methodology is used in other similar studies as well [2]. Furthermore, to examine the relationship between the variables of our interest, a log-linear specification is applied by using an error-corrected autoregressive distributed lag (ARDL). This is a common methodology proposed in similar studies [11,40]. The regression part of the analysis was carried out by using Stata Corp's STATA V.13 software.

The abovementioned log-linear specification investigates the relationship between variables concerning the dependent variable of gross domestic product (GDP) and the independents of RES and Non-RES energy consumption, gross fixed capital formation (GFCF), and labor force in the long-run. Such models, where GDP is set to be predicted by RES and Non-RES' consumption, GFCF and labor force are used in other studies of the same scientific field as well [4–11,40].

4. Results and Discussion

Initially, we focussed on energy consumption of the examined countries for the 10-year period of 2007–2016. Figure 1 shows that there is an ongoing shift from fossil fuels to RESs, which depicts the implementation of the European strategy for gradual independence from fossil fuels.

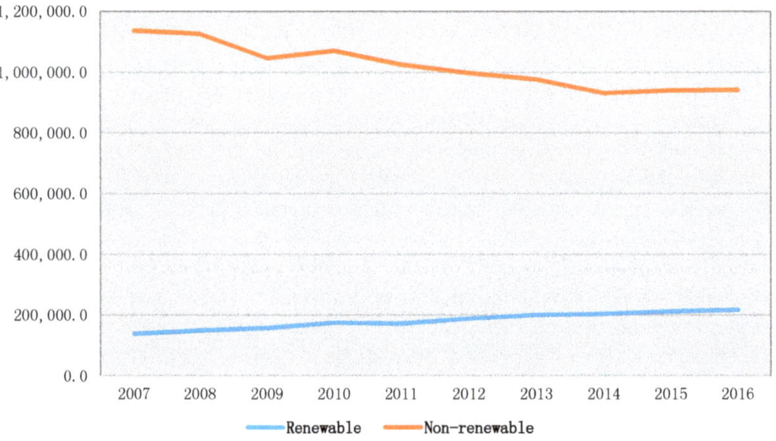

Figure 1. Energy usage in European countries (in thousand tons of oil equivalent (TOE)). Data source: Eurostat [39].

The next step of the analysis was to categorize the countries based on their GDP and energy produced from RESs. According to Equation (1), the countries are clustered by using the Ward's method for hierarchical cluster analysis, which considers that the distance between two clusters is how much the sum of squares will increase [41]:

$$\Delta(,) = \sum_{i \in A \cup B} \|\vec{x}_i - \vec{m}_{A \cup B}\|^2 - \sum_{i \in A} \|\vec{x}_i - \vec{m}_A\|^2 - \sum_{i \in B} \|\vec{x}_i - \vec{m}_B\|^2 \tag{1}$$

where $\vec{m}$ denotes the cluster's centers. This method minimizes the variances within clusters as much as possible [3,41]. Based on Ward's method dendrogram, two clusters were identified, as shown in Figure 2.

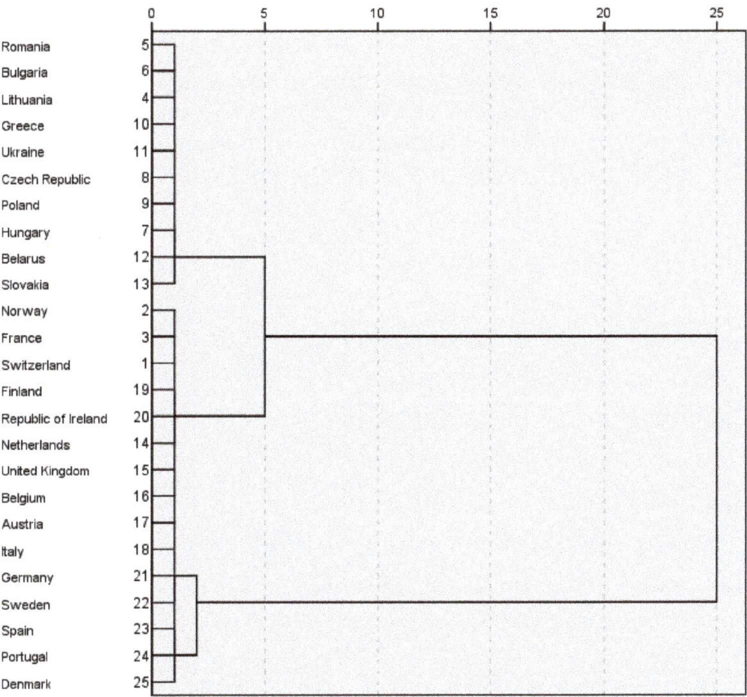

Figure 2. Dendrogram using Ward's method.

As we can see in Table 1, high GDP countries with high consumption of RESs were classified in cluster 1 (60% of countries); countries with low GDP and low consumption of RESs were classified in cluster 2.

Table 1. Countries clusters according to GDP and RES consumption.

	Cluster Distribution		
		Number of Countries	**% of Combined**
	1	10	40.0%
Cluster	2	15	60.0%
	Combined	25	100.0%

As we can note at Table 2, the average GDP for countries belonging in cluster 1 is 34,548 Euros per capita; in the second cluster there is an average of 16,526 Euros. The per capita consumption of renewable energy of the first cluster is 0.254 tons of oil equivalent (TOE), while the second cluster consumes 0.049 TOE. It is remarkable that the countries of the first cluster have a considerable deviation while the countries of the second cluster appear to be more concentrated around their mean.

Table 2. Clusters according to GDP per capita (in thousand Euros) and renewable energy consumption per capita (in tons of oil equivalent).

		GDP per Capita (in Thousand Euros)		RES Consumption per Capita (in TOE)	
		Mean	Std. Deviation	Mean	Std. Deviation
Cluster	1	34.548	7.430	0.254	0.160
	2	16.526	5.760	0.049	0.039

The above results show a potential correlation between RES consumption and economic growth. This obtained correlation between economic growth and RES consumption is depicted in the following scatter diagram, Figure 3, below. More specifically, it is obvious that most of the countries of higher GDP (cluster 1) had higher RES consumption than those of lower GDP (cluster 2).

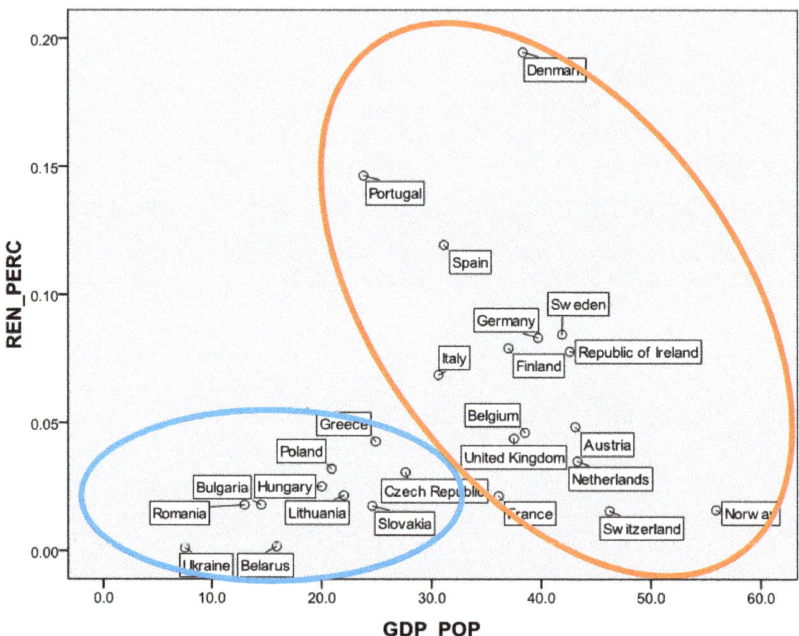

Figure 3. European countries according to RES consumption and GDP. Data source: Eurostat [39].

To attest and further analyze the observed correlation between RES consumption and economic growth in the long-run, we used a production function (Equation (2)), as proposed in other studies [4–11,40]:

$$GDP = f(GFCF, LF, RES, NONRES) \tag{2}$$

where:

- GDP denotes the gross domestic product;
- GFCF denotes the gross fixed capital formation;

- LF denotes the total labor force;
- RES denotes the renewable energy sources consumption; and
- NONRES denotes the non-renewable energy consumption

To examine the possible correlation between the abovementioned variables during the examined period, a log-linear specification was used (Equation (3)). The applied log-linear specification was of the following form:

$$lnGDP_t = a_0 + a_1 lnGFCF_t + a_2 lnLF_t + a_3 lnRES_t + a_4 lnNONRES_t + u_t \tag{3}$$

The above log-linear specification was examined using an autoregressive distributed lag (ARDL) as according to many authors it has more advantages over other techniques [11]. The main ARDL model's equation (Equation (4)) is of the following form [42]:

$$y_t = c_0 + c_1 t + \sum_{i=1}^{p} \varphi_i y_{t-i} + \sum_{i=0}^{q} \beta_i' x_{t-i} + u_t \tag{4}$$

The above model was reparametrized as follows (Equation (5)) to be error-corrected [42–44]:

$$\Delta y_t = c_0 + c_1 t - \left(1 - \sum_{j=1}^{p} \varphi_i\right)\left(y_{t-1} - \frac{\sum_{j=0}^{q} \beta_j}{a} x_{t-1}\right) + \sum_{i=1}^{p-1} \psi_{yi} \Delta y_{t-i} + \omega' \Delta x_t + \sum_{i=1}^{q-1} \psi_{xi}' \Delta x_{t-i} + u_t \tag{5}$$

Based on the above analysis, the data of Table 3 was obtained. This table summarizes the results for the log-linear specification models. More specifically, it concerns the impact of predictors for the examined countries' economic growth as it is measured based on their GDP.

Table 3. Variables and coefficients of the error-corrected log-linear specification models.

Model	Variables in Model	Coefficient	Std. Error	F-Statistic	*p*-Value
	Constant	15,979.230	9517.702	6.978	0.000
	lnGFCF$_t$	1.760	2.514	4.929	0.006
	lnLF$_t$	0.722	0.444	6.413	0.000
Model for cluster 1 countries.	lnRES$_t$	0.603	0.231	20.070	0.001
Dependent: lnGDP$_t$	lnNONRES$_t$	0.310	0.504	7.245	0.000

	Diagnostic Tests	Test Values	*p*-Value
	F-Statistic	22.380	0.000
	Jarque-Bera	0.194	0.812
	Adjusted R-squared	0.680	

Model	Variables in Model	Coefficient	Std. Error	F-Statistic	*p*-Value
	Constant	8848.629	3367.971	6.210	0.000
	lnGFCF$_t$	2.932	0.258	9.070	0.002
	lnLF$_t$	0.152	1.074	7.032	0.008
Model for cluster 2 countries.	lnRES$_t$	0.477	0.497	16.0766	0.000
Dependent: lnGDP$_t$	lnNONRES$_t$	0.718	0.586	6.649	0.000

	Diagnostic Tests	Test Values	*p*-Value
	F-Statistic	20.070	0.000
	Jarque-Bera	0.208	0.753
	Adjusted R-squared	0.738	

The first critical step was to find evidence that there was a correlation between GDP and its predictors for the examined period. This evidence was deduced from the F-Statistic values as shown in Table 3. Specifically, the F-Statistic values were greater than their upper critical bounds as they were proposed for the analysis of level relationship tests [44] and cointegration tests for less than 30 observations [42,45]. Thus, we can confirm that there was a statistically significant correlation between GDP and its predictors in the long-run concerning the countries of both the examined clusters.

Furthermore, in all the examined cases, the p-values were less than 1%, which means that there was a positive correlation between the examined GDPs and their predictors at 1% level of significance.

For the countries belonging to cluster 1, RES's coefficient was equal to 0.603; thus, if RES consumption increases by 1%, and the other predictors of economic growth remain constant (ceteris paribus), the countries' GDP will increase by 0.603%. Similarly, as far as the countries of cluster 2 were concerned, when RES consumption increases by 1%, ceteris paribus, the countries' GDP will increase by 0.477%.

Based on the adjusted R-squared values, the independent variables explained 68% of the first cluster countries' economic growth; as far as the second cluster's countries are concerned, the independent variables explained 74% of their economic growth. Moreover, the diagnostic tests' results show that both the proposed models had a reasonable good fit; there was no serial correlation, while skewness and kurtosis matching normal distribution and homoscedasticity were shown.

Consequently, RES consumption seemed to be correlated with the economic growth of both clusters' countries in the long-run. However, this correlation was higher with the economic growth of countries with higher GDP, while countries with lower economic growth were more dependent on Non-RESs.

5. Conclusions

The aim of this study was to investigate the relationship between RES consumption and countries' economic growth. To examine this relationship, data concerning 25 European countries' GDP, RES consumption, Non-RES consumption, GFCF, and labor force were statistically analyzed using the Ward's method for hierarchical cluster analysis. This method revealed the existence of two countries' clusters; a cluster of high GDP and high RES consumption countries and another cluster of low GDP and low RES consumption countries.

To sharpen the above results and further examine the correlation between RES consumption and a country's GDP, a log-linear specification by applying an error-corrected autoregressive distributed lag (ARDL) was employed. The results of this analysis for the countries of the two clusters suggest that there was a statistically significant correlation between GDP and its predictors in the long-run. More specifically, as far as energy consumption is concerned, the applied models indicated that RES's correlation was higher with high GDP countries; on the other hand, countries with lower GDP rely more on Non-RESs.

The results also discovered a prominent level of heterogeneity concerning the deployment of RESs in the examined countries' energy policy. For the case of low RES usage between several high GDP countries, the results are likely to indicate a low level of know-how on their effective usage. Furthermore, this could be the result of the lack of financial resources for a higher level of RES inclusion to these countries' energy mix. A two-sided interrelation between RES consumption and GDP growth is possible to exist as well [16]. This interrelation is the subject of further analysis.

The above findings show that policy-makers of both high and low GDP countries should take all the needed measures to increase RES contribution to the energy mix. Such measures involve the creation of a friendly environment for large-scale investments in RESs [46] and the provision of other incentives such as financial facilities to citizens [47–50]. Financial support is necessary since the richest countries can afford to use RESs. Furthermore, the creation of a common EU-wide energy production balance is critical as even countries that cannot produce a type of renewable energy due to their restricted natural resources (e.g., northern countries do not have the adequate sunlight to produce solar energy) can use RESs by relying to imports from other countries. Moreover, the development of know-how and the removal of economic and political barriers are some of the important steps towards the further development and deployment of RESs [4].

Based on the subject of the present study, there are several avenues for future research. First, it may be useful to apply various statistical techniques such as Augmented Dickey-Fuller (ADF) or Kwiatkowski-Phillips-Schmidt-Shin (KPSS) tests to better understand the relationship between renewable energy consumption and economic growth. Moreover, since carbon emissions reduction is a core feature of EU's environmental strategy, it would be useful to add it as an additional parameter to the statistical analysis [2]. Lastly, a comparative analysis between the EU and other countries could highlight the potential differences in the use of RESs and, at the same time, their global contribution to economic development.

Author Contributions: S.N., M.S., and M.C. gathered the data and carried out the implementation, and performed the calculations and the computer programming. G.K., G.A., S.G., and A.B. gathered and implemented all the theoretical background of the paper, and having the input from the experimental development, they reviewed and discussed the results of the study as well. A.K. performed the overall grammar check based on the narrative and its smoothing throughout the text content.

Funding: This research received no external funding.

Conflicts of Interest: The authors declare no conflict of interest.

References

1. Rodríguez-Monroy, C.; Mármol-Acitores, G.; Nilsson-Cifuentes, G. Electricity generation in Chile using non-conventional renewable energy sources—A focus on biomass. *Renew. Sustain. Energy Rev.* **2018**, *81*, 937–945. [CrossRef]
2. Ntanos, S.; Ziatas, T.; Merkouri, A. Renewable energy consumption, carbon dioxide emissions and economic growth: Evidence from Europe and Greece. In Proceedings of the e-RA 10 International Scientific Conference, Piraeus, Greece, 23–25 September 2015; pp. 46–56.
3. Papageorgiou, A.; Skordoulis, M.; Trichias, C.; Georgakellos, D.; Koniordos, M. Emissions trading scheme: Evidence from the European Union countries. In *Communications in Computer and Information Science*; Kravets, A., Shcherbakov, M., Kultsova, M., Shabalina, O., Eds.; Springer International Publishing: Cham, Switzerland, 2015; pp. 222–233.
4. Svenfelt, Å.; Engström, R.; Svane, Ö. Decreasing energy use in buildings by 50% by 2050—A backcasting study using stakeholder groups. *Technol. Forecast. Soc. Chang.* **2011**, *78*, 785–796. [CrossRef]
5. Alper, A.; Oguz, O. The role of renewable energy consumption in economic growth: Evidence from asymmetric causality. *Renew. Sustain. Energy Rev.* **2016**, *60*, 953–959. [CrossRef]
6. Apergis, N.; Danuletiu, D.C. Renewable energy and economic growth: Evidence from the sign of panel long-run causality. *Int. J. Energy Econ. Policy* **2014**, *4*, 578–587.
7. Destek, M.A.; Aslan, A. Renewable and non-renewable energy consumption and economic growth in emerging economies: Evidence from bootstrap panel causality. *Renew. Energy* **2017**, *111*, 757–763. [CrossRef]
8. Pao, H.T.; Fu, H.C. Renewable energy, non-renewable energy and economic growth in Brazil. *Renew. Sustain. Energy Rev.* **2013**, *25*, 381–392. [CrossRef]
9. Koçak, E.; Şarkgüneşi, A. The renewable energy and economic growth nexus in Black Sea and Balkan countries. *Energy Policy* **2017**, *100*, 51–57. [CrossRef]
10. Sasana, H.; Ghozali, I. The impact of fossil and renewable energy consumption on the economic growth in Brazil, Russia, India, China and South Africa. *Int. J. Energy Econ. Policy* **2017**, *7*, 194–200.
11. Shahbaz, M.; Loganathan, N.; Zeshan, M.; Zaman, K. Does renewable energy consumption add in economic growth? An application of auto-regressive distributed lag model in Pakistan. *Renew. Sustain. Energy Rev.* **2015**, *44*, 576–585. [CrossRef]
12. Salim, R.A.; Hassan, K.; Shafiei, S. Renewable and non-renewable energy consumption and economic activities: Further evidence from OECD countries. *Energy Econ.* **2014**, *44*, 350–360. [CrossRef]
13. Apergis, N.; Payne, J.E. Renewable and non-renewable energy consumption-growth nexus: Evidence from a panel error correction model. *Energy Econ.* **2012**, *34*, 733–738. [CrossRef]
14. Bhattacharya, M.; Paramati, S.R.; Ozturk, I.; Bhattacharya, S. The effect of renewable energy consumption on economic growth: Evidence from top 38 countries. *Appl. Energy* **2016**, *162*, 733–741. [CrossRef]
15. Inglesi-Lotz, R. The impact of renewable energy consumption to economic growth: A panel data application. *Energy Econ.* **2015**, *53*, 58–63. [CrossRef]
16. Omri, A.; Nguyen, D.K. On the determinants of renewable energy consumption: International evidence. *Energy* **2014**, *72*, 554–560. [CrossRef]

17. Al-Mulali, U.; Fereidouni, H.G.; Lee, J.Y.; Sab, C.N. Examining the bi-directional long run relationship between renewable energy consumption and GDP growth. *Renew. Sustain. Energy Rev.* **2013**, *22*, 209–222. [CrossRef]

18. Jebli, M.B.; Youssef, S.B. Output, renewable and non-renewable energy consumption and international trade: Evidence from a panel of 69 countries. *Renew. Energy* **2015**, *83*, 799–808. [CrossRef]

19. Sadorsky, P. Renewable energy consumption and income in emerging economies. *Energy Policy* **2009**, *37*, 4021–4028. [CrossRef]

20. Yildirim, E.; Saraç, Ş.; Aslan, A. Energy consumption and economic growth in the USA: Evidence from renewable energy. *Renew. Sustain. Energy Rev.* **2012**, *16*, 6770–6774. [CrossRef]

21. Fang, Y. Economic welfare impacts from renewable energy consumption: The China experience. *Renew. Sustain. Energy Rev.* **2011**, *15*, 5120–5128. [CrossRef]

22. Lin, B.; Moubarak, M. Renewable energy consumption—Economic growth nexus for China. *Renew. Sustain. Energy Rev.* **2014**, *40*, 111–117. [CrossRef]

23. Dogan, E. The relationship between economic growth and electricity consumption from renewable and non-renewable sources: A study of Turkey. *Renew. Sustain. Energy Rev.* **2015**, *52*, 534–546. [CrossRef]

24. Cambini, C.; Franzi, D. Assessing the EU Pressure for Rules Change: The Perceptions of Southern Mediterranean Energy Regulators. *Mediterr. Politics* **2014**, *19*, 59–81. [CrossRef]

25. Pang, X.; Mörtberg, U.; Brown, N. Energy models from a strategic environmental assessment perspective in an EU context—What is missing concerning renewables? *Renew. Sustain. Energy Rev.* **2014**, *33*, 353–362. [CrossRef]

26. Dogan, E.; Sebri, M.; Turkekul, B. Exploring the relationship between agricultural electricity consumption and output: New evidence from Turkish regional data. *Energy Policy* **2016**, *95*, 370–377. [CrossRef]

27. Sutter, D. Propagandistic Research and the U.S. Department of Energy: Energy Efficiency in Ordinary Life and Renewables in Electricity Production. *Econ. J. Watch* **2017**, *14*, 103–120.

28. Cherukupalli, N. Renewables Can Help Transform Lives in Rural Areas. *Proc. IEEE* **2015**, *103*, 862–867. [CrossRef]

29. Haerer, D.; Pratson, L. Employment trends in the U.S. Electricity Sector, 2008–2012. *Energy Policy* **2015**, *82*, 85–98. [CrossRef]

30. DeLlano-Paz, F.; Calvo-Silvosa, A.; Antelo, S.I.; Soares, I. The European low-carbon mix for 2030: The role of renewable energy sources in an environmentally and socially efficient approach. *Renew. Sustain. Energy Rev.* **2015**, *48*, 49–61. [CrossRef]

31. Knopf, B.; Nahmmacher, P.; Schmid, E. The European renewable energy target for 2030—An impact assessment of the electricity sector. *Energy Policy* **2015**, *85*, 50–60. [CrossRef]

32. Bolük, G.; Mert, M. Fossil and renewable energy consumption, GHGs (greenhouse gases) and economic growth: Evidence from a panel of EU (European Union) countries. *Energy* **2014**, *74*, 439–446. [CrossRef]

33. Caraiani, C.; Lungu, C.; Dascălu, C. Energy consumption and GDP causality: A three-step analysis for emerging European countries. *Renew. Sustain. Energy Rev.* **2015**, *44*, 198–210. [CrossRef]

34. Borkowska, B.; Klimczak, M. From a monopoly towards an imperfectly competitive electricity market in Poland. *Transform. Bus. Econ.* **2011**, *10*, 463–474.

35. Tellidou, A.C.; Bakirtzis, A.G. Agent-based analysis of monopoly power in electricity markets. In Proceedings of the International Conference on Intelligent Systems Applications to Power Systems, Kaohsiung, Taiwan, 5–8 November 2007.

36. Philippou, M.; Tassoulis, A.; Katsigiannakis, G. The liberalization process of the electricity market in Greece: The design of the transmission system operator's IT infrastructure. In Proceedings of the International Conference on Electric Power Engineering, PowerTech, Budapest, Hungary, 29 August–2 September 1999; p. 15.

37. Andrianesis, P.; Biskas, P.; Liberopoulos, G. An overview of Greece's wholesale electricity market with emphasis on ancillary services. *Electr. Power Syst. Res.* **2011**, *81*, 1631–1642. [CrossRef]

38. Chatzisideris, M.D.; Laurent, A.; Christoforidis, G.C.; Krebs, F.C. Cost-competitiveness of organic photovoltaics for electricity self-consumption at residential buildings: A comparative study of Denmark and Greece under real market conditions. *Appl. Energy* **2017**, *208*, 471–479. [CrossRef]

39. Eurostat. Supply, Transformation and Consumption of Renewable Energies—Annual Data Supply, Transformation and Consumption of Renewable Energies—Annual Data. Available online: http://ec.europa.eu/eurostat/data/database (accessed on 1 June 2018).
40. Taeyoung, J.; Jinsoo, K. Coal Consumption and Economic Growth: Panel Cointegration and Causality Evidence from OECD and Non-OECD Countries. *Sustainability* **2018**, *10*, 660. [CrossRef]
41. Punj, G.; Stewart, D.W. Cluster analysis in marketing research: Review and suggestions for application. *J. Mark. Res.* **1983**, *20*, 134–148. [CrossRef]
42. Kripfganz, S.; Schneider, D.C. ARDL: Stata module to estimate autoregressive distributed lag models. In Proceedings of the Stata Conference, Chicago, IL, USA, 28–29 July 2016.
43. Goh, S.K.; Wong, K.N. Could inward FDI offset the substitution effect of outward FDI on domestic investment? Evidence from Malaysia. *Prague Econ. Pap.* **2014**, *23*, 413–425.
44. Pesaran, M.H.; Shin, Y.; Smith, R.J. Bounds testing approaches to the analysis of level relationships. *J. Appl. Econ.* **2001**, *16*, 289–326. [CrossRef]
45. Narayan, P.K. The saving and investment nexus for China: Evidence from cointegration tests. *Appl. Econ.* **2005**, *37*, 1979–1990. [CrossRef]
46. Azam, M.; Khan, A.Q.; Zafeiriou, E.; Arabatzis, G. Socio-economic determinants of energy consumption: An empirical survey for Greece. *Renew. Sustain. Energy Rev.* **2016**, *57*, 1556–1567. [CrossRef]
47. Ntanos, S.; Kyriakopoulos, G.; Chalikias, M.; Arabatzis, G.; Skordoulis, M.; Galatsidas, S.; Drosos, D. A Social Assessment of Renewable Energy Sources Usage and Contribution to Life Quality: The Case of an Attica Urban Area in Greece. *Sustainability* **2018**, *10*, 1414. [CrossRef]
48. Ntanos, S.; Kyriakopoulos, G.; Chalikias, M.; Arabatzis, G.; Skordoulis, M. Public Perceptions and Willingness to Pay for Renewable Energy: A Case Study from Greece. *Sustainability* **2018**, *10*, 687. [CrossRef]
49. Menanteau, P.; Finon, D.; Lamy, M.L. Prices versus quantities: Choosing policies for promoting the development of renewable energy. *Energy Policy* **2003**, *31*, 799–812. [CrossRef]
50. Garcia, A.; Alzate, J.M.; Barrera, J. Regulatory design and incentives for renewable energy. *J. Regul. Econ.* **2012**, *41*, 315–336. [CrossRef]

Article

Multi-Port High Voltage Gain Modular Power Converter for Offshore Wind Farms

Sen Song [1], Yihua Hu [1], Kai Ni [1,*], Joseph Yan [1], Guipeng Chen [2], Huiqing Wen [3] and Xianming Ye [4]

[1] Department of Electrical Engineering and Electronics, The University of Liverpool, Liverpool L69 3BX, UK; sgssong2@student.liverpool.ac.uk (S.S.); y.hu35@liverpool.ac.uk (Y.H.); Yaneee@liverpool.ac.uk (J.Y.)
[2] College of Electrical Engineering, Zhejiang University, Hangzhou 310027, China; cgp2017@xmu.edu.cn
[3] Department of Electrical Engineering and Electronics, Xi'an Jiaotong-Liverpool University, Suzhou 215123, China; Huiqing.Wen@xjtlu.edu.cn
[4] Department of Electrical, Electronic and Computer Engineering, University of Pretoria, Pretoria 0084, South Africa; xianming.ye@up.ac.za
* Correspondence: k.ni@student.liverpool.ac.uk; Tel.: +44-751-920-3720

Received: 30 May 2018; Accepted: 25 June 2018; Published: 26 June 2018

Abstract: In high voltage direct current (HVDC) power transmission of offshore wind power systems, DC/DC converters are applied to transfer power from wind generators to HVDC terminals, and they play a crucial role in providing a high voltage gain, high efficiency, and high fault tolerance. This paper introduces an innovative multi-port DC/DC converter with multiple modules connected in a scalable matrix configuration, presenting an ultra-high voltage step-up ratio and low voltage/current rating of components simultaneously. Additionally, thanks to the adoption of active clamping current-fed push–pull (CFPP) converters as sub-modules (SMs), soft-switching is obtained for all power switches, and the currents of series-connected CFPP converters are auto-balanced, which significantly reduce switching losses and control complexity. Furthermore, owing to the expandable matrix structure, the output voltage and power of a modular converter can be controlled by those of a single SM, or by adjusting the column and row numbers of the matrix. High control flexibility improves fault tolerance. Moreover, due to the flexible control, the proposed converter can transfer power directly from multiple ports to HVDC terminals without bus cable. In this paper, the design of the proposed converter is introduced, and its functions are illustrated by simulation results.

Keywords: high voltage direct current (HVDC); power transmission; DC/DC converter; high voltage gain; modular; multi-port

---

1. Introduction

The global number of offshore wind farms has increased in recent years [1,2]. In Europe, 560 new offshore wind turbines were built in 17 wind farms in 2017 with a total generation capacity of 3148 MW, which is about 20% of the total offshore generation capacity [3]. The power generated offshore is typically transmitted over an average distance of 41 km (in 2017) through submarine cables before reaching a connection point with the existing onshore grid [3]. For example, Hornsea wind farm is located around 40 km from the onshore station. Compared with high voltage alternating current (HVAC) transmission, high voltage direct current (HVDC) transmission is the preferred method to transfer power over a long distance (>40 km) in terms of the factors of economy and power efficiency [4,5]. When delivering the same amount of power, the purchase price for the bipolar HVDC cable is lower than that of two parallel 3-core HVAC ones [6]. Additionally, HVDC has a higher transmission efficiency than HVAC since no inductance-reactive power exists within DC transmission cables.

Figure 1 presents three HVDC configurations. The hybrid HVDC system illustrated in Figure 1a uses a medium voltage (MV)—high voltage (HV) AC/DC converter to obtain high voltage DC power. However, line-frequency (50/60 Hz) AC/AC transformers for low-voltage (LV)—MV conversion still occupy a significant portion of the substation space. In this topology, the transformer is replaced by a converter as shown in Figure 1b, which significantly reduces the system size and weight [7]. These two configurations use two-stage conversion to meet the voltage level of HVDC transmission. However, Reference [8] points out that the configuration of two DC/DC conversion stages has the highest power loss, which is around four times that of a one conversion stage configuration presented in Figure 1c taking into consideration the winding and core losses of transformers and the given datasheets of the semiconductors. The converter applied in Figure 1c not only provides a high voltage gain but also transfers power directly from multiple generators to HVDC terminals without bus cable. The converter design is a technical challenge for boosting LV directly to HV due to the conflict between the required high voltage level, e.g., ±800 kV [9] and the restricted voltage ratings of semiconductor components, e.g., 22 kV for SiC thyristors, and 15 kV for SiC transistors [10]. Fortunately, with the development of semiconductors and converter topologies, possible solutions are provided [11–13].

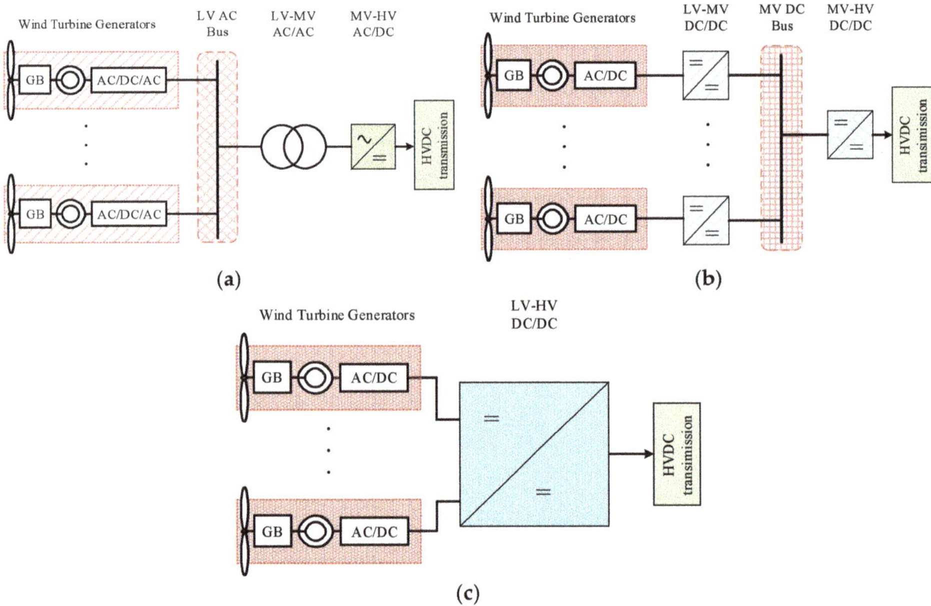

Figure 1. High voltage direct current (HVDC) configurations for wind power transmission: (**a**) DC-based connection with two-stage hybrid conversion; (**b**) DC-based connection with two-stage DC/DC conversion; (**c**) DC-based connection with the proposed modular converter.

To address the challenges, DC converters with high voltage gains [14–17], modular multilevel converters (MMCs) [18–22] and multi-module converters [23–25] are studied extensively. Although dual-active bridge (DAB) converters [14,15] and resonant converters [16,17] can obtain high voltage step-up ratios, the voltage stress on their semiconductor components is high, which can be reduced by applying MMCs. By adding sub-modules (SMs), a high output voltage is achieved without increasing the voltage stress. However, the MMC topologies based on half-bridge (HB) or full-bridge (FB) [18,19] and resonant MMC [20] cannot provide electrical isolation. The isolated MMCs in References [21,22] are presented with DC/AC/DC configuration, where medium-frequency high turns-ratio transformers are employed, resulting in a vast volume. Although resonant MMCs [26] achieve galvanic isolation and a small volume of transformers at the same time, its conversion ratio only satisfies MV applications. Alternatively, transformers can be decentralized into multi-module converters, enabling the installation of high-frequency transformers, which reduces the sizes of transformers and reactive components. However, by adopting active-clamping flyback–forward converters as SMs, the currents of different SMs are unbalanced because of the non-ideal factors such as unequable leakage inductances of transformers, and only half of the power switches can achieve zero voltage switching (ZVS), resulting in high switching losses [23].

In this paper, based on the multi-modular converter in Reference [23], active-clamping current-fed push–pull (CFPP) converters are adopted to replace the flyback–forward SMs. The currents of modules in series-connection are auto-balanced, and all power switches can achieve soft-switching. Therefore, control complexity and switching loss are reduced. Furthermore, the matrix configuration brings about high control flexibility, which improves the fault tolerance capability. Additionally, thanks to the independent operation of each port, the converter can collect power from multiple sources without bus cable.

The paper is organized as follows: The basic cell and two interleaved working modes are analyzed in Section 1; the scalable topology design is discussed in Section 3; Section 4 depicts the fault tolerance strategies of the topology; simulation results are presented in Section 5 to demonstrate the effectiveness and efficiency of the converter, and finally, conclusions are drawn in Section 4.

2. Analysis of the Basic Cell and Working Strategies of Matrix Configuration

2.1. Operation of the Basic Cell

The basic cell topology based on CFPP converter shown in Figure 2 has similar operation principles and characteristics as the converter presented in Reference [27]. S_1–S_2 are the two main switches. S_{c1}–S_{c2} are the two active clamping switches. C_c is the clamp capacitor. L_1 is the input transistor. The tri-winding transformer has a turns ratio of $N_{p1}:N_{p2}:N_s = 1:1:n$ and leakage inductance L_k at the secondary side. Furthermore, the secondary circuit consists of four rectifier diodes D_1–D_4, one output capacitor C_{out} and a load resistor R.

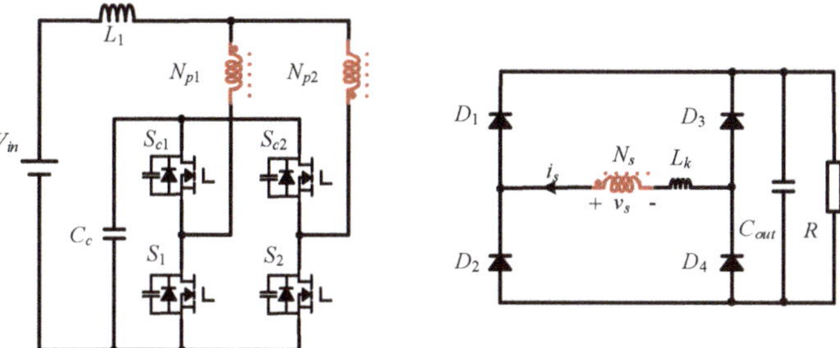

Figure 2. Topology of the basic cell based on active-clamping fed push–pull (CFPP) converter.

The key operating waveforms of CFPP cell are depicted in Figure 3. V_{gs1}–V_{gs2} are the control signals for the two main switches S_1–S_2, which have the phase shift angle of 180°. V_{gsc1}–V_{gsc2} are the control signals for the two clamp switches S_{c1}-S_{c2}. The control signals of the main switches and clamping switches are complementary. v_{ds1}–v_{ds2} and v_{dsc1}–v_{dsc2} are the drain-to-source voltages of the main switches and clamping switches respectively. i_{L1} is the current of input inductor L_1. v_s and i_s are the secondary voltage and current of the transformer respectively. The following assumptions are made to simplify the analysis.

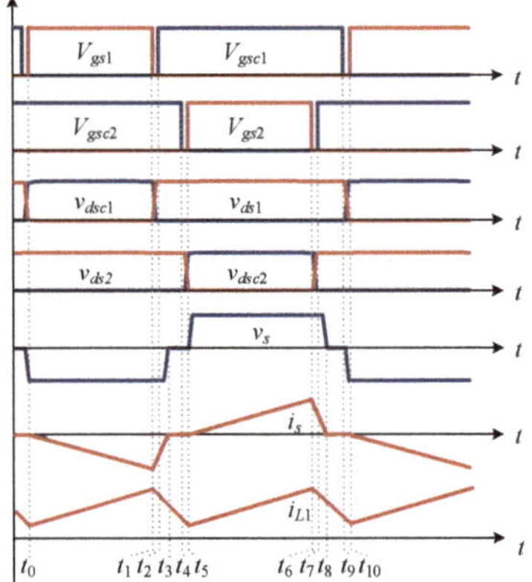

Figure 3. Operating waveforms of the basic cell.

- All switches and diodes are identical.
- The capacitance of clamp capacitor is large enough so that its voltage ripple can be ignored. Due to the symmetrical operation, a brief introduction of the operation during t_0–t_5 when $D \leq 0.5$ is presented in this part.

Mode 1 (t_0–t_1): In this mode, the main switch S_1 and the clamping switch S_{c2} are on. The power is transferred to the output. The diodes D_2 and D_3 are forward biased, and the secondary current i_s decreases.

Mode 2 (t_1–t_2): At t_1, the main switch S_1 is turned off. The leakage inductances L_k resonate with the parasitic capacitances of S_1 and S_{c1}. Then, the voltage of S_{C1} drops to zero at t_2 to achieve ZVS turn-on. At the same time, capacitance C_{S1} is charged.

Mode 3 (t_2–t_3): At t_2, the clamping switch S_{C1} is turned on with zero voltage. Because both the clamping switches S_{C1} and S_{C2} are on, the primary sides of the transformer are short-circuited. Then the power is transferred to the input inductor L_1, and the secondary current i_s rises rapidly.

Mode 4 (t_3–t_4): At t_3, the secondary current reaches zero. All four diodes are reverse biased. Additionally, the secondary voltage recovers to zero within a short time.

Mode 5 (t_4–t_5): At t_4, the clamping switch S_{C2} is turned off. The leakage inductances L_k resonate with parasitic capacitances of S_2 and S_{C2}. The voltage across S_2 drops to zero at t_5 so that ZVS turn-on of S_2 is obtained.

The operation in intervals (t_0–t_5) and (t_5–t_{10}) is symmetrical. The power is transferred to the output load R when one main switch and one clamping switch are on, and then the power flows from the input to inductor L_1 when both clamping switches are on. All switches can obtain ZVS turn-on and the energy stored in L_k is recycled by the parasitic capacitances of clamping switches, which contributes to a higher conversion efficiency.

The voltage of clamp capacitor V_{Cc} can be obtained according to the flux balance of L_1:

$$V_{Cc} = \frac{V_{in}}{1 - D} \tag{1}$$

Additionally, with the turns ratio as 1:1:n, the output voltage of basic cell $V_{o,BC}$ can be determined:

$$V_{o,BC} = \frac{n}{2} V_{Cc} = \frac{n}{2} \frac{V_{in}}{(1 - D)} \tag{2}$$

The voltage stress of all four power switches V_{ds} can be obtained by:

$$V_{ds} = V_{Cc} = \frac{V_{in}}{1 - D} \tag{3}$$

2.2. Working Strategies of Matrix Configuration

A fundamental 2×2 modular topology is presented in Figure 4. The primary circuits of power cells are parallel connected so that they have an equal secondary voltage value with the same duty cycle of main switches. The secondary side of each cell is connected with four rectifier diodes for power regulation.

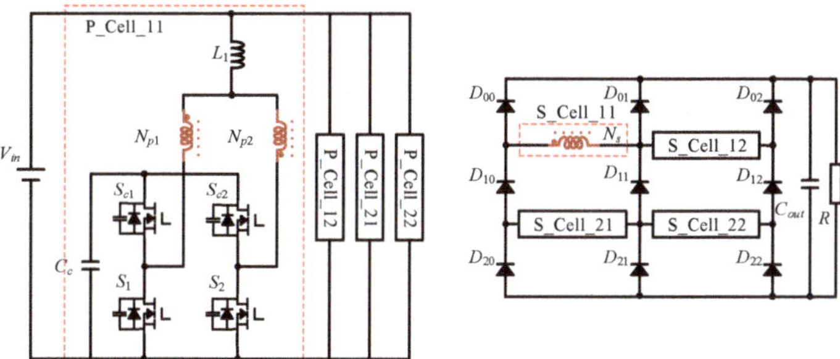

Figure 4. 2×2 topology of the isolated high voltage gain DC/DC converter with basic cells.

As illustrated in Figure 5a,b, adjacent cells have opposite polarities in column interleaved modes. For example, the polarity of cell 11 is opposite to those of cell 21 and 12. In this case, cells in the same column are connected in series, and the adjacent columns are in parallel connection. The voltage ratings of diodes in the first and last rows, D_{00}–D_{02} and D_{20}–D_{22}, are equal to the voltage of power cells v_s. While the voltage ratings of other diodes D_{10}–D_{12} have twice the value of v_s since they are connected with two cells. The current stresses of diodes in the first and last columns, D_{00}–D_{20} and D_{02}–D_{22} have the same value as the secondary current of one column. Diodes D_{01}–D_{21} connect with two columns so that they have double the current stress of the others. Additionally, the sum of the average diode currents in all columns is equal to the output current. Therefore, in column interleaved modes, the voltage and current ratings of diodes in an expanded topology as shown in Figure 6 composed of s rows and p columns can be obtained:

$$V(D_{ij}) = \begin{cases} \frac{1}{s}V_{o,s\times p} = V_{o,BC} = \frac{n}{2(1-D)}V_{in} & i = 0, s \\ \frac{2}{s}V_{o,s\times p} = 2 \times V_{o,BC} = \frac{n}{(1-D)}V_{in} & i = 1, 2, \cdots, (s-1) \end{cases} \tag{4}$$

$$I(D_{ij}) = \begin{cases} \frac{1}{2}I_{o,BC} = \frac{1}{2p}I_{o,s\times p} & j = 0, p \\ I_{o,BC} = \frac{1}{p}I_{o,s\times p} & j = 1, 2, \cdots, (p-1) \end{cases} \tag{5}$$

$$\sum_{j=0}^{p} I(D_{ij}) = I_{o,sp} \tag{6}$$

where $I_{o,BC}$ is the average output current of a single basic cell; $V_{o,s\times p}$ and $I_{o,s\times p}$ are the output voltage and current of $s \times p$ topology. In this case, all semiconductor components have low voltage and current ratings.

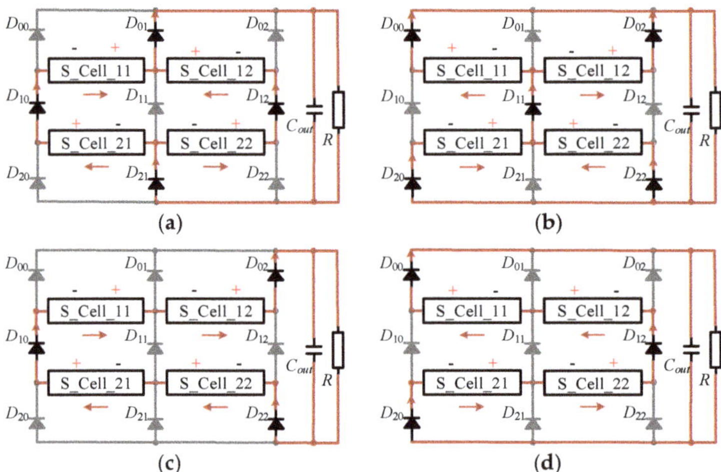

Figure 5. 2×2 topology with different interleaved strategies: (**a**) Column interleaved mode 1; (**b**) Column interleaved mode 2; (**c**) Series interleaved mode 1; (**d**) Series interleaved mode 2.

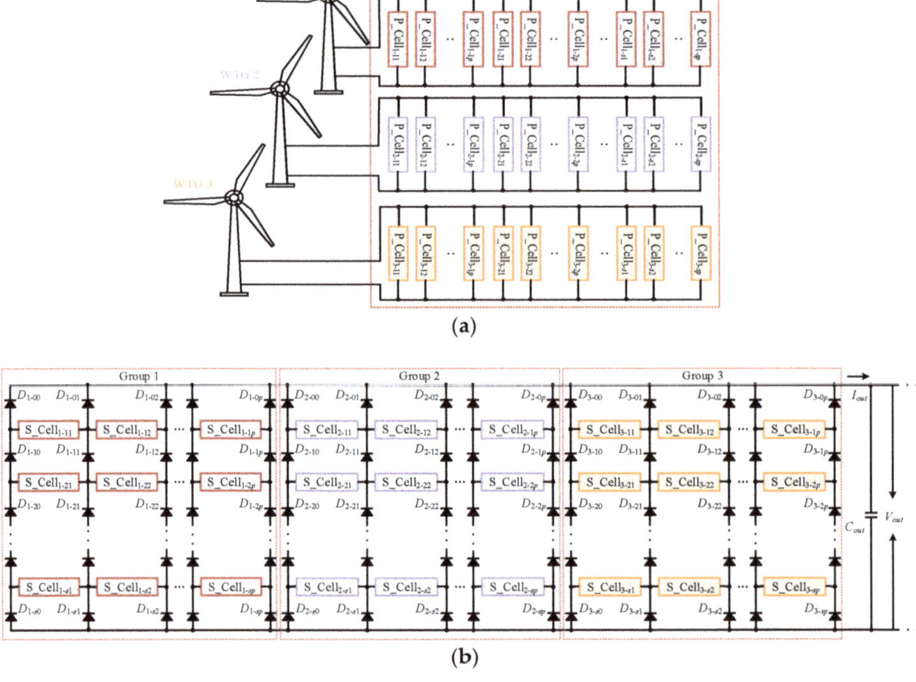

Figure 6. Topology of the proposed converter with three input-ports: (**a**) primary circuits with three power sources; (**b**) secondary circuits collecting power and delivering it to load.

From Figure 5c,d, in series interleaved modes, all cells are series-connected. The cells in adjacent rows have the opposite polarities while the cells in the same row have the same polarity. Similar to Equation (4), diodes D_{10} and D_{12} have twice the voltage rating higher than that of diodes in the first and last rows because they connect with two rows. The currents of all operating diodes have the same value since they are in series-connection. Hence, the voltage and current ratings of diodes in series interleaved modes can be calculated as:

$$V(D_{ij}) = \begin{cases} \frac{1}{s}V_{o,s\times p} = p \times V_{o,BC} = \frac{p\times n}{2(1-D)}V_{in} & i = 0, s \\ \frac{2}{s}V_{o,s\times p} = 2p \times V_{o,BC} = \frac{p\times n}{(1-D)}V_{in} & i = 1,2,\cdots,(s-1) \end{cases} \tag{7}$$

$$I(D_{ij}) = \frac{1}{2}I_{o,BC} = \frac{1}{2}I_{o,s\times p} \qquad j = 0, p \tag{8}$$

According to Equation (8), for series interleaved modes, the current rating of diodes is increased with the increase of output power. Besides which, the diodes D_{01}–D_{21} are blocked, which benefits the fault tolerance operation to be described in Section 4.

3. Analysis of the Proposed Converter with Multi-Input Ports

3.1. Scalable Topology

Thanks to modularity, multi-module converters can be easily expanded by increasing its row number and column number to attain a high voltage gain and the desired high power level. In the normal scenario, the proposed converter operates with column interleaved modes to keep a low voltage/current rating of components. Three wind-turbine-generators, WTGs 1–3, are connected to the proposed converter as illustrated in Figure 6a. Figure 6b shows the secondary circuits that are divided into three independent, Groups 1–3, by diodes $D_{1\text{-}0p}$–$D_{1\text{-}sp}$, $D_{2\text{-}00}$–$D_{1\text{-}s0}$ and $D_{2\text{-}0p}$–$D_{2\text{-}sp}$, $D_{3\text{-}00}$–$D_{3\text{-}s0}$ on the basis of input ports. The output power of each group consisting of the $s \times p$ expanded topology can be controlled individually, and the bus cable is eliminated. With the same input voltage and power of each port, the voltages and currents of all basic cells are identical.

The cells in the same column are in series-connection. Hence, every column has the same terminal voltage which is the sum of the voltages of cells in the same column. For a converter with multi-ports, output voltage V_{out} equals to the identical terminal voltage of columns and the output current I_{out} is the sum of secondary currents of all columns. Therefore, the row number s determines the output voltage, and the group number x with column number p determines the output power.

$$\begin{cases} V_{out} = V_{o,s\times p} = s \times V_{o,BC} = \frac{s\times n}{2(1-D)}V_{in} \\ I_{out} = x \times I_{o,s\times p} = x \times p \times I_{o,BC} \end{cases} \tag{9}$$

3.2. Current Balance with Column Interleaved Mode

The control complexity is reduced by the auto-balanced currents of cells in the same column. According to the currents through diodes D_{10}–D_{13} and D_{20}–D_{23}, as shown in Figure 7, the relationships among all cells can be obtained as:

$$\begin{cases} I_{o,BC11+} + I_{o,BC12+} = I_{o,BC21+} + I_{o,BC22+} \\ I_{o,BC31+} = I_{o,BC21+} \\ I_{o,BC23+} = I_{o,BC13+} \\ I_{o,BC32+} + I_{o,BC33+} = I_{o,BC22+} + I_{o,BC23+} \end{cases} \tag{10}$$

$$\begin{cases} I_{o,BC11-} = I_{o,BC21-} \\ I_{o,BC31-} + I_{o,BC32-} = I_{o,BC21-} + I_{o,BC22-} \\ I_{o,BC12-} + I_{o,BC13-} = I_{o,BC22-} + I_{o,BC23-} \\ I_{o,BC33-} = I_{o,BC23-} \end{cases} \tag{11}$$

where $I_{o,BC+}$ is the average current of secondary circuit in the interval (t_0–t_5) and $I_{o,BC-}$ is that in the interval (t_5–t_{10}). According to the symmetrical operation of the basic cell between the two intervals, it can be derived that $I_{o,BC+} = I_{o,BC-}$. Therefore, Equation (12) is obtained with $I_{o,BC} = I_{o,BC+} + I_{o,BC-}$.

$$\begin{cases} I_{o,BC11} = I_{o,BC21} = I_{o,BC31} \\ I_{o,BC12} = I_{o,BC22} = I_{o,BC32} \\ I_{o,BC13} = I_{o,BC23} = I_{o,BC33} \end{cases} \tag{12}$$

According to Equation (12), the average currents for cells in the same column are auto-balanced. Therefore, as shown in Figure 8, only the control of output voltage and current sharing in different columns is employed to achieve the required output voltage and power in the $s \times p$ topology, where $v_{o,ref}$ is the desired output voltage and $i_{L,i1}$–$i_{L,ip}$ are the currents collected from columns 1–p.

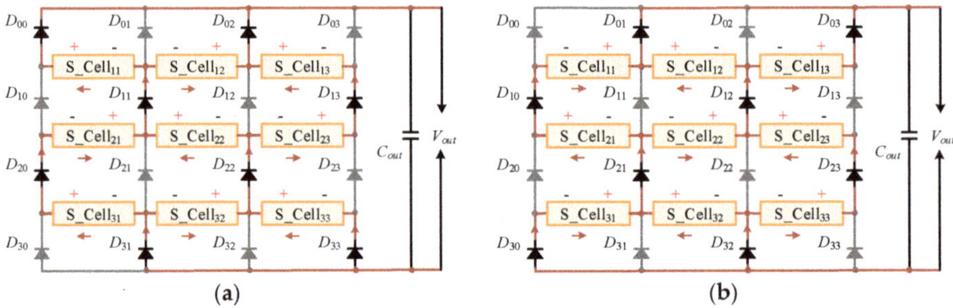

(a)   (b)

Figure 7. Auto-balanced currents of cells in the same column with column interleaved working strategy: (**a**) Column interleaved working mode 1; (**b**) Column interleaved working mode 2.

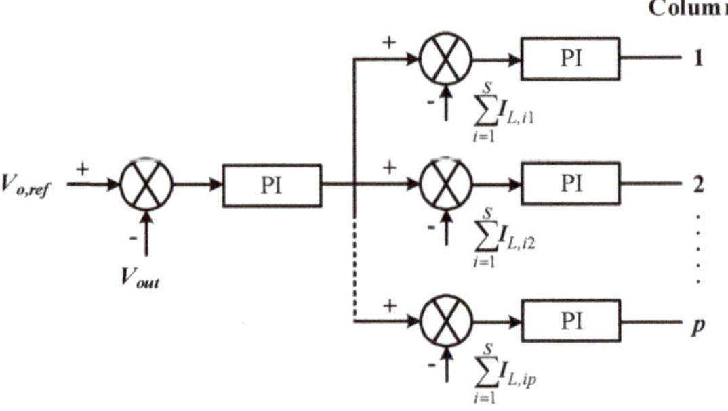

Figure 8. Control scheme of the $s \times p$ topology.

4. Fault Tolerance

4.1. Fault Tolerance for WTGs with Different Output Power

When disturbances occur, proper control strategies of WTGs and converters should be applied to ensure system protection and high power efficiency [28,29]. For the proposed converter, the output power of one group is determined not only by the column number of the secondary circuits but also, by the duty-cycles of main switches. According to Equation (8), the output power of one group is obtained as:

$$P_{group} = (p - m) \times V_{out} \times I_{o,BC} \tag{13}$$

where m is the number of idle column in the corresponding group.

The variation of a duty-cycle will cause the change of currents through the cells in one column as:

$$I_{sub} = I_{o,BC} = \frac{2}{T_s} \cdot \int_0^{t_{on}} i_{out} \approx i_{max} \cdot D \tag{14}$$

Hence the output power is:

$$P_{group} = p \cdot V_{out} \cdot I_{sub} = p \cdot D \cdot V_{out} \cdot i_{max} \tag{15}$$

where i_{max} is the maximum output current of power cells.

4.2. Fault Tolerance for Semiconductor Components

Semiconductor components are vulnerable components in a converter [30]. For offshore HVDC stations, faulty components take a long time for maintenance, resulting in high cost and loss [31].

Figure 9 shows the fault tolerance operation derived by installing redundant power cells. To maintain normal operation, the column in a red dotted box containing the damaged Cell 11 is replaced by one redundant column that is in the other red dotted box. For the faulty diodes D_{00}–D_{s0}, they require only one redundant column since they connect with one column. However, for other diodes, two redundant columns are demanded. For example, when diode D_{11} fails, the columns 1–2 in the blue dotted box are idle, and the redundant columns $p1$–$p2$ are applied.

$$V_{o,FBC} = \frac{V_{o,s \times p}}{2 \times s} = \frac{1}{2} V_{o,BC} \tag{16}$$

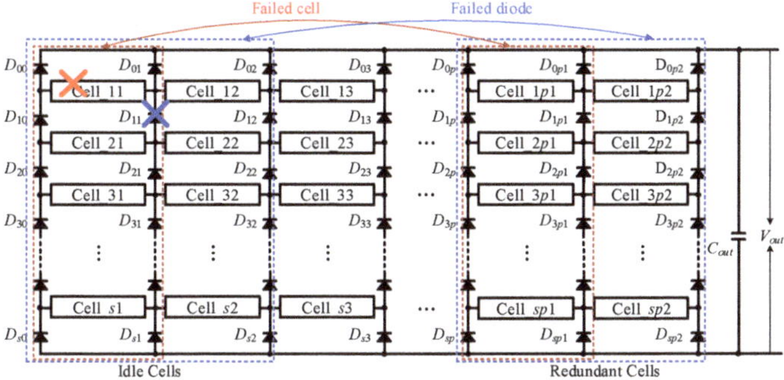

Figure 9. Fault tolerance with redundancy.

The proposed topology can also obtain fault tolerant operation of diodes and power switches without redundancy. In Figure 10a,b, when diode D_{11} is short-circuited, the cells in rows 1–2 are inactive to block the faulty components, while the fault tolerance group still operates with the column interleaved strategy. Additionally, to maintain normal operation, the output voltages of the faulty group are controlled according to Equations (2) and (9) to ensure the output voltage is the same as that in the normal case. When diode D_{11} is open-circuited as illustrated in Figure 10c,d, the fault tolerant operation group consisting of two columns works under the series interleaved modes to block D_{11}. Compared with the normal operation group, the number of cells in series connection in the faulty group is doubled. The voltages of cells in the fault group are adjusted as illustrated in Equation (15) to ensure they have the same terminal voltage.

Moreover, the fault tolerant operation of damaged cells without redundancy is similar to that when diodes are short-circuited. For instance, when cell 11 fails, cells 11-1p and 21-2p are idle so that the faulty one is blocked.

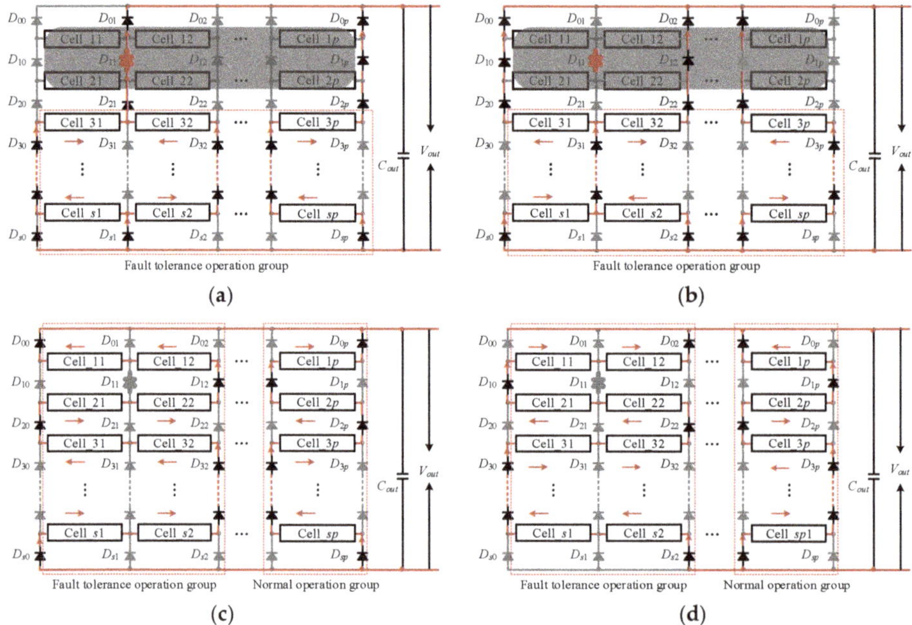

Figure 10. Fault tolerance without redundancy: (**a**) Working mode 1 under diode short-circuit; (**b**) Working mode 2 under diode short-circuit; (**c**) Working mode 1 under diode open-circuit; (**d**) Working mode 2 under diode open-circuit.

5. Simulation Results and Discussion

To illustrate the functionality of the proposed power converter, a simulation model consisting of 3 groups × 4 rows × 5 columns with 2 redundant columns is built by software *PSIM*, which is similar to the model shown in Figure 6. The cells in columns 1–3 are active, while those in columns 4–5 are inactive.

Table 1 presents the initial values for the simulation. It is noted that to verify the auto-balanced current characteristic, the leakage inductances of cells in column 1 are set as 70, 75, 80, 85 µH.

Table 1. Initial values of simulation.

System Parameters	Values	Components	Values
Input Voltage	650 V	Turns ratio 1:1:n	1:1:2
Output Voltage	6400 V	Leakage inductance	80 µH
Switching Frequency	5 kHz	Input inductance	5 mH
Output Power	40 kW	Clamp capacitor	20 µF

The steady-state waveforms of the converter with the same input power from WTGs are shown in Figure 11. All groups work under the column interleaved strategy, and the voltages of the adjacent cells in the same column have opposite polarities. Meanwhile, the currents of cells with diverse leakage inductances in column 1 are almost equal. The voltage stress of power switches is 1/4 of the output voltage, and all diodes have the voltage and current stresses as low as 1/4 or 1/2 of the output voltage and current, respectively.

For Figure 12, every group has different numbers of rows. Only one row operates in Group 1; in Group 2, there are two rows; Group 3 has four rows. The duty-cycle D of cells in each group is regulated to obtain the same terminal voltage, and the voltages of cells are shown in Figure 12 as: $V_{s\_cell\ 1\text{--}11} = 2 \times V_{s\_cell\ 2\text{--}11} = 4 \times V_{s\_cell\ 3\text{--}11} = V_{out}$.

The peak current values of different columns are almost equal so that the duty-cycle power transmission control presented in Equation (15) is verified.

Figure 13 shows the voltages of diodes in Group 1–2. The voltage of diodes in Group 3 has the same waveforms as those presented in Figure 10. It can be found that the voltage value is increased as the number of idle rows increases.

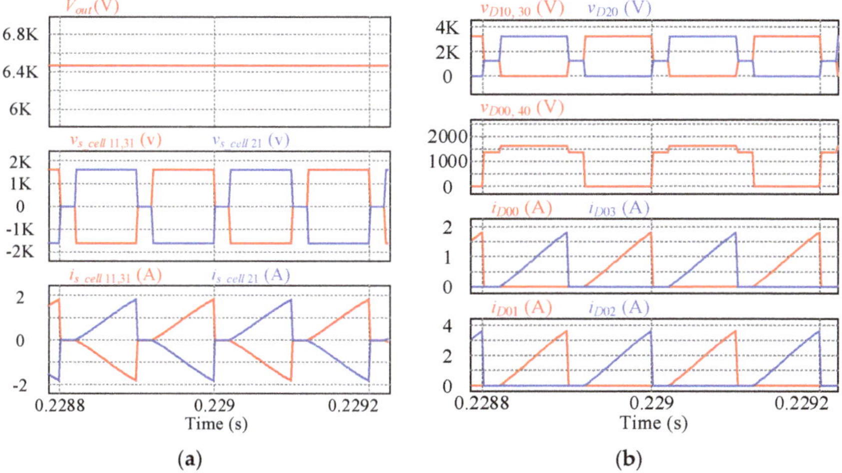

Figure 11. Steady-state waveforms for cells and diode: (**a**) voltage/current of basic cells; (**b**) voltage/current of diodes.

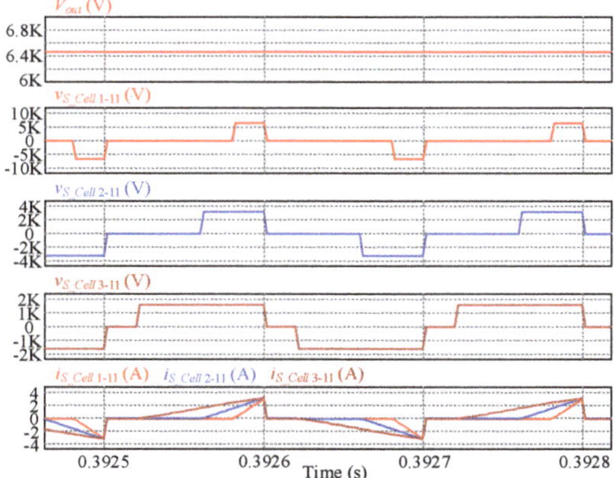

Figure 12. Waveforms of cells when Groups 1–3 have different output power.

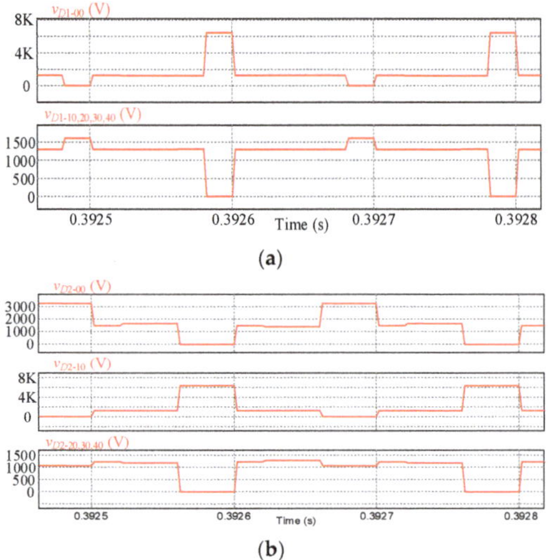

Figure 13. Voltages of diodes in each Group when Groups 1–3 have different output power: (**a**) Voltages of diodes in Group 1; (**b**) Voltages of diodes in Group 2.

As illustrated in Figure 14, before turning on switches S_1 or S_{C1}, the drain-to-source current flows through the parasitic diode of the switch to achieve ZVS operation. Similarly, the switches S_2 and S_{C2} can also obtain soft-switching. Therefore, the switching loss is significantly reduced.

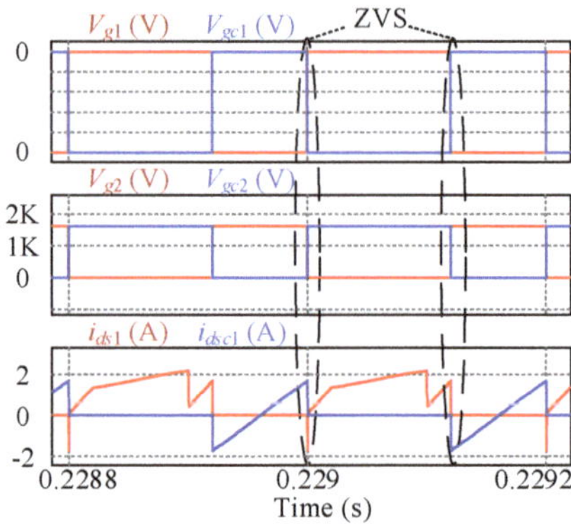

Figure 14. Zero voltage switching (ZVS) of main switches and clamping switches.

Figure 15a shows the fault tolerance operation with redundancy. At 0.07 s, fault cells 1–2 are idle, and the redundant columns 4–5 start to work to guarantee normal operation. Figure 15b,c presents the fault tolerance without redundancy, where only columns 1–3 are active. In Figure 15b, when the diode D_{11} is short-circuited, the cells in rows 1–2 are blocked. The voltages of cells in row 3–4 are doubled to achieve the same terminal voltage. The diode D_{11} is open in Figure 15c. The faulty group consisting of columns 1–2 works in the series interleaved mode by changing the polarities of cells. Moreover, the voltages of cells in the faulty group are reduced to half of that in the normal operation group to obtain the same output voltage.

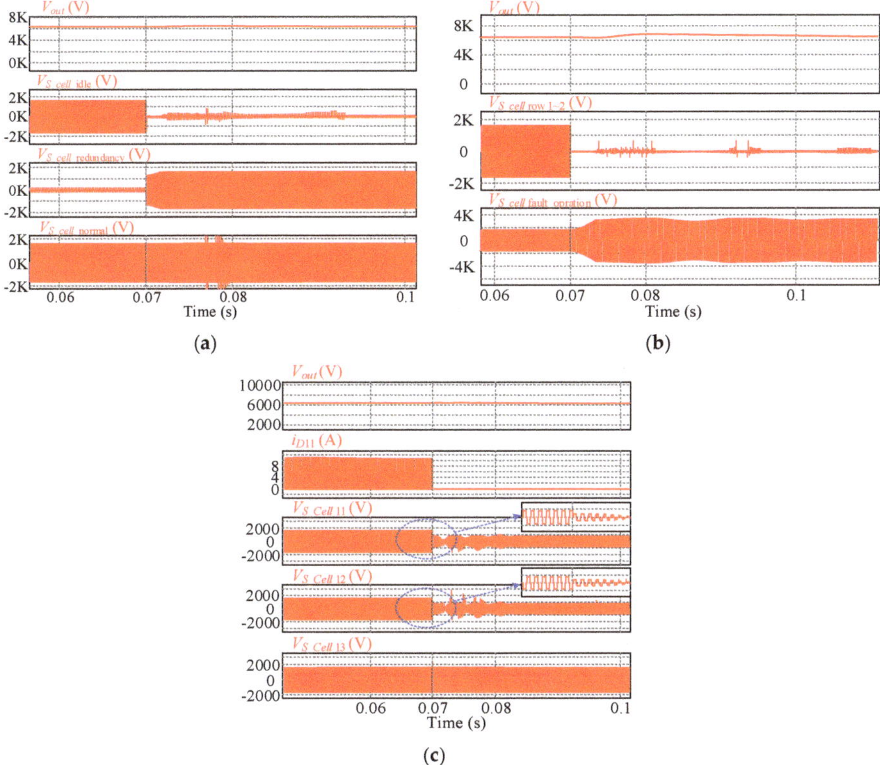

Figure 15. Fault tolerance operation: (**a**) With redundancy; (**b**) Without redundancy for D_{11} short-circuit; (**c**) Without redundancy for D_{11} open-circuit.

Where N is the number of SMs or power cells; n is the turns ratio of transformers. Table 2 shows a performance comparison among several literatures. Due to hard switching, References [21,23] have higher switching losses. Multi-module converters can achieve a higher voltage step-up ratio by increasing the power cells number and the voltage gain of power cells. For MMCs, fault tolerance is achieved by redundancy. However, the proposed converter has an improved fault tolerance capability which can maintain normal operation without redundancy. Therefore, to achieve the same performance, a multi-module converter with CFPP cells requires fewer components.

Table 2. Performance comparison.

	MMCs		Multi-Module Converters	
	HB/FB [21,32]	**Resonant MMC [20]**	**Flyback–Forward [23]**	**CFPP Converter**
Soft-switching	✗	✓	✗	✓
Conversion ratio	$\propto N, n$	$\propto N$	$\propto N, \frac{1}{1-D}, n$	$\propto N, \frac{1}{1-D}, n$
Transformer	Large	✗	Small	Small
Fault tolerance	Low	Low	Medium	High

6. Conclusions

A multi-port high voltage gain modular DC/DC power converter applied in offshore wind farms is proposed in this paper. Thanks to the modularity, the high output voltage and power is achieved by adding power cells. With the independent operation of each port and high control flexibility, the converter can collect power from multi-sources without bus cable. Additionally, the CFPP cells reduce the switching losses and control complexity.

The performances of MMC with galvanic isolation, resonant MMC, multi-module converter with flyback–forwarding cells and CFPP cells are compared. The proposed model appears to be more efficient and reliable, including fewer switching losses, higher conversion ratio, fewer components, smaller volume, and higher reliability.

The simulation results verify the advantages of the proposed converter are the soft-switching of all power switches, flexible control, and improved fault tolerance operation.

Author Contributions: Conceptualization, Y.H. and G.C.; Formal analysis, S.S. and Y.H.; Investigation, S.S.; Methodology, S.S., Y.H. and G.C.; Software, S.S. and G.C.; Supervision, Y.H.; Validation, S.S.; Writing—original draft, S.S.; Writing—review and editing, K.N., J.Y., H.W. and X.Y.

Funding: This study is supported by the State Key Laboratory of Alternate Electrical Power System with Renewable Energy Sources under Grant LAPS17022.

Conflicts of Interest: The authors declare no conflict of interest.

References

1. Zervos, A.; Kjaer, C. *Pure Power. Wind Energy Scenarios up to 2030*; The European Wind Energy Association (EWEA): Brussels, Belgium, 2006.
2. Council, G.W.E. Global Wind Report 2016. Available online: http://files.gwec.net/files/GWR2016.pdf?ref=Website (accessed on 25 June 2018).
3. Remy, T.; Mbistrova, A. Offshore Wind in Europe-Key trends and statistics 2017. Available online: https://windeurope.org/wp-content/uploads/files/about-wind/statistics/WindEurope-Annual-Offshore-Statistics-2017.pdf (accessed on 25 June 2018).
4. Oni, O.E.; Davidson, I.E.; Mbangula, K.N. A review of LCC-HVDC and VSC-HVDC technologies and applications. In Proceedings of the 2016 IEEE 16th International Conference on Environment and Electrical Engineering (EEEIC), Florence, Italy, 7–10 July 2016.
5. Oates, C. Modular multilevel converter design for VSC HVDC applications. *IEEE J. Emerg. Sel. Top. Power Electron.* **2015**, *3*, 505–515. [CrossRef]
6. Van Eeckhout, B.; Van Hertem, D.; Reza, M.; Srivastava, K.; Belmans, R. Economic comparison of VSC HVDC and HVAC as transmission system for a 300 MW offshore wind farm. *Eur. Trans. Electr. Power* **2009**, *20*, 661–671. [CrossRef]
7. Chen, W.; Huang, A.Q.; Li, C.; Wang, G.; Gu, W. Analysis and comparison of medium voltage high power DC/DC converters for offshore wind energy systems. *IEEE Trans. Power Electron.* **2013**, *28*, 2014–2023. [CrossRef]
8. Meyer, C.; Hoing, M.; Peterson, A.; De Doncker, R.W. Control and design of DC grids for offshore wind farms. *IEEE Trans. Ind. Appl.* **2007**, *43*, 1475–1482. [CrossRef]
9. Wang, X.; Cao, C.; Zhou, Z. Experiment on fractional frequency transmission system. *IEEE Trans. Power Syst.* **2006**, *21*, 372–377.
10. Vechalapu, K.; Bhattacharya, S.; Van Brunt, E.; Ryu, S.-H.; Grider, D.; Palmour, J.W. Comparative evaluation of 15-kv sic mosfet and 15-kv sic igbt for medium-voltage converter under the same dv/dt conditions. *IEEE J. Emerg. Sel. Top. Power Electron.* **2017**, *5*, 469–489. [CrossRef]
11. Nami, A.; Liang, J.; Dijkhuizen, F.; Demetriades, G.D. Modular Multilevel Converters for HVDC Applications: Review on Converter Cells and Functionalities. *IEEE Trans. Power Electron.* **2015**, *30*, 18–36. [CrossRef]
12. Ghat, M.B.; Shukla, A. A New H-Bridge Hybrid Modular Converter (HBHMC) for HVDC Application: Operating Modes, Control, and Voltage Balancing. *IEEE Trans. Power Electron.* **2018**, *33*, 6537–6554. [CrossRef]
13. Gowaid, I.; Adam, G.; Massoud, A.M.; Ahmed, S.; Williams, B. Hybrid and Modular Multilevel Converter Designs for Isolated HVDC–DC Converters. *IEEE J. Emerg. Sel. Top. Power Electron.* **2018**, *6*, 188–202. [CrossRef]

14. Soltau, N.; Stagge, H.; De Doncker, R.W.; Apeldoorn, O. Development and demonstration of a medium-voltage high-power dc-dc converter for dc distribution systems. In Proceedings of the 2014 IEEE 5th International Symposium on Power Electronics for Distributed Generation Systems (PEDG), Galway, Ireland, 24–27 July 2014.

15. Engel, S.P.; Soltau, N.; Stagge, H.; De Doncker, R.W. Dynamic and balanced control of three-phase high-power dual-active bridge DC–DC converters in DC-grid applications. *IEEE Trans. Power Electron.* **2013**, *28*, 1880–1889. [CrossRef]

16. Chen, W.; Wu, X.; Yao, L.; Jiang, W.; Hu, R. A step-up resonant converter for grid-connected renewable energy sources. *IEEE Trans. Power Electron.* **2015**, *30*, 3017–3029. [CrossRef]

17. Meyer, C.; De Doncker, R.W. Design of a three-phase series resonant converter for offshore DC grids. In Proceedings of the 42nd IAS Annual Meeting. Conference Record of the 2007 IEEE on Industry Applications Conference, New Orleans, LA, USA, 23–27 September 2007.

18. Kish, G.J.; Ranjram, M.; Lehn, P.W. A modular multilevel DC/DC converter with fault blocking capability for HVDC interconnects. *IEEE Trans. Power Electron.* **2015**, *30*, 148–162. [CrossRef]

19. Martinez-Rodrigo, F.; Ramirez, D.; Rey-Boue, A.B.; de Pablo, S.; Herrero-de Lucas, L.C. Modular Multilevel Converters: Control and Applications. *Energies* **2017**, *10*, 1709. [CrossRef]

20. Zhang, X.; Xiang, X.; Green, T.C.; Yang, X. Operation and Performance of Resonant Modular Multilevel Converter with Flexible Step Ratio. *IEEE Trans. Ind. Electron.* **2017**, *64*, 6276–6286. [CrossRef]

21. Kenzelmann, S.; Rufer, A.; Dujic, D.; Canales, F.; De Novaes, Y.R. Isolated DC/DC structure based on modular multilevel converter. *IEEE Trans. Power Electron.* **2015**, *30*, 89–98. [CrossRef]

22. Gowaid, I.; Adam, G.; Massoud, A.M.; Ahmed, S.; Holliday, D.; Williams, B. Quasi two-level operation of modular multilevel converter for use in a high-power DC transformer with DC fault isolation capability. *IEEE Trans. Power Electron.* **2015**, *30*, 108–123. [CrossRef]

23. Hu, Y.; Zeng, R.; Cao, W.; Zhang, J.; Finney, S.J. Design of a Modular, High Step-Up Ratio DC–DC Converter for HVDC Applications Integrating Offshore Wind Power. *IEEE Trans. Ind. Electron.* **2016**, *63*, 2190–2202. [CrossRef]

24. Chen, G.; Deng, Y.; He, X.; Hu, Y.; Jiang, L. Analysis of high voltage gain DC-DC converter with active-clamping current-fed push-pull cells for HVDC-connected offshore wind power. In Proceedings of the 42nd Annual Conference of the IEEE Industrial Electronics Society (IECON 2016), Florence, Italy, 23–26 October 2016.

25. Mohammadpour, A.; Parsa, L.; Todorovic, M.H.; Lai, R.; Datta, R.; Garces, L. Series-input parallel-output modular-phase dc–dc converter with soft-switching and high-frequency isolation. *IEEE Trans. Power Electron.* **2016**, *31*, 111–119. [CrossRef]

26. Xiang, X.; Zhang, X.; Chaffey, G.P.; Green, T.C. An Isolated Resonant Mode Modular Converter with Flexible Modulation and Variety of Configurations for MVDC Application. *IEEE Trans. Power Deliv.* **2017**, *33*, 508–519. [CrossRef]

27. Nome, F.J.; Barbi, I. A ZVS clamping mode-current-fed push-pull DC-DC converter. In Proceedings of the IEEE International Symposium on Industrial Electronics (ISIE'98), Pretoria, South Africa, 7–10 July 1998.

28. Duong, M.Q.; Leva, S.; Mussetta, M.; Le, K.H. A Comparative Study on Controllers for Improving Transient Stability of DFIG Wind Turbines during Large Disturbances. *Energies* **2018**, *11*, 480. [CrossRef]

29. Wang, H.; Wang, Y.; Duan, G.; Hu, W.; Wang, W.; Chen, Z. An Improved Droop Control Method for Multi-Terminal VSC-HVDC Converter Stations. *Energies* **2017**, *10*, 843. [CrossRef]

30. Yang, Z.; Chai, Y. A survey of fault diagnosis for onshore grid-connected converter in wind energy conversion systems. *Renew. Sustain. Energy Rev.* **2016**, *66*, 345–359. [CrossRef]

31. Zhang, W.; Xu, D.; Enjeti, P.N.; Li, H.; Hawke, J.T.; Krishnamoorthy, H.S. Survey on fault-tolerant techniques for power electronic converters. *IEEE Trans. Power Electron.* **2014**, *29*, 6319–6331. [CrossRef]

32. Zhang, X.; Green, T.C. The modular multilevel converter for high step-up ratio DC–DC conversion. *IEEE Trans. Ind. Electron.* **2015**, *62*, 4925–4936. [CrossRef]

Article

Day-Ahead Probabilistic Model for Scheduling the Operation of a Wind Pumped-Storage Hybrid Power Station: Overcoming Forecasting Errors to Ensure Reliability of Supply to the Grid

Jakub Jurasz [1,*] and Alexander Kies [2,*]

[1] Department of Engineering Managment, AGH University of Science and Technology, 30-059 Kraków, Poland
[2] Frankfurt Institute for Advanced Studies, Goethe University Frankfurt, 60438 Frankfurt am Main, Germany
*   Correspondence: jakubamiljurasz@gmail.com (J.J.); kies@fias.uni-frankfurt.de (A.K.);
    Tel.: +48-792-612-485 (J.J)

Received: 19 May 2018; Accepted: 11 June 2018; Published: 13 June 2018

Abstract: Variable renewable energy sources (VRES), such as solarphotovoltaic (PV) and wind turbines (WT), are starting to play a significant role in several energy systems around the globe. To overcome the problem of their non-dispatchable and stochastic nature, several approaches have been proposed so far. This paper describes a novel mathematical model for scheduling the operation of a wind-powered pumped-storage hydroelectricity (PSH) hybrid for 25 to 48 h ahead. The model is based on mathematical programming and wind speed forecasts for the next 1 to 24 h, along with predicted upper reservoir occupancy for the 24th hour ahead. The results indicate that by coupling a 2-MW conventional wind turbine with a PSH of energy storing capacity equal to 54 MWh it is possible to significantly reduce the intraday energy generation coefficient of variation from 31% for pure wind turbine to 1.15% for a wind-powered PSH The scheduling errors calculated based on mean absolute percentage error (MAPE) are significantly smaller for such a coupling than those seen for wind generation forecasts, at 2.39% and 27%, respectively. This is even stronger emphasized by the fact that, those for wind generation were calculated for forecasts made for the next 1 to 24 h, while those for scheduled generation were calculated for forecasts made for the next 25 to 48 h. The results clearly show that the proposed scheduling approach ensures the high reliability of the WT–PSH energy source.

Keywords: variable renewable energy sources; hybrid energy sources; scheduling; non-dispatchability; reliability

---

1. Introduction

The randomness and high variability of wind power strongly limits its potential to be a reliable energy source. However, its availability in certain parts of the world, its economic superiority over other energy sources, and government-driven financial incentives have all led to it acquiring a significant role in several existing power systems. Due to the limited predictability of wind generation and legal regulations which give priority to wind sources, the operation of conventional power plants must be adjusted not only to varying energy demand, but now also to the fluctuating energy yield from wind turbines.

The problem of forecasting energy demand has been addressed in a great number of scientific papers [1,2] and over time forecasting errors have been significantly reduced. However, with the advent of larger-scale VRES generation this problem has escalated on the supply side of the energy market. The problem of VRES integration has recently been scrutinised by [3–5], whereas [6,7] have summarised approaches which can be used to ease and overcome this problem. Those include, inter alia: energy storage; spatial and temporal complementarity of energy sources; demand side management (DMS); hybrid energy sources; vehicle-to-grid (V2G); power generation capacity oversizing; and transmission lines connecting energy systems in different locations. This paper, to some extent, builds upon those concepts and aims to overcome the problem of the inherent variability of wind generation by coupling it with almost the only bulk (and certainly the most mature) form of energy storage in the form of PSH, and by proposing a novel scheduling formula for the next 25 to 48 h.

1.1. WT–PSH Hybrid Energy Sources

A great number of scientific papers has been dedicated to the concept of wind turbines operating together with an energy storage facility in the form of PSH. Due to the popularity of this concept as a solution for supplying isolated communities with electricity, the majority of authors focus on WT–PSH hybrids on islands. At least five papers [8–12] have been dedicated to the WT–PSH system on Lesbos Island (the Aegean Sea), with the authors of those papers claiming that such hybrids can significantly increase the utilisation of wind energy and lead to greater penetration by RES, translating into lower consumption of the oil so commonly used in off-grid systems. A paper written by Karsaprakakis et al. [13] points out that Greek law defines the WT–PSH hybrid as a conventional power plant with full dispatchability. In those authors' opinions, such a hybrid energy sources does not have to provide power all the time and its use can be limited to specific periods in order to facilitate the operation of other energy sources. Two more papers [14,15] investigated the WT–PSH concept on Ikaria Island (the Aegean Sea). Similarly, as in our research, the authors of the aforementioned papers have assumed that the available energy from the WT will first be pumped into the upper reservoir and then dispatched when needed. Additionally, the PSH upper reservoir can be, to some extent, supplied by the excess flow from the additional reservoir which is located above it and used for irrigation purposes.

Three papers [16–18] investigated the possibility of covering the energy needs of a local community inhabiting a small island near Hong Kong. The analyses presented in these papers are part of a larger project aiming to ensure total energy autarky by means of a WT–PSH hybrid cooperating with a PV installation. Results note, for example, the environmental aspects of large scale battery banks and the superiority of PSH in this regard.

Some articles have also been dedicated to mainland WT–PSH hybrids. Hessami and Bowly [19] investigated a 190-MW wind farm connected to various forms of energy storage and found the optimal one to guarantee maximal revenue. Of the three considered, namely PSH, compressed air energy storage (CAES), and thermal energy storage (TES), CAES seems to be the most promising in terms of the rate of return (ROR) criterion. Hedegaard and Meilbom [20] analyzed the WT–PSH concept from the perspective of the Danish energy system, where wind generation plays an important role in covering energy needs. According to the authors' results, wind-powered PSHs can both be used as a first-tier reserve power plant with very swift reaction times, and also be applied to compensate for the seasonal variability of WT generation. Canales et al. [21] and Murage and Anderson [22] analyzed the WT–PSH hybrid from the Brazilian and Kenyan points of view, respectively. Canales et al. [21] point out that PSH facilities have larger upfront costs than traditional hydropower plants with pumping installations, but that their operational costs and environmental impact are significantly smaller.

Scheduling WT–PSH Operation

To the best of the author's knowledge, the issue of scheduling the operation of wind-powered PSH has been the subject of the following papers:

- Varkani et al. [23] proposed the concept of a self-scheduling strategy which is based on stochastic programming techniques and aims to minimise the impact of wind power generation's uncertainty. The strategy presented by the authors can be applied for day-ahead energy generation offers. The uncertainty of the wind generation forecasts is modelled by means of an artificial neural network (ANN);

- Tan et al. [24] developed a two-stage scheduling optimisation model for the operation of a wind power and energy storage system based on day-ahead and ultra-short-term wind power forecasting. The results indicate that synergies of demand response and energy storage can be used to overcome wind generation variability and lead to an improvement in utilisation of wind energy, as well as decreasing coal consumption. Similarly, as in [23], the uncertainty of wind power was simulated, in this case being based on a scenario analysis;

- Jurasz and Mikulik [25] suggested a model for scheduling the operation of WT–PSH based on the occupancy of the upper reservoir. It was assumed that the sets of pumps and generators of the PSH can operate simultaneously. The results indicate that WT–PSH operating on such a formula can be a reliable energy source operating as a baseload (with potential modifications regarding when the energy from the upper reservoir should be dispatched). Additionally, the scheduling formula eliminates the occurrence of wind energy rejection. However, the schedule of the WT–PSH operation for the next 1 to 24 h is known only one hour ahead;

- Castronuovo nd Lopes [26] proposed a model for improving wind parks' operational economic gains by means of PSH facilities. Using an hourly-discretised algorithm the authors identified an optimal daily operation strategy assuming that wind power forecasts were available; and

- Papaefthimiou et al. [27] suggested several possible operation modes for WT–PSH focusing on their verbal description. In general: the WT–PSH can submit an energy offer for the next 24 h considering the state of charge of its upper reservoir; the system operator dispatches the PSH operation mainly during peak load hours or to shave the peaks; the energy generated by the WT is stored in the upper reservoir and the PSH pumps balance the varying WT energy yield—basically, using variable-speed pumps the PSH is able to constantly adjust to the changes in wind generation; the system operator requires the WT–PSH to deliver power in certain periods in order to fill deficits resulting from the limited capacity of the conventional power sources; and the WT–PSH is designed to deliver guaranteed power, but if the upper reservoir state of charge is not sufficient and/or the wind generation forecasts are unfavourable then it is admissible to use electricity from the grid and dispatch it when required. According to their results, the authors state that the proposed operating policies enable the integration of wind generation on a larger scale.

Clearly this problem is investigated in the literature, but the number of papers and approaches is limited. The authors of the aforementioned papers have investigated the problem of WT–PSH operation from different perspectives. Therefore, the ability to draw general conclusions is constrained. The common denominator of the majority of those papers is that the authors point to the relevance of the accuracy of wind speed forecasts and the increasing role of wind generation in today's power systems.

1.2. Wind Speed Forecasting

As mentioned by [6,7] and the authors of the papers presented in Section 1.1 [23–28], forecasting techniques are a possible tool for smoothing the integration of variable renewable energy sources into the power system. Additionally, accurate forecasts with low absolute and percentage errors can make the process of scheduling the operation of conventional power plants more accurate and economically feasible.

Soman et al. [29] provided an overview of current wind power and wind speed forecasting methods with different time horizons. Depending on the application, the wind forecasting can be divided into four time horizon categories: very short term forecasting (from seconds to 30 min ahead), short term forecasting (from 30 min to 6 h ahead), medium term forecasting (from 6 h to 1 day ahead), and long term forecasting (ranging from one day ahead to one week ahead). The ranges of individual categories are not fixed and can be modified. From the above we conclude that the scheduling formula proposed in Section 2.2. would be based on a combination of short and medium-term forecasting.

Advances in computing power, as well as deeper and broader understanding of various phenomena, have led to the rapid development of various forecasting tools and approaches [30]. The majority of them are also being used in the area of wind speed forecasting. The most commonly-applied techniques are based on: persistence methods/naïve predictors, which are usually used as benchmarks; physical approaches, which use numerical weather predictors (NWP); statistical approaches based on time series analysis or artificial neural networks (ANN); and emerging methods based on wavelet transformation or fuzzy logic (hybrid approaches which develop new hybrid tools which utilise various combinations of, for example, ANN + wavelet transformation or ANN + fuzzy logic).

Recently published papers have mostly concentrated on various hybrid approaches [31–33] and investigated the impact of various phenomena, such as the spatial distribution on forecasts' error correlation [34]. The abundance of literature and the presence of various concepts and tools for forecasting wind speed or wind power shows the complexity of this problem and proves its importance in the context of the VRES integration into the power system. As an example of papers which clearly show the complexity of VRES integration which applied dynamic programing to overcome the problems of models non-linearity and non-convexity we refer the reader to the works of Korkas et al. [35] and Baldi et al. [36]. Other examples of wind-solar or solar integration by means of hydropower to the national power system can be found in [37–39].

In this paper we do not aim to develop a new method or apply the existing ones. The potential user of our scheduling approach can arbitrarily select the forecasting methods which best suits his needs or are the best available solution for the location of a given wind farm. Such aspects of wind speed forecasting and the role of their accuracy in RES-based hybrid energy sources has been investigated in [40]. For the purpose of this paper the forecasting errors have been generated based on the assumption of their normal distribution and percentage value increasing with an elongating time horizon—detailed information on this is provided in Section 3.2.

1.3. WT–PSH

The main idea behind the scheduling concept introduced in this paper is that it is possible to overcome the inherent variability of wind energy by using the storage capacity of the upper reservoir to offset the inaccuracy of wind speed forecasts. The conceptual design of the wind powered PSH has been presented in Figure 1.

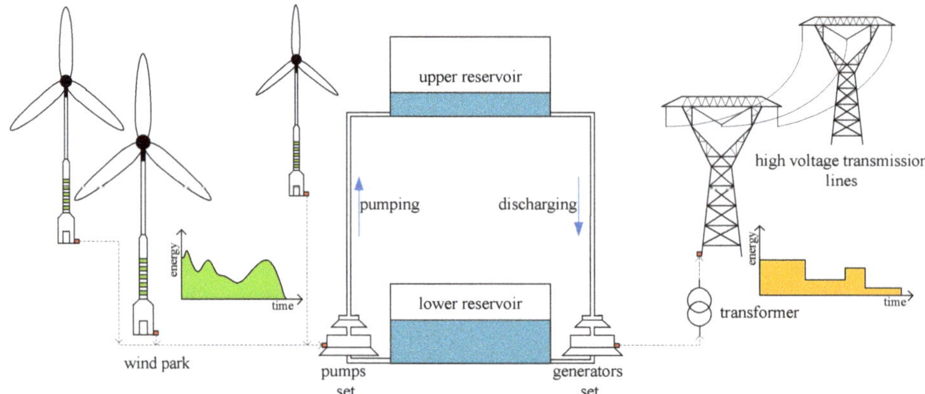

Figure 1. Conceptual design of WT–PSH hybrid along with energy and water flows and schematic charts visualising the initial and resulting variability of energy generation.

The operation of such a system is based on day-ahead schedules created to determine the volume of energy which should be discharged over the next 25 to 48 h. The energy available from the WTs is always firstly stored in the upper reservoir. Naturally, the situation may occur in which the reservoir is full and some portion of WT-derived energy will have to be fed to the grid. Such situations, as well as partially realised schedules, should be avoided because they introduce undesirable disruptions into the operation of the energy market.

1.4. Paper Objectives

Considering the aspects and potential role of WT–PSH hybrid energy sources described above, this paper aims to realise the following goals:

- to develop a MINLP model for simulating the operation of the WT–PSH hybrid and scheduling its operation;
- to establish measures for estimating the model's accuracy; and
- to investigate the impact of the scheduling formula parameters and upper reservoir capacity on WT–PSH performance.

2. Mathematical Model

In this paper we have used a mathematical model which has been used for a discretised hourly wind speed time series and can be applied in various simulation and optimisation software packages.

2.1. Energy Yield from Wind Turbines

The energy generated by a wind turbine can be estimated based on the known wind speed and the power curve of a given type of wind generator. Here, a commonly-applied Gamesa G90 [41] wind turbine has been selected for simulation purposes. Figure 2 visualises its power curve for wind speeds ranging from 3 m/s (v^{cut-in}) to 12 m/s (v^{rated}), with the operation of WT being stopped when the wind speed exceeds 25 m/s ($v^{cut-off}$). In the model introduced in Section 2.2, the equation is used to calculate the actual (EWT) and forecasted (EWT*) energy yield from wind. For the forecasted values, predicted wind speeds (v^*) were used as shown in Section 3.2.

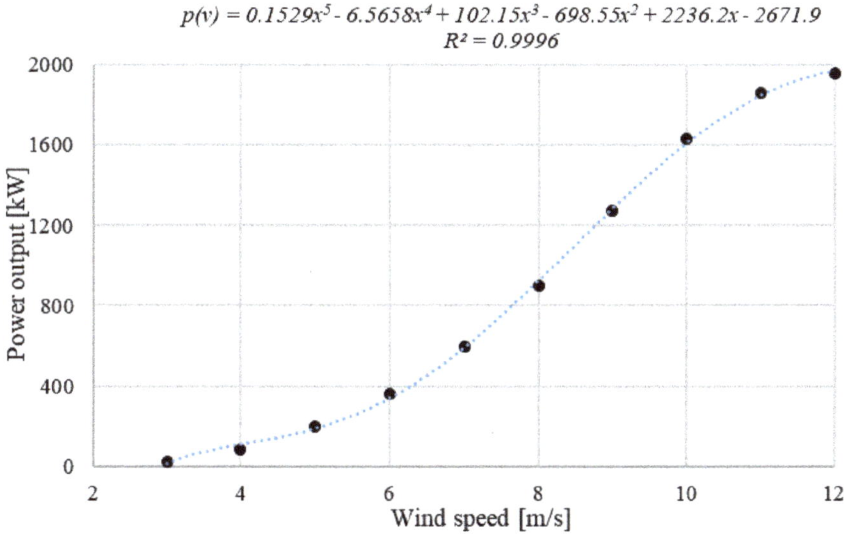

Figure 2. The Gamesa G90 power curve.

$$E^{WT} = \begin{cases} P^{WT} \times t \times n \ for \ v \in \left\langle v^{rated}; v^{cut-off} \right) \\ p(v) \times t \times n \ for \ v \in \left\langle v^{cut-in}; v^{rated} \right) \\ 0 \ otherwise \end{cases} \qquad (1)$$

where E^{WT} is the energy yield from wind turbine (kWh), P^{WT} is the nominal capacity of wind turbine (kW), t is the time (1 h), n is the number of wind turbines (-), v^{rated} is the wind speed at which the wind turbine starts to operate with its nominal capacity (m/s), $v^{cut-off}$ is the wind speed at which turbine operation is stopped for safety reasons (m/s), $p(v)$ is the polynomial approximating wind turbine power curve (kW), and v^{cut-in} is the wind speed at which wind turbine starts to generate electricity (m/s).

2.2. WT–PSH Operation

In this section we introduce and describe the approach used to schedule the operation of the WT–PSH. In the following paragraphs it is visualised in Figure 3, a verbal algorithm and a mathematical formulation.

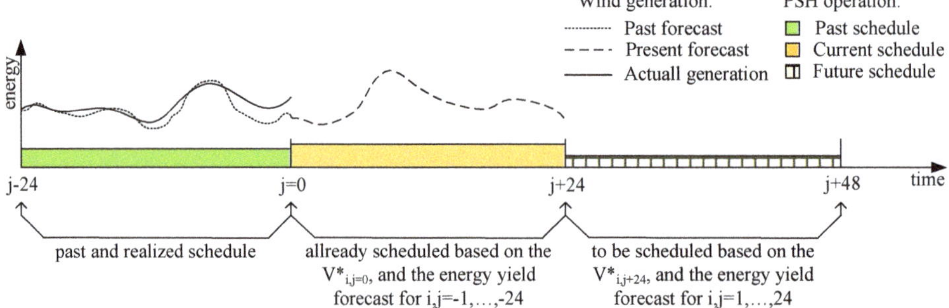

Figure 3. Chart depicting the WT–PSH scheduling method.

WPSH Operation

Generally speaking, the schedule creation procedure can be summarised in the five steps listed hereunder:

Step 1. Obtain/generate wind turbine energy yield forecast for the next 24 h.

Step 2. Determine the upper reservoir state of filling at the end of the upcoming day ($j = 24$) based on the already scheduled discharge covering the next 24 h ($j = 1, \ldots, 24$) and possible energy generation from wind turbines calculated in the first step.

Step 3. Considering the estimated reservoir occupancy of the upper reservoir in step 2, generate a uniform discharge schedule for the next 25 to 48 h ($j = 25, \ldots, 48$).

Step 4. After realising the previously-planned schedule for the period ($j = 1, ..., 24$) juxtapose the scheduled and actual operation of the PSH as well as the wind turbine generation and upper reservoir occupancy. In case of any discrepancies replace the estimated occupancy of the upper reservoir calculated in step 2 by the actual one.

Step 5. Return to the step 1 and repeat the whole procedure.

The verbal description of the procedure described above can also be expressed by means of several mathematical formulas and equations. As mentioned earlier the first step is to determine the volume of energy stored in the upper reservoir. Here, an assumption has been made that the initial reservoir occupancy ($i = 1, j = 0$) is half its maximal capacity. It is important to emphasise that, for simulation purposes, we have been simultaneously calculating the reservoir occupancy for a scheduled PSH operation and forecasted wind power generation (Equation (2)), as well as for actual PSH operation and actual WT energy yield (Equation (3)). This was done in order to assess the accuracy and quality of the proposed scheduling approach. Such an approach is necessary due to the intermittent nature of wind generation and the non-ideal accuracy of wind generation forecasts. The idea behind Equations (2) and (3) is as follows: within each unit of time (hour) a certain volume of energy is being discharged from the upper reservoir and, simultaneously, some energy from wind generation is being pumped upwards. The calculating procedure must ensure that the characteristics of the PSH will not be exceeded, meaning that the calculated momentary volume of energy stored in the upper reservoir is not smaller than zero or greater than its maximal storage capacity. Additionally, the momentary energy storage capacity of the PSH is limited by the throughput of its pumps; therefore, from the available wind generation only a volume ranging from 0 to E^{Pump} can be considered as possible for storing:

$$V_{i,j}^* = \min(\max(V_{i,j-1}^* - E_{i,j}^C \frac{1}{\eta^{PSH}} + \min(E_{i,j}^{WT*}\eta^{PSH}; E^{Pump}); 0); V^M) \tag{2}$$

$$V_{i,j} = \min(\max(V_{i,j-1} - E_{i,j}^C \frac{1}{\eta^{PSH}} + \min(E_{i,j}^{WT}\eta^{PSH}; E^{Pump}); 0); V^M) \tag{3}$$

where: $V_{i,j}^*$ is the forecasted volume of energy stored in the upper reservoir (kWh), $V_{i,j}$ is the actual volume of energy stored in the upper reservoir (kWh), $E_{i,j}^C$ is the scheduled energy generation from the PSH (kWh), η^{PSH}—pumping or generating efficiency of the PSH [%], $E_{i,j}^{WT*}$ is the forecasted energy yield from the wind generation (kWh), E^{Pump} is the maximal energy throughput of the PSH pumps (kWh), and V^M is the PSH upper reservoir energy storing capacity (kWh).

In Equations (2) and (3) we have used the scheduled PSH generation value ($E_{i,j}^C$). This can be calculated based on Equation (4) which is the most important part of the presented scheduling method. The schedule for the next 25 to 48 h ($i + 2, j = 1, \ldots, 24$) is generated based on the estimated occupancy of the upper reservoir and the predicted energy yield from wind turbines. Consequently, the exact volume of the available energy for the next considered scheduling period is uncertain. Therefore, to the proposed formula we have introduced two additional parameters (β, α) whose potential values are from 0 to 1. Their main task is to compensate for the inaccuracy of the estimated available energy in the upper reservoir (β) and the energy yield from wind turbines (α). It is important to highlight that the scheduling formula introduced in Equation (4) leads to a uniform (uniform in theory only, due to forecasting errors) energy generation from the PSH over the next 25 to 48 h. Those generation values will be known a day ahead. Equation (4) also ensures that the scheduled energy discharge from the PSH will not be greater than its energy generating capacity (E^{Gen}).

Naturally, those energy generation patterns can be modified in such a manner that the PSH will generate electricity only during demand peak hours, only during daylight, or over periods when the energy price is highest, thereby increasing the owner's revenue:

$$E_{i+2, j = 1,\ldots,24}^C = \min\left(\min\left(\frac{V_{i+1,j=24}^*}{24}\eta^{PSH}; E^{Gen}\right)\beta + \min\left(\frac{\sum_{j=1}^{24} E_{i+1,j}^{WT*}}{24}\eta^{PSH}; E^{Gen}\right)\alpha; E^{Gen}\right) \tag{4}$$

As already mentioned, the scheduling formula is prone to forecasting errors and periods may consequently occur during which the WT–PSH hybrid is unable to realise the created energy generation schedule. Therefore, to assess the quality of our approach we have introduced Equation (5), which determines whether the schedule was realised or not and what the actual electricity generation ($E_{i,j}^{CR}$) was in comparison to that scheduled ($E_{i,j}^C$):

$$E_{i,j}^{CR} = \begin{cases} E_{i,j}^C \ for \ V_{i,j-1} \geq E_{i,j}^C \frac{1}{\eta^{PSH}} \\ V_{i,j-1}\eta^{PSH} \ otherwise \end{cases} \tag{5}$$

In the case of non-dispatchable energy sources coupled to the national power system (NPS) or any other type of grid, two distinctive situations may occur. Namely, energy generation is less than demand (in this paper the scheduled generation is as shown in Equation (5)) and an energy deficit occurs which should be covered by any other energy source or, alternatively, available energy is greater than demand and an energy surplus appears. In the case of the WT–PSH some part of the excess energy which has not been used in the scheduled generation can be stored in the upper reservoir. However, this is limited by the maximal capacity of the reservoir and, consequently, sometimes some volume of energy coming from WT may be neither used in the PSH generation schedule nor stored. This energy is rejected and its volume ($E_{i,j}^{WT\_R}$) can be calculated based on Equation (6). By rejection we mean a direct transfer of that energy to the grid—a situation which is not desirable because it increases variability and unpredictability on the energy market. The problem of an unexpected energy flow between hybrid energy sources based on PV and WT coupled with the PSH and the NPS has been investigated in [42] where it is revealed that, to some extent, they can be quite accurately forecasted:

$$E_{i,j}^{WT\_R} = \begin{cases} E_{i,j}^{WT} \eta^{PSH} - E^{Pump} \text{ for } E_{i,j}^{WT} \eta^{PSH} > E^{Pump} \\ E_{i,j}^{WT} - \left(V^M - \left(V_{i,j-1} - E_{i,j}^{C} \frac{1}{\eta^{PSH}}\right)\right) \text{ for } V_{i,j-1} - E_{i,j}^{C} \frac{1}{\eta^{PSH}} + E_{i,j}^{WT} \eta^{PSH} > V^M \\ 0 \text{ otherwise} \end{cases} \quad (6)$$

The situations described by Equations (5) and (6) are both undesirable events which should be avoided. Therefore, their volumes and occurrences will be the subject of the optimisation model introduced in Section 3.2.

3. Input Data

3.1. Wind Data

For simulation purposes, an hourly time series of wind speed data covering the years 2014–2016 for the Koszalin measuring station (Northern Poland—10 km straight-line distance from the Baltic Sea) was downloaded from [43]. Figure 4 visualises the variability of considered wind speeds.

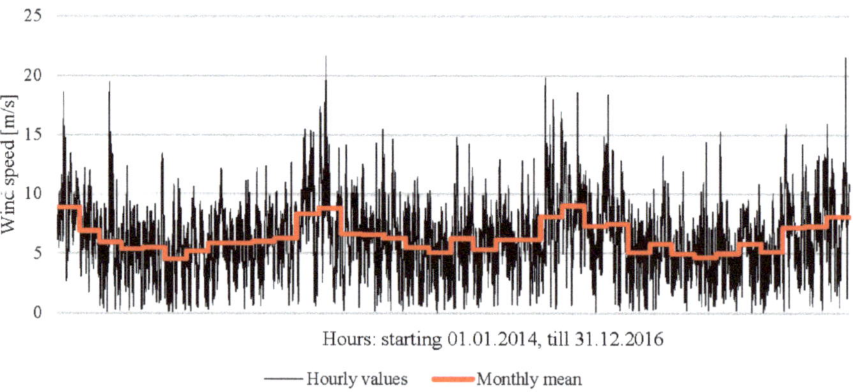

Figure 4. Wind speed variability over the 2014–2016 period.

3.2. Wind Speed Forecast Accuracy

Since no historical records on wind speed forecast quality were available, synthetic forecasts were generated based on the assumption that with the extending prediction horizon the accuracy of the forecast deteriorates. It is important to emphasise that the approach used here was made only for demonstration purposes and the assumptions of forecast mean error values and standard deviation of errors for each prediction horizon were determined arbitrarily. Naturally, in reality, better quality forecasts can be employed and, as a consequence, the proposed approach can be used as a benchmark. Figure 5 visualises the MAPE (mean absolute percentage error) values of errors for various time horizons. When the simulations were run, the average MAPE error over the whole considered period for forecasts ranging from 1 to 24 h ahead was 27%.

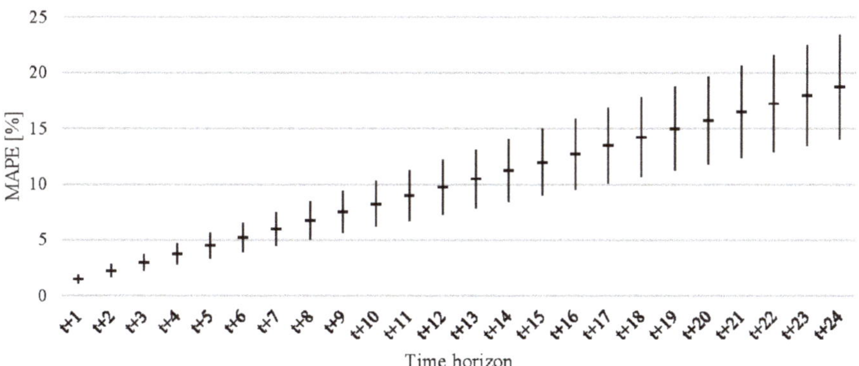

Figure 5. Forecast errors and their range for various prediction time horizons. The errors are normally distributed.

3.3. WT-PSH Parameters

For simulation purposes we have assumed that the WT–PSH hybrid will consist of a single wind turbine with a rated capacity of 2 MW coupled to a PSH facility which has a set of pumps capable of accommodating the maximal hourly yield from the WT. The power output of the water turbines used to generate electricity is equal to the set of pumps and is 2 MW. The energy storage capacity of the PSH (the volume of the upper reservoir) will be the subject of sensitivity analysis, with the minimal storage capacity calculated as $V^M = 24 \times 1/2 \times \eta^{PSH} \times P^{TW}$, or in words, half of the energy generated by the WT operating for 24 h at nominal capacity and taking into account the efficiency of the PSH pumps. The considered cycle efficiency of the PSH was 81%.

4. Results and Discussion

For the considered wind speed conditions, the 2 MW Gamesa G90 wind turbine should generate 5.33 GWh of electricity annually, which translates into a capacity factor of 30%. For the forecast quality presented in Section 3.2 the mean absolute forecast error was 131 kWh, while the lowest and highest were 0 kWh and 2000 kWh, respectively. Considering above in our analysis we have considered a WT-PSH hybrid which is characterized by the following parameters: a 2 MW Gamesa G90 wind turbine will be directly coupled with the PSH pump which is characterized by a 90% efficiency and installed capacity of 2 MW; the PSH generates electricity by means of hydro unit of 2 MW rated capacity and efficiency of 90%; the PSH upper reservoir storage capacity is subject of change and varies from 21.6 MWh to 172.8 MWh (with an increment of 10.8 MWh); the capacity of the lower reservoir is in the considered case not subject of analysis and was considered as infinite (or significantly larger than the upper reservoir).

4.1. WT-PSH Parameters

The accuracy of the introduced scheduling approach was assessed based on the MAPE criterion, whose value can be calculated from Equation (7). This criterion was further used in the optimisation model:

$$\text{MAPE} \;=\; \frac{1}{mn} \sum_{i=1}^{m} \sum_{j=1}^{n} \left| \frac{E_{i,j}^{C} - E_{i,j}^{CR}}{E_{i,j}^{C}} \right| 100\% \tag{7}$$

The precision of day-ahead predictions of the available quantity of produced energy was optimised by finding the minimal value of Equation (7) by changing the values of α and β parameters, which are part of the scheduling formula (Equation (4)) while also ensuring that the volume of the rejected energy from wind generation will not be greater than the constant value R (Equation (8)). In this case, R was arbitrarily equal to 5%:

$$\sum_{i=1}^{m} \sum_{j=1}^{n} E_{i,j}^{WT\_R} \leq R. \tag{8}$$

To find the optimal solution a *brute-force* method was applied and the subs of possible pairs of α and β parameters was limited by assuming that their values would range from $\langle 0; 1 \rangle$ with an increment of 0.1. Optimal and acceptable solutions were found for all considered upper reservoir capacities, which ranged from 21.6 to 172.8 MWh with an increment of 10.8 MWh. In total, 1815 various configurations of these three variables have been considered and 1356 (almost 75%) satisfied the imposed constraint given in Equation (8). Additional calculations revealed that by considering $R = 2.5\%$ this number dwindles to 1280 and to 1218 for $R = 1\%$. From the configurations which satisfied the constraint imposed in Equation (8) only one configuration for each capacity of the upper reservoir which exhibited the lowest objective function value (MAPE) was selected. The results are given in Table 1.

First of all, it is important to highlight that the coefficient of variation (CV) calculated for the wind turbine energy generation hourly time series is 101%, whereas for the intraday values it is 38%. Here we understand the intraday variability as the changes in energy generation which occurred within a single day. It is important to emphasise that such variability will have to be considered by the power system operator in scheduling the operation of the conventional power plants. However, by introducing the approach proposed in this paper those variations, and especially the intraday ones, can be significantly reduced.

Analysis of the results presented in Table 1 reveals that for relatively small capacities of the upper reservoir (S1–S4) the imposed constraint ($R < 5\%$) resulted in an increase in the observed discrepancies between the scheduled and realised operation of the PSH (this is indicated by the value of the MAPE criterion). In scenario S1, which considered the smallest capacity of the upper reservoir (equal to the half of the daily energy generation of the WT with its maximal capacity), we observed an increase in CV values for the hourly generation values (by 8.5% relative to the purely wind turbine generation) and a significant reduction in the intraday CV. The CV values calculated for the hourly time series over the whole considered period describe the general variability of the observed phenomena, whereas the intraday values point to the quality of the scheduling formula. In the best-case scenario, the CV values for the intraday energy generation should be equal to or close to zero. Such can be observed in scenarios S5–S15 which considered greater capacities of the upper reservoir. Therefore, the general conclusion is quite straightforward and common in the area of variable renewable energy sources: with an increasing upper reservoir capacity, the accuracy/realisability of generated schedules increases. This is indicated by the decreasing values of the MAPE and CV criteria.

In the scheduling formula the α parameter dictates to what extent wind energy generation forecasts should be considered in the scheduling procedure, whereas the β parameter refers to the forecasted upper reservoir state of filling. Interestingly (as shown in Table 1), for the small upper reservoir capacities (S1–S4) the optimal scheduling formula is almost entirely based on the value of the predicted upper reservoir state of filling, whereas the role of the WT forecasts for the next 24 h is neglected for $\alpha \leq 0.2$. This indicates that the upper reservoir is, to the same extent, capable of offsetting the inaccuracies in the wind generation forecasts. However, its limited capacity will often lead to situations in which an unreasonable (to bulk energy generation) schedule will ultimately deplete the reservoir or, conversely, the wind generation forecasting errors will lead to the inability of the upper reservoir to store excess energy which has not been considered in the schedule.

When the capacity of the upper reservoir is greater than 54 MWh, this tendency changes, and for scenarios S5–S9 the α is equal to 0.9 and for the remaining ones (S10–S15) it is 0.7. The greater alpha parameter and smaller beta (for S5–S9, $\beta = 0$ and for S10–S15, $\beta = 0.1$) mean that the upper reservoir capacity assumed in those scenarios is sufficient to compensate for the variability and inaccuracy of the wind generation forecasts. From the results presented in Table 1 we can also observe that increasing the upper reservoir capacity above 75.6 MWh does not significantly decrease the values of the MAPE and CV parameters. In general, by introducing the proposed scheduling formula we were able to reduce:

- the unpredictability of the wind generation from 27% (MAPE error for the wind generation forecasts) to less than 1.5% for scenarios S6–S15;
- the hourly variability of the wind generation time series by at least 10%, as in scenarios S2–S9, or almost 23%, as in scenarios S10–S15; and
- the intraday variability from 31% to less than 1%, as in scenarios S5–S10.

The last result is the most important contribution of the concept introduced in this paper. By combining WTs with the PSH, and applying an appropriate scheduling approach to them, we have obtained a dispatchable energy source with a known energy output for the next 25 to 48 h.

Table 1. Optimal values of α and β parameters for various upper reservoir capacities and $R \leq 5\%$.

Scenario	Variables			Mape [%]	CV [%]	
	V^M	α	β		Hourly	Intraday
S1	21.6	0.2	1	17.63	109.48	15.84
S2	32.4	0.1	1	4.85	90.78	2.31
S3	43.2	0.3	1	3.20	90.05	1.86
S4	54.0	0.1	1	2.39	90.02	1.15
S5	64.8	0.9	0	1.73	88.36	0.76
S6	75.6	0.9	0	1.37	88.22	0.61
S7	86.4	0.9	0	1.15	88.05	0.43
S8	97.2	0.9	0	1.15	88.05	0.43
S9	108.0	0.9	0	1.15	88.05	0.43
S10	118.8	0.7	0.1	1.10	78.37	0.27
S11	129.6	0.7	0.1	1.10	78.37	0.27
S12	140.4	0.7	0.1	1.10	78.37	0.27
S13	151.2	0.7	0.1	1.10	78.37	0.27
S14	162.0	0.7	0.1	1.10	78.37	0.27
S15	172.8	0.7	0.1	1.10	78.37	0.27

The varying capacity of the upper reservoir not only has an impact on the accuracy of the scheduling formula, but also determines the mean scheduled volume of energy. Figure 6 visualises this phenomena and indicates that for reservoir capacities equal to or greater than 54 MWh the mean scheduled generation ranges from 489 to 491 kWh. For the scenarios with smaller upper reservoir capacity (S1–S3) those values were smaller by 20 to 30 kWh. This shows that an increase in reservoir capacity above a certain value (here, 54 MWh) does not contribute to an increase in mean scheduled generation (values ranging from 489 to 491 kWh are exactly equal to the mean generation of the wind turbines which, for the considered location, were equal to 608 kWh decreased by the efficiency of the PSH) meaning that a limit for such a WT–PSH configuration has been reached. Therefore, it seems that there is no need to oversize the proposed system in terms of its energy storage capacities. This is supported by the mean upper reservoir state of filling depicted in Figure 6, which decreases with an increasing maximal reservoir storing capacity. Such a situation indicates the underutilisation of the available storing potential.

In Figure 7 we have visualised the operation of the WT–PSH energy source in January 2014, 2015, and 2016. As can be seen, the greater volume of scheduled generation from the PSH is usually preceded by an increase in wind generation, indicating that the scheduling formula correctly interprets both wind generation forecasts and the predicted upper reservoir state of filling, and makes correct decisions when it comes to the energy generation schedule.

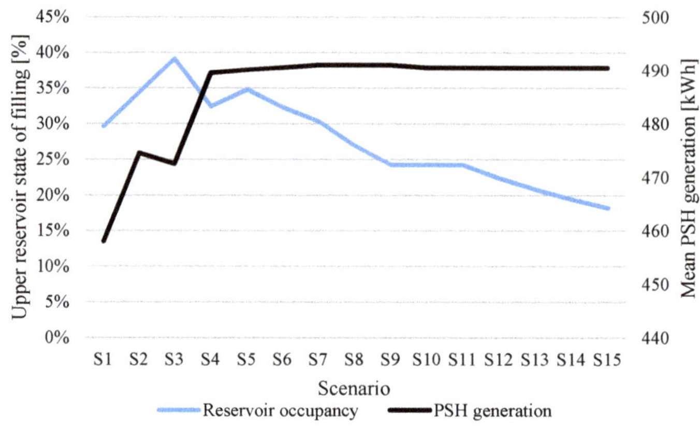

Figure 6. Mean values of the upper reservoir state of filling and the PSH generation over the years 2014–2016.

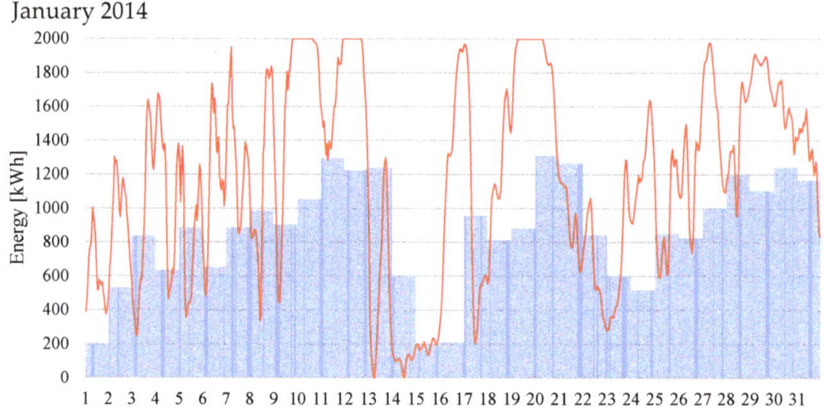

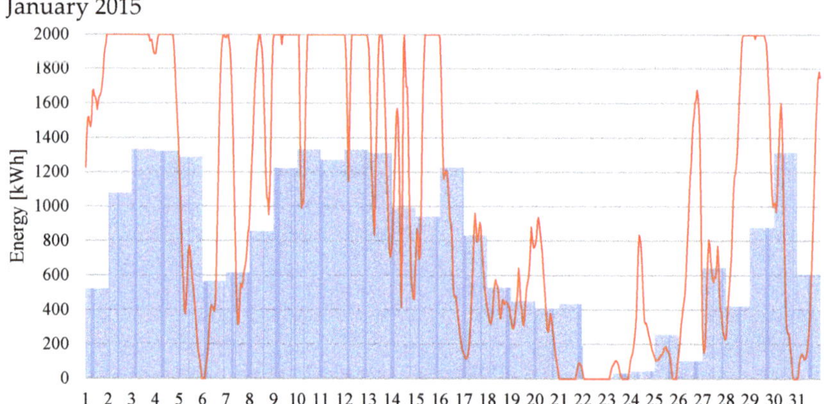

Figure 7. *Cont.*

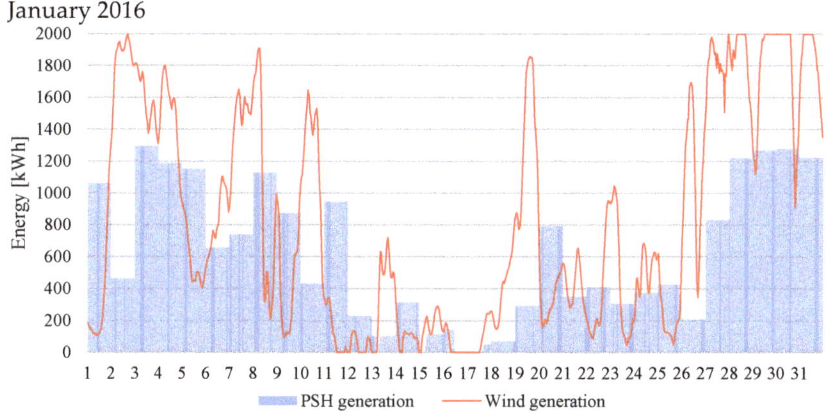

Figure 7. Operation of the WT–PSH hybrid in scenario S4. *X* axis: days.

4.2. Sensitivity Analysis

In the optimisation model we have considered three parameters which may impact the operation of the WT–PSH energy source. All of them have a potential impact on the accuracy of the scheduling formula as well as the volume of the rejected energy from the wind generation. Figure 6 summarises that impact. On each of the charts we have visualised all possible combinations of those parameters with a stress on the individual parameter given on the *X* axis.

As can be seen, an increase in the α parameter generally leads to an increase in the MAPE error and a decrease in the rejected volume of energy from wind generation. This results from the fact that this parameter in the scheduling formula refers to the volume of energy from the forecasted wind generation which should be considered in the schedule for the next 25 to 48 h. Low values of α make it possible to avoid errors in scheduling (MAPE) whereas those close to 1.0 will be prone to errors. At the same time, α close to zero will neglect some portion of the forecasted wind generation and, in some circumstances (e.g., upper reservoir is almost full), that energy will be rejected.

In the case of the beta parameter, which determined the usage of the predicted upper reservoir state of filling, one can observe similar dependencies, as in the case of the alpha parameter. However, it seems that regardless of the remaining parameters (α, V^M) for values of β ranging from 0 to 0.3 the error of the forecasting approach is relatively low and remains below 10%. An increase in the β value leads to a simultaneous increase in the MAPE criterion. This results from the fact that the upper reservoir state of filling is a consequence of the previously created schedule and the forecasted energy yield from wind turbines. It is important to note that by increasing the utilisation of the upper reservoir in the scheduling formula the volume of rejected energy from wind generation dwindles. This results from the fact that the scheduling formula with β close to 1 will always tend to discharge almost all available energy in the next considered period and, as a consequence, there will almost always be a possibility to store there the energy originating from wind generation.

The upper reservoir capacity (V^M) seems not to have any significant impact on the value of the MAPE criterion. However, a slight decrease in energy rejected from the wind generation is observed when the upper reservoir energy storage capacity increases. Naturally, it is not observed in all configurations of α and β. For some, regardless of the V^M value, the share of rejected energy is still above 10%, or even close to 90%. This occurred for very low values of both of these parameters. Regression analysis conducted for the V^M impact on the share of the rejected energy from WT generation shows that, on average, with every increase of 1 MWh in the upper reservoir capacity one should expect the volume of the rejected energy to decrease by 0.06%.

The correlation analysis between the objective function and the volume of the rejected energy from wind generation revealed a rather loose negative interdependence. The average coefficient of correlation (CC) for all considered combinations of input variables was −0.38 (Figure 8). Interestingly, the highest were observed for the combinations with relatively low upper reservoir capacities. For V^M = 21.6 MWh the CC was −0.76 and tended to decrease with increasing reservoir capacity. This results from the fact that the upper reservoir energy storage capacity has the potential to compensate for forecasting errors and minimises the relation between the objective function and the imposed constraint—the volume of rejected energy from wind generation.

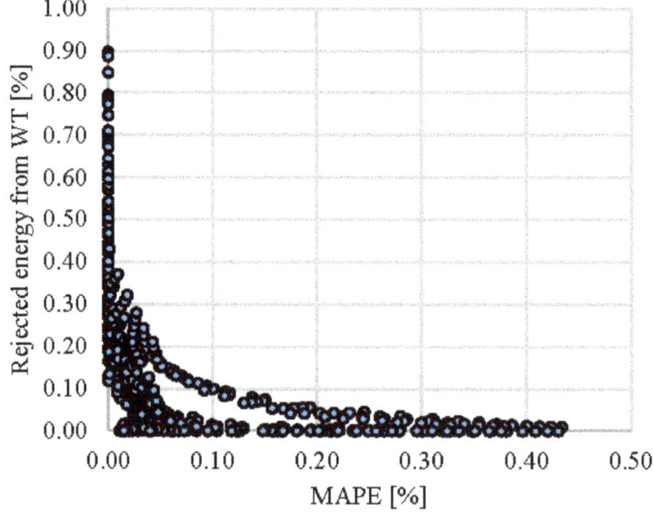

Figure 8. Relation between objective function (MAPE) and the constraint (percentage of rejected energy).

A similar correlation analysis has been conducted for the individual parameters and the aforementioned objective function and the volume of rejected energy from WTs. The obtained CC values are summarised in Table 1. Once again the resulting CC are rather low and, in the case of upper reservoir capacity (V^M), rather insignificant when it comes to their relation with the objective function and the constraint. Greater values are observed for the α and β parameters and they are coherent with the analysis made based on the scatter plots in Figure 9. An increase in α will generally lead to a decrease in the volume of rejected energy from WT generation and an increase in scheduling inaccuracy. Meanwhile, in the case of β (CC = 0.85) the higher the value, the higher the schedule error (MAPE) and the lower the volume of rejected energy (CC = −0.35, REJECTED), Table 2.

Table 2. Correlation coefficients between variables (α, β, V^M), objective function (MAPE), and constraint (REJECTED).

	MAPE	Rejected
α	0.3915	−0.5186
β	0.8456	−0.3568
V^M	−0.0215	−0.1545

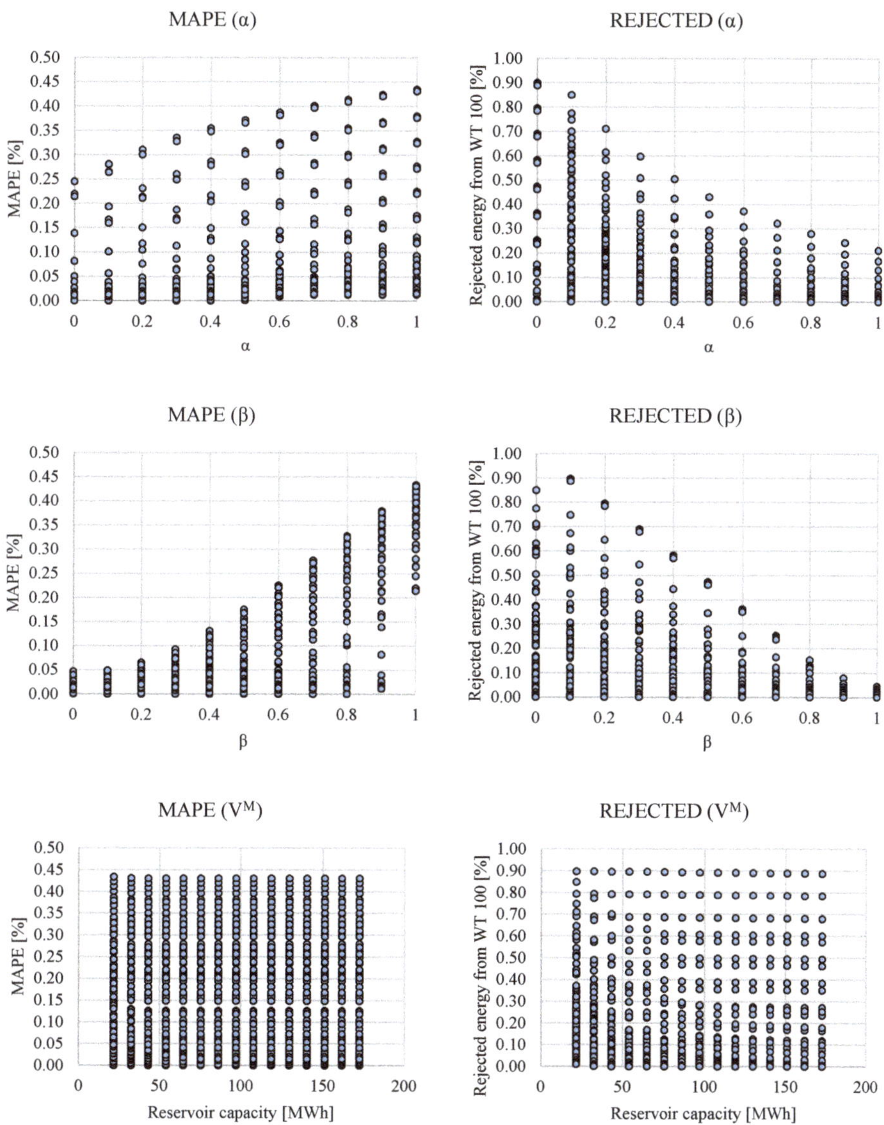

Figure 9. Objective function and the main constraint in the function of individual variables.

4.3. Analysis of the Dispatch Errors

As observed efigurearlier, the proposed scheduling/dispatching procedure is not error-free. Hereunder, we provide a concise analysis of observed absolute errors in the case of the considered scenarios. Their ordered values (from largest to smallest) are depicted in Figure 10. Please note that in scenarios S1 and S2 some errors occurred during 20% of the hours considered in our analysis, whereas in all the remaining scenarios it is significantly less than 2% or, in other words, there were no dispatch errors over more than 98% of the considered time period. This shows that for larger storage capacity the proposed scheduling method is very efficient, but a question arises whether it is, from an economic point of view, cost effective to invest in larger storage or accept potential penalties for unrealized dispatch.

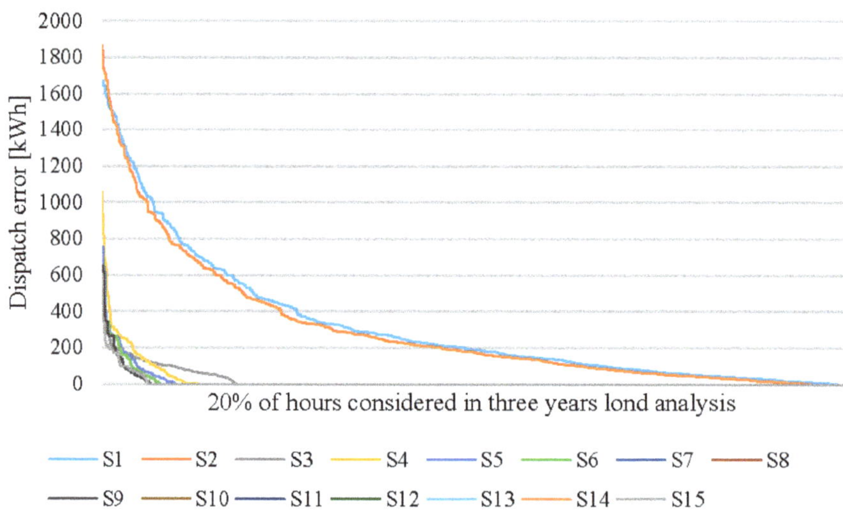

Figure 10. Ordered values of dispatch errors for 15 considered scenarios of upper reservoir capacity.

Further analysis of the dispatch errors focused on their statistical parameters. As can be observed in Table 3 there is a visible trend of decreasing values of dispatching errors for larger upper reservoir capacities. Not only is the average value significantly reduced, but also the observed standard deviation and, as a consequence, the coefficient of variation. The analysis of kurtosis and skewness shows that the error distributions are leptokurtic (with fatter tails) and are positively skewed (the distribution is concentrated on the left tail).

Table 3. Basic statistical parameters of observed dispatch errors.

Scenario	S1	S2	S3	S4	S5	S6	S7–S9	S10–S15
AVG (kWh)	61.3	57.1	4.0	4.2	2.8	2.3	1.8	1.5
STD (kWh)	199.3	191.7	25.9	38.3	28.0	25.4	23.0	16.2
CV [%]	30.7%	29.8%	15.5%	10.8%	10.0%	9.2%	8.0%	9.2%
Skeweness	4.6	4.9	9.4	14.0	14.7	15.1	17.9	14.8
Kurtosis	23.7	27.5	123.1	250.1	272.7	280.1	390.7	285.7

5. Conclusions

The advent of variable renewable energy sources made a significant change on the energy market. In NPSs with a high penetration of VRES generation, not only does the demand side of the energy market start to vary, but so too does the supply side. In general, based on the coefficient of variation, the variability of wind and solar generation is greater than that of energy demand. So far various approaches have been proposed to overcome this problem. In this paper a novel strategy for scheduling the operation of a wind turbine/park and pumped-storage hydroelectricity is presented. The proposed approach takes into account the forecasted energy generation from the wind turbines for the next 24 h and the estimated upper reservoir state of filling at the end of the upcoming day. Based on those estimates an energy generation schedule is formulated for the next 25 to 48 h.

The developed strategy was tested based on a three-year hourly time series of wind speed and synthetic wind speed forecasts generated for the purpose of this paper. The results indicate that a scheduling formula for a 2-MW wind turbine and a PSH with an energy storing potential equal to 54 MWh is capable of reducing the intraday variability (calculated based on the coefficient of variation) of energy generation from 31% to 1.15%. The low value of CV for the WT–PSH energy source indicated the good quality of the scheduling approach and the fact that the inherently variable energy from wind generation becomes manageable. For this configuration, and the α and β parameter values assumed in the scheduling formula, the MAPE criterion for the schedule accuracy was 2.39%, whereas that considered for wind generation was 27%.

The results also show that an increase in upper reservoir capacity above a certain volume does not have a significant impact on the performance of the considered energy source. However, the larger the reservoir, the lower the value of the MAPE criterion and the volume of rejected energy will be. This, however, comes at the cost of underutilised reservoir storage capacity. Therefore, the α and β parameters, which are used in the scheduling formula, remain the most relevant.

There are several ways in which such an energy source may be used in the power system. First and foremost, it can be applied to increase the share of wind generation in covering the energy demand without increasing the variability of the supply side of the energy market. Secondly, the scheduling formula may be modified in such a way that the WT–PSH energy source will generate electricity only during peak hours. As a consequence it will operate as a conventional PSH power plant, but the generally greater energy prices during the peak demand periods will increase the economic potential of the wind energy. Finally, in systems with an already high penetration of VRES, this type of energy source can be used as a conventional power plant whose energy generation can be scheduled for the next 25 to 48 h. Its reliability and availability mainly during the summer period (which has less beneficial wind speed conditions at the considered location) can be increased by coupling it with a PV installation.

In this paper we have made a new contribution by proposing a novel mathematical model for scheduling the operation of a wind-powered pumped-storage hydroelectric hybrid power source. Additionally, we have introduced a measure to assess the accuracy and quality of the scheduled generation and, based on the sensitivity analysis, we revealed the impact of the individual parameters on the performance of the WT–PSH. It is expected, by applying more accurate wind speed/generation forecasts, that the accuracy of the schedule generation will also increase.

Naturally, not all aspects of such an energy source operating in the very complex environment of the energy market have been investigated. From the author's perspective, the most important future research direction would be to investigate the economic side of such an energy source, as well as its potential in energy systems which depend on a share of VRES in covering energy demand.

Author Contributions: J.J. has retrieved the data, designed the experiment, performed the calculations and written the draft of the manuscript. Both authors have discussed the findings and agreed on the final form of the manuscript.

Funding: A. Kies is financially supported by the national R&D project NetAllok, Methoden und Anwendungen der Netzkostenallokation (Bundesministerium für Wirtschaft und Energie, FKZ 03ET4046A).

Conflicts of Interest: The authors declare no conflict of interest.

Nomenclature

Abbreviations

AVG	Average
ANN	Artificial Neural Network
CAES	Compressed Air Energy Storage
CC	Coefficient of Correlation
CV	Coefficient of Variation
DMS	Demand Side Management
NPS	National Power System
NWP	Numerical Weather Predictors
MAPE	Mean Absolute Percentage Error
PSH	Pumped Storage-Hydroelectricity
PV	Photovoltaics
ROR	Rate of Return
STD	Standard Deviation
VRES	Variable Renewable Energy Sources
V2G	Vehicle to Grid

Indexes

i	Index of days ($i = 1, \ldots , 1095$)
j	Index of hours ($j = 1, \ldots , 24$)

Parameters and constants

η^{PSH}	Efficiency of the PSH pumps and generators (%)
E^C	Scheduled energy generation of the PSH (kWh)
E^{CR}	Scheduled and realised generation of the PSH (kWh)
E^{Gen}	Maximal energy generating capacity of the PSH per unit of time (kWh)
E^{Pump}	Maximal capacity of the PSH pumps to accommodate and store the energy in the upper reservoir per unit of time (kWh)
E^{WT}	Energy yield from wind turbine (kWh)
E^{WT*}	Forecasted energy yield from wind turbine (kWh)
E^{WT_R}	Energy yield from wind turbine which has not been used in the PSH generation and/or cannot be stored in the upper reservoir (kWh)
n	Number of wind turbines with rated power capacity equal to P^{WT} (-)
$p(v)$	Polynomial approximating the wind turbine power curve (kW)
p^{WT}	Rated power output of the selected type of wind turbine (kW)
t	Time (hour)
v	Wind speed (m/s)
v^*	Forecasted wind speed (m/s)
V	Upper reservoir state of filling (kWh)
V^*	Estimated/forecasted upper reservoir state of filling (kWh)
$v^{cut\text{-}in}$	Wind speed at which wind turbine starts to generate electricity (m/s)
$v^{cut\text{-}off}$	Wind speed at which wind turbine operation is interrupted to prevent it being damaged (m/s)
v^{rated}	Wind speed at which wind turbine generates electricity at its rated capacity (m/s)

Variables

α	Parameter used to define the share of potential wind turbine energy yields used in the PSH generation schedule
β	Parameter used to define the contribution of the estimated upper reservoir occupancy to the PSH generation schedule
V^M	Maximal energy storage capacity of the PSH (kWh)

References

1. Suganthi, L.; Samuel, A.A. Energy models for demand forecasting—A review. Renew. *Sustain. Energy Rev.* **2014**, *16*, 1223–1240. [CrossRef]
2. Jung, J.; Broadwater, R.P. Current status and future advances for wind speed and power forecasting. *Renew. Sustain. Energy Rev.* **2014**, *31*, 762–777. [CrossRef]
3. Kyritsis, E.; Andersson, J.; Serletis, A. Electricity prices, large-scale renewable integration, and policy implications. *Energy Policy* **2017**, *101*, 550–560. [CrossRef]
4. Santos, S.F.; Fitiwi, D.Z.; Cruz, M.R.; Cabrita, C.M.; Catalão, J.P. Impacts of optimal energy storage deployment and network reconfiguration on renewable integration level in distribution systems. *Appl. Energy* **2017**, *185*, 44–55. [CrossRef]
5. Jones, L.E. *Renewable Energy Integration: Practical Management of Variability, Uncertainty, and Flexibility in Power Grids*; Academic Press: Cambridge, MA, USA, 2014.
6. Jacobson, M.Z.; Delucchi, M.A. Providing all global energy with wind, water, and solar power, Part I: Technologies, energy resources, quantities and areas of infrastructure, and materials. *Energy Policy* **2011**, *39*, 1154–1169. [CrossRef]
7. Delucchi, M.A.; Jacobson, M.Z. Providing all global energy with wind, water, and solar power, Part II: Reliability, system and transmission costs, and policies. *Energy Policy* **2011**, *39*, 1170–1190. [CrossRef]
8. Kaldellis, J.K.; Kapsali, M.; Kavadias, K.A. Energy balance analysis of wind-based pumped hydro storage systems in remote island electrical networks. *Appl. Energy* **2010**, *87*, 2427–2437. [CrossRef]
9. Kapsali, M.; Kaldellis, J.K. Combining hydro and variable wind power generation by means of pumped-storage under economically viable terms. *Appl. Energy* **2010**, *87*, 3475–3485. [CrossRef]
10. Kapsali, M.; Anagnostopoulos, J.S.; Kaldellis, J.K. Wind powered pumped-hydro storage systems for remote islands: A complete sensitivity analysis based on economic perspectives. *Appl. Energy* **2012**, *99*, 430–444. [CrossRef]
11. Kavadias, K.A.; Kapsali, M.; Kaldellis, J.K. An integrated computational method for the optimum sizing of a wind-based pumped hydro storage system. In Proceedings of the European Wind Energy Conference, Marseille, France, 16–19 March 2009.
12. Papaefthymiou, S.V.; Papathanassiou, S.A. Optimum sizing of wind-pumped-storage hybrid power stations in island systems. *Renew. Energy* **2014**, *64*, 187–196. [CrossRef]
13. Katsaprakakis, D.A.; Christakis, D.G.; Pavlopoylos, K.; Stamataki, S.; Dimitrelou, I.; Stefanakis, I.; Spanos, P. Introduction of a wind powered pumped storage system in the isolated insular power system of Karpathos–Kasos. *Appl. Energy* **2012**, *97*, 38–48. [CrossRef]
14. Katsaprakakis, D.A. Hybrid power plants in non-interconnected insular systems. *Appl. Energy* **2016**, *164*, 268–283. [CrossRef]
15. Anagnostopoulos, J.S.; Papantonis, D.E. Pumping station design for a pumped-storage wind-hydro power plant. *Energy Convers. Manag.* **2007**, *48*, 3009–3017. [CrossRef]
16. Ma, T.; Yang, H.; Lu, L. Feasibility study and economic analysis of pumped hydro storage and battery storage for a renewable energy powered island. *Energy Convers. Manag.* **2014**, *79*, 387–397. [CrossRef]
17. Ma, T.; Yang, H.; Lu, L.; Peng, J. Pumped storage-based standalone photovoltaic power generation system: Modeling and techno-economic optimization. *Appl. Energy* **2015**, *137*, 649–659. [CrossRef]
18. Ma, T.; Yang, H.; Lu, L.; Peng, J. Optimal design of an autonomous solar–wind-pumped storage power supply system. *Appl. Energy* **2015**, *160*, 728–736. [CrossRef]
19. Hessami, M.A.; Bowly, D.R. Economic feasibility and optimisation of an energy storage system for Portland Wind Farm (Victoria, Australia). *Appl. Energy* **2011**, *88*, 2755–2763. [CrossRef]
20. Hedegaard, K.; Meibom, P. Wind power impacts and electricity storage—A time scale perspective. *Renew. Energy* **2012**, *37*, 318–324. [CrossRef]
21. Canales, F.A.; Beluco, A.; Mendes, C.A.B. A comparative study of a wind hydro hybrid system with water storage capacity: Conventional reservoir or pumped storage plant? *J. Energy Storage* **2015**, *4*, 96–105. [CrossRef]

22. Murage, M.W.; Anderson, C.L. Contribution of pumped hydro storage to integration of wind power in Kenya: An optimal control approach. *Renew. Energy* **2014**, *63*, 698–707. [CrossRef]

23. Varkani, A.K.; Daraeepour, A.; Monsef, H. A new self-scheduling strategy for integrated operation of wind and pumped-storage power plants in power markets. *Appl. Energy* **2011**, *88*, 5002–5012. [CrossRef]

24. Tan, Z.F.; Ju, L.W.; Li, H.H.; Li, J.Y.; Zhang, H.J. A two-stage scheduling optimization model and solution algorithm for wind power and energy storage system considering uncertainty and demand response. *Int. J. Electr. Power Energy Syst.* **2014**, *63*, 1057–1069. [CrossRef]

25. Jurasz, J.; Mikulik, J. Scheduling opearation of wind powered pumped-storage hydroelectricity. In Proceedings of the 13th International Conference on Industrial Logistics, Zakopane, Poland, 28 September–1 October 2016; AGH University of Science and Technology: Kraków, Poland, 2016; pp. 74–83.

26. Castronuovo, E.D.; Lopes, J.P. On the optimization of the daily operation of a wind-hydro power plant. *IEEE Trans. Power Syst.* **2004**, *19*, 1599–1606. [CrossRef]

27. Papaefthimiou, S.; Karamanou, E.; Papathanassiou, S.; Papadopoulos, M. Operating policies for wind-pumped storage hybrid power stations in island grids. *IET Renew. Power Gener.* **2009**, *3*, 293–307. [CrossRef]

28. Helseth, A.; Gjelsvik, A.; Mo, B.; Linnet, U. A model for optimal scheduling of hydro thermal systems including pumped-storage and wind power. *IET Gener. Transm. Distrib.* **2013**, *7*, 1426–1434. [CrossRef]

29. Soman, S.S.; Zareipour, H.; Malik, O.; Mandal, P. A review of wind power and wind speed forecasting methods with different time horizons. In Proceedings of the North American Power Symposium (NAPS), Arlington, TX, USA, 26–28 September 2010; IEEE: Piscataway, NJ, USA, 2010; pp. 1–8.

30. Makridakis, S.; Wheelwright, S.C.; Hyndman, R.J. *Forecasting Methods and Applications*; John Wiley & Sons: Hoboken, NJ, USA, 2008.

31. Zhao, W.; Wei, Y.M.; Su, Z. One day ahead wind speed forecasting: A resampling-based approach. *Appl. Energy* **2016**, *178*, 886–901. [CrossRef]

32. Dong, L.; Wang, L.; Khahro, S.F.; Gao, S.; Liao, X. Wind power day-ahead prediction with cluster analysis of NWP. *Renew. Sustain. Energy Rev.* **2016**, *60*, 1206–1212. [CrossRef]

33. Azimi, R.; Ghofrani, M.; Ghayekhloo, M. A hybrid wind power forecasting model based on data mining and wavelets analysis. *Energy Convers. Manag.* **2016**, *127*, 208–225. [CrossRef]

34. Miettinen, J.J.; Holttinen, H. Characteristics of day-ahead wind power forecast errors in Nordic countries and benefits of aggregation. *Wind Energy* **2017**, *20*, 959–972. [CrossRef]

35. Korkas, C.D.; Baldi, S.; Michailidis, I.; Kosmatopoulos, E.B. Occupancy-based demand response and thermal comfort optimization in microgrids with renewable energy sources and energy storage. *Appl. Energy* **2016**, *163*, 93–104. [CrossRef]

36. Baldi, S.; Karagevrekis, A.; Michailidis, I.T.; Kosmatopoulos, E.B. Joint energy demand and thermal comfort optimization in photovoltaic-equipped interconnected microgrids. *Energy Convers. Manag.* **2015**, *101*, 352–363. [CrossRef]

37. Jurasz, J.; Mikulik, J.; Krzywda, M.; Ciapała, B.; Janowski, M. Integrating a wind-and solar-powered hybrid to the power system by coupling it with a hydroelectric power station with pumping installation. *Energy* **2018**, *144*, 549–563. [CrossRef]

38. Jurasz, J.; Ciapała, B. Integrating photovoltaics into energy systems by using a run-off-river power plant with pondage to smooth energy exchange with the power gird. *Appl. Energy* **2017**, *198*, 21–35. [CrossRef]

39. Yang, Z.; Liu, P.; Cheng, L.; Wang, H.; Ming, B.; Gong, W. Deriving operating rules for a large-scale hydro-photovoltaic power system using implicit stochastic optimization. *J. Clean. Prod.* **2018**, *195*, 562–572. [CrossRef]

40. Moustris, K.P.; Zafirakis, D.; Alamo, D.H.; Medina, R.N.; Kaldellis, J.K. 24-h Ahead Wind Speed Prediction for the Optimum Operation of Hybrid Power Stations with the Use of Artificial Neural Networks. In *Perspectives on Atmospheric Sciences*; Springer International Publishing: Basel, Switzerland, 2017; pp. 409–414.

41. Siemens Gamesa Renewable Energy. Available online: http://www.gamesacorp.com/ (accessed on 20 January 2017).

42. Jurasz, J. Modeling and forecasting energy flow between national power grid and a solar–wind–pumped-hydroelectricity (PV–WT–PSH) energy source. *Energy Convers. Manag.* **2017**, *136*, 382–394. [CrossRef]

43. National Aeronautics and Space Administration Goddard Space Flight Center. Available online: https://gmao.gsfc.nasa.gov/reanalysis/MERRA-2/ (accessed on 20 January 2017).

 *sustainability*

Article

Environmental Behavior of Secondary Education Students: A Case Study at Central Greece

Stamatios Ntanos [1,*], **Grigorios L. Kyriakopoulos** [2], **Garyfallos Arabatzis** [1], **Vasilios Palios** [3] and **Miltiadis Chalikias** [4]

[1] Department of Forestry and Management of the Environment and Natural Resources, School of Agricultural and Forestry Sciences, Democritus University of Thrace, 68200 Orestiada, Greece; garamp@fmenr.duth.gr

[2] School of Electrical and Computer Engineering, National Technical University of Athens, 15780 Zografou, Greece; gregkyr@chemeng.ntua.gr

[3] Faculty of Pure and Applied Sciences, Open University of Cyprus, 12794, 2252 Latsia, Cyprus; bpalios@yahoo.gr

[4] Department of Tourism Management, School of Business, Economics and Social Sciences, University of West Attica, 12244 Aigaleo, Greece; mchalikias@hotmail.com

* Correspondence: sdanos@ath.forthnet.gr or sdanos@puas.gr; Tel.: +30-210-538-1275

Received: 3 April 2018; Accepted: 18 May 2018; Published: 21 May 2018

Abstract: During the last three decades, human behavior has been becoming energy alarming towards environmental sustainability. One of the most influential initiatives towards environmental protection and increased environmental consciousness is the solidification of primary and secondary environmental education. The purpose of this paper is to investigate different environmental profiles amongst secondary education students, in light of a multi-parametric analysis that involved the contributive role of school and family towards environmental awareness and participation. By reviewing relevant studies, the benefits offered by environmental education are presented. Accordingly, a questionnaire survey was deployed using a sample of 270 secondary education students, from schools situated in the prefecture of Larissa, central Greece. The statistical methods included factor analysis and cluster analysis. Particularly, four groups of different environmental characteristics are identified and interviewed. Results suggest that most students are environmental affectionate, although there is a need for more solidified environmental education and motivation from out-of-school societal opportunities, such as in the contexts of family and public socialization. The deployed research method and analysis can be proven supportive in adopting and scheduling school environmental programs after an initial identification of the various environmental attitudes among the student population.

Keywords: environmental education; renewable energy sources; cluster analysis; factor analysis; survey

---

1. Introduction

In recent years, severe environmental problems are associated with rising energy consumption due to economic development and population growth, while a simultaneous imperative need for a sustainable environment for humans necessitates the scientific and technological research to be concentrated on energy preservation and the abiding role of renewable energy sources (RES) [1–7]. In the Brundtland Report (1987), issued under the World Commission on Environment and Development, the need to value the obligations for future generations in balance to the needs of present generations was introduced, setting the foundations for "sustainable development" [8]. In the environment-based literature production, the need for a safe, environmentally sound and economically viable energy pathway that supports human progress into future generations was recognized, in line with efficient energy use and the development of renewables [9–11]. Implementing the context of sustainability in the real world, the EU set a target of increasing RES share up to 20% upon total energy consumption by the year 2020, while Greece has committed to achieving RES penetration up to 18% of its total national energy consumption [12]. European countries, China and Greece have set additional national targets for achieving sustainable development under national strategies for sustainable development [13–16].

In response to global alarm for current environmental depletion, many studies have been published, especially during the last decade. Several studies focus on the effective production, distribution, and exploitation of biomass energy, especially within residential areas, where diverse RES exists [17–20]. These studies also signify an increasing awareness of energy-focused education programs. It is reported that shifts in energy consumption and energy behavior can be achieved by changing cultural beliefs that are also abiding to the need of continuing and life-long education.

In a normal societal context, energy behavior is promptly instilled inside the family environment and is subsequently developed by the citizens' cultural and societal behavior out of the core-household environment. The existing research production on energy behavior is mainly focused on energy consumption in residential areas, attempting to determine the factors that simultaneously define human behavior and energy use [21–40]. Households consume energy not only to satisfy subsistence needs of their members, but also towards a wider spectrum of societal activities, such as transport, work, leisure, and entertainment [41]. In residential areas, domestic energy consumption can widely vary within households, since it is directly associated with households' behavior and sociodemographic characteristics [42,43]. Relevant studies highlight the impact of individuals' energy behavior and conclude that around 1/3 of the residential energy consumption is attributed to diversified variations that are prevailing in each household [44,45].

An essential step towards changing energy behavior is the notional development of an "environmental culture". In the relevant literature production, the concept of environment in sustainable development (ESD) is not new, dating over a decade ago, both for North Europe and for countries worldwide [46,47]. Environmental education aims at formulating an ecological awareness from early ages [48]. In this respect, programs implemented as early as in preschool age constitute a much different educational approach in comparison to the traditional educational methods, which have been implemented in past educational curricula. Such programs are characterized by their student-centered nature, the active role attributed to the person undergoing education and, among others, the mutual collaboration between the children and the educator for the desired outcome to be achieved for children of preschool age [48].

In the context of energy education, there is a growing interest for research in the dynamic linkage between energy-based issues and their educational context of applicability. The pronounced role of science education adaptation to the taught educational curriculum in primary schools is highlighted [49]. Such educational reforms enable students to be scientifically and technologically literate citizens and shapes their thinking and behavior towards the importance of sustainable development.

Therefore, it can be noted that environmental behavior change is a dynamic but low-pace process that can be proven achievable through proper education.

Within the above theoretical framework, the sampling methodology suggested in this study is pooled from secondary education students, taking as its premise that, if properly educated, they are vital and effective stakeholders and future decision makers in promoting environmentally responsible behavior at their family and socializing environment [50,51]. According to the proposed methodology, the profile of secondary education students concerning environmental awareness and environmental behavior is captured and used to classify them into clusters, to facilitate decision makers for the application of efficient environmental education programs. Furthermore, a theoretical review on the benefits that environmental education offers, with implications for secondary education, are presented in the theoretical section.

2. Benefits of Environmental Education

Several studies manipulate the interdisciplinary issues of Environmental education/environmental issues/energy behavior, in an integrated manner [52–54]. Energy-oriented curricula of environmental education must be an integral and inseparable part of contemporary education for all citizens. The key-entity of "sustainability" can develop environmentally conscious citizens from their early phases of education. The purpose of a "sustainable school" is not only to help children acquire environmental literacy and understand the dimensions of sustainability. The main goal of such an educational reform is to change the school itself for promoting sustainability and helping today students to follow sustainable practices in their everyday life. Indeed, this educational reform is the tool for achieving sustainable development and instilling energy behavior into all levels of educational operation: teaching, social/organizational, and technical/financial [55–66]. Educational activities play a dominant role to educate students on how to take initiatives, set goals, make decisions, handle information data, and be engaged in creative argumentation and productive criticism [67]. Energy-educated children can stand on an equal footing to the adult members of the family, participate, take initiatives, and rationalize their argumentation in taking household decisions and priorities concerning energy consumption. Environmental education in schools can also promote the agenda of energy wastes reduction in households [68,69].

A cognitive gap between attitudes and behavior has been attributed to a contradiction between a generalized interest for the environment and a feeling of hopelessness and incapability in applying this interest to specific actions [70]. Under this framework, environmental education can play a determining role, enabling students to develop sustainability-conscious behavior and make decisions in favor of a sustainable environment. Students are vital contributors to the development of environmental consciousness and the promotion of environmentally responsible behavior [50]. Furthermore, it has been reported that secondary school students are prone to rapid intellectual development, with the ability of abstracted thinking. This period of students' psychosomatic development is significant because students of this group-age develop their personality values of socialization and shape their attitudes towards environment protection [71].

Teaching about climate change in environmental secondary education curricula is considered crucial for developed countries [72]. Particularly, teachers in the US are willing to teach climate change topics with their students, taking this as an opportunity to teach them the nature of sciences and strengthen their skills upon data analysis, systematic thinking, and critical argumentation [72]. Such a climate-driven curriculum offers the opportunity to make climate change topic familiar to local communities, by enriching the knowledge upon local ecosystem impacts and opening policies' argumentation upon the relevant taught curricula.

In the case of developing countries, environmental education is crucial for primary education because in many countries this educational level is the only formal education that children may receive throughout their life-span. In this respect, energy-related school courses may facilitate students and parents with energy-oriented information, especially in those countries in which there is a cultural trait for parents to be actively involved in their children's schoolwork [73]. In a relevant study, it has been pointed out that many secondary education students in Turkey were introduced to the renewable energy entity during their secondary school education [74]. The contribution of greenhouse gas emissions to environmental degradation, as well as the pronounced role of education driven public awareness upon environment pollution and petroleum exploration has been pointed out in a study for Uganda [54].

In the Greek context, environmental sustainability in higher education students has been investigated in a study on Aegean University [75]. The authors reported important constraints concerning the promotion of environmental sustainability. Particularly, according to the interviewed students, the most important constraint is lack of environmental knowledge among the academic community members. Another critical issue is lack of environmental projects funding from private-based ownership, since Greek university funding is mainly state-dependent [75]. Furthermore, a study for Greek elementary students in the Prefecture of Evros reveals that students are sensitive to energy management issues and recycling [76].

Conclusively, a successful adaptation and implementation of an environmental education to current curricula of secondary education worldwide should reflect the accomplishment of:

- Critical thinking skills [72]: Creative thinking in environmental education remains a greatly under-researched topic, thus connections must be opened to broader approaches that are currently narrowing creativity of the relevant researches undertaken to handle environmental problems [77]. Developing contextual thinking in students seems to be a key-parameter for the perspectives of educational reforms [78]. Under an environmental education context, the need for active engagement with environmental topics and the active participation in monitoring and problem-solving activities should prevail over the need for obtaining individual knowledge [78]. Educational activities play a dominant role to educate students how to take initiatives, set goals, make decisions, handle information data, and be engaged to creative argumentation and productive criticism [67].

- The entities of "sustainability" and "environmental clubs" that reflect how these notions can help both teachers and learners to develop all skills and positive attitudes' needed towards the environmental sustainability [79]. In a relevant study, it was pointed out that biodiversity loss has encouraged scientists to begin promoting the idea of ecosystem services that can be offered to humans to be active supporters of conservation policies. To this end, the concept of ecosystem services can be designed to communicate societal needs. This societal viewpoint is perfectly served by the school system because it plays a key role in educating students to be active and responsible citizens [80]. In another study, the link between the "sustainable development" concept and development was investigated [81]. This author proposed some examples of sustainability integration in new textbooks for primary education. Under this research framework, it was illustrated the extent of curricular materials that could contribute to developing skills, values, and attitudes aligned with sustainable development perspectives [81].

- The elevation of the environment as a vital natural asset that strengthens the proactive and active inquiry to promote problem-based learning approaches in real environment-degradation issues. Under this framework, technology can also be used as a supportive tool that induces a new educational approach, not just to be used in reproducing traditional teaching roles that are based in conservative repetition of environmental concepts [82]. Students and citizens familiar with the methodological tools of web-repositories and web-quests for environmental education can assist in promoting environmental affection [83].

- Key-project stakeholders to explore the challenges of project sustainability and promote participation. Applying the principles of interdisciplinary learning theory, the contradictions that emerge from the interaction between different project stakeholders can be rationalized and taught within the school environment, as powerful sources for learning [84].

3. Materials and Methods

The purpose of this study was the classification of students according to their profile on environmental awareness. Data collection was performed using a structured questionnaire including various dichotomous type questions and Likert type questions on the topics of school and family role toward environmental behavior, students' environmental consciousness and environmental knowledge.

The statistical methods used in this study included descriptive statistics, reliability analysis, Principal Component Analysis (PCA) and cluster analysis by using the K-means algorithm. K-means clustering is applicable because of the large sample size [85]. This statistical process initiates by assigning temporary k-centroids. If the distance of an observation from the center is higher than the shortest distance between that center and all other centers, this observation replaces the nearest cluster center. The centers are continually re-evaluated, based on the above criteria through a loop process.

The process is repeated until there is no change in the centers or the convergence criterion has been achieved [85–87].

In the Results Section, the socio-demographic profile and respondents' opinion on environmental questions are presented. Affirmative responses to the dichotomous (yes/no) variables are expressed as percentages of the total sample, whereas responses to Likert-type questions are presented by using the mean and standard deviation. Subsequently, reliability analysis by using Cronbach's alpha is deployed. Principal Component Analysis (PCA) method is afterwards used to identify factors (components). The same methodology was applied in an analogous study for the energy behavior of secondary education Greek students in Grevena [88]. In this method, each one of the identified factors (components) interprets a percentage of the variance that has not been captured by previous factors (components). The final number of factors (components) is determined according to the Kaiser criterion, where a factor is created when the calculated eigenvalue is greater than 1. Finally, by using the variables created from the PCA method, k-mean analysis estimated the number of students included in each cluster, sustaining common environmental characteristics.

This research was conducted during 2014 within Larissa Regional Unit (Figure 1). The research area is in the northeastern part of Thessaly, where Larissa is largest regional unit at about 5381 km^2 occupying 38.3% of the total area of the Region [89]. The population is 284,420 inhabitants according to the 2011 National Statistical Service census. The economic character of Larissa Regional Unit is outlined under the three main economic sectors (primary, secondary and tertiary). The primary sector is a key activity for the area and is characterized by: (a) low field resting rate; (b) the dominance of arable crops; and (c) minor contribution of forestry and fishery products. The secondary economic sector is focused on processing of agricultural products. There are also activities such as wood processing, textiles, garment production, food, paper, engineering and machining. The tertiary economic sector is mainly dominated by commercial activity and industry services owing to the area's strategic geographical position on the county map.

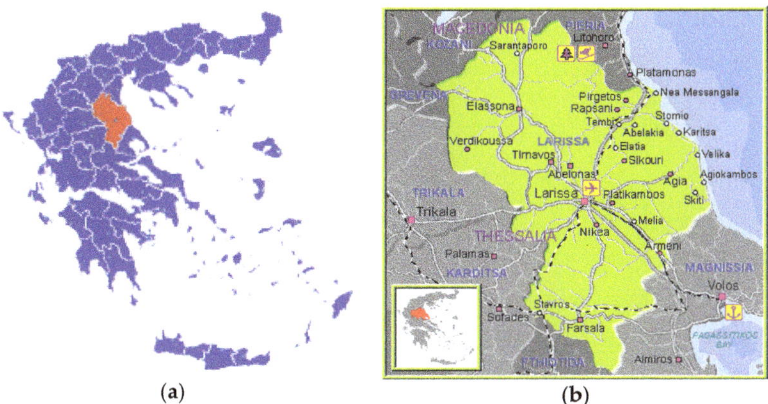

(a)   (b)

Figure 1. (a) Map of Greece depicting the regional unit of Larissa citation upon the map [90]; and (b) detailed research area map [91].

Under the research framework of this study, students attending the third grade of Secondary education schools constituted the research population. The total number of students (15 year old) attending the third grade, at the Regional Unit of Larissa at the time of the survey, was 2589 [92]. The sample of this research consisted of 270 students, accounting for 10.4% of the total student population attending the third Grade of secondary education (Table 1). Concerning sample size calculation, according a formula provided by the CheckMarket survey company [93], for a confidence level of 95%, we expected an error margin of 5.65% for our analysis. The sample was taken from four representative schools of the research area: the 1st Secondary school of Larissa, the 1st High School of Elassona, the 1st Secondary School of Farsala, and the 2nd Secondary school of Tyrnavos (Table 1). The selection of those specific school units was undertaken under the basis that they capture the characteristics of the typical school units operating in the region. The research area included urban areas (Larissa and Tyrnavos), semi-mountainous areas with large livestock activities (Elassona) and suburban areas (Farsala). These four sampled cities are the largest ones throughout the research, while the selected school units gather students from all their surrounding areas. Another fundamental precondition for the sampling stratification was the representation of all social strata and all occupational professions of the local inhabitants. The sample selection of students in the third grade of Secondary school was considered appropriate since there exist taught courses upon environmental education. This approach matches the scopes of the conducted research, since under the educational curricula and the deliverable learning outcomes, students have already engaged in energy issues under the taught courses of Science and Technology. Therefore, in the context of the interdisciplinary teaching of courses and consequently of the wider acquisition of knowledge and skills, it is argued that students sustain appropriate cognitive capacities and knowledge background to be familiarized with the set of the surveyed questions. Indeed, students at the age of 15 years who have undertaken the formal educational curriculum, have already developed formal thinking which enables them to theoretically deepen their thought and manipulate information in such way that they can clearly and accurately express their personal views and worldview experiences [71].

Table 1. Socio-geographic features of the schools and corresponding number of students who participated in the survey.

	School Unit (All Units Are in Central Greece)	Total Number of Students in Third Grade	Number of Students Participated in the Survey	School Location Parameters
1	1st Secondary school of Larissa	65	65	School in urban area
2	2nd Secondary school of Tyrnavos	56	56	School in urban area with high agricultural activity
3	1st Secondary school of Elassona	119	83	School in semi-urban and semi-mountainous areas rich in livestock activity
4	1st Secondary school of Farsala	66	66	School in semi-urban area with developed agricultural and livestock activities
	Total	309	270	

We cannot overlook that people in a survey do not only represent theories, words, and numbers. Every participant in a survey must be "treated" with respect and courtesy, especially in the case of underaged participants. The ethics of research with children needs to balance between different conditions. On the one hand, it is necessary to avoid or minimize damage caused by research and to ensure the protection of children and young people and, on the other hand, research needs to be alert to the dangers and harm caused by the concealment of children's opinions and experiences by excluding them. Consensus, confidentiality and anonymity were the three ethical parameters of this research. In addition, this study was about human emotions. Following the suggestion of Seale and Filmer [94], there is a moral "obligation" not to take advantage of the time and confidence of the participants. In this research we followed articles 3 and 12 on the Rights of the Child of the United Nations [95] (UNCRC), which also applies to the Greek legal context. Article 3 provides that the child's interests are a fundamental prerequisite for all actions relating to the child. Article 12 requires that children have, depending on their age and maturity, the ability to form opinions, the right to freely express themselves and their views on all matters concerning them. Permission to conduct the survey was initially granted by the School Directors, as proposed by relevant literature for research in education [96,97]. A visit to the students' classroom was afterwards carried out with the escort of the teacher, briefly informing the students about the process of conducting the research. Students were informed by the researcher about the purpose of the visit and the process of completing the questionnaire. The participants in the survey were asked to complete the questionnaire as reliably as possible by honestly answering the questions in an anonymous manner, without the physical presence of their educators. The completion of the process required one teaching hour for each class. Research conditions could be characterized as excellent since there was a climate of collaboration between students, teachers and researchers.

The questionnaire was written in a way to be understandable, to be complemented by all participants, to minimize potential errors and not be boring or tiring. It consists of four sections: Section A is divided into two subsections. Subsection A1 includes the socio-demographics, such as place of residence, parents' occupation and personal details. Under Subsection A2, there are questions about students' level of knowledge related to ecological awareness, information on ecological and environmental terms and student habits towards environment. Section B includes questions concerning the role of school and family towards environmental motivation, as well as some conceptual questions about students' environmental sensitivity. Section C consists of questions on students' self-evaluation concerning environmental knowledge, environmental education and willingness to participate in ecological activities. In conclusion, the content of the questionnaire included information and data related to actions concerning energy use and "green" behavior taken inside and outside the classroom.

4. Results

The sample consists of 270 students attending the third grade of secondary education, with an average age of 15 years, while the students' place of residence is the same as the place of the school (Table 1). Regarding gender, 49.6% of the students are boys and 50.4% are girls. About the overall school performance of the sample, 40.4% of students have an average score of 18.1 to 20 (with a maximum of 20), while 35.6% achieved a performance from 16.1 to 18. The remaining 24% of the sample achieved a graduation grade lower than 16. In response to family statues, 13.7% live in a three-member family, 53% live in a four-member (two parents and two children), 25% live in five-member family and 9% live in large families of six or more members. Regarding type of residence, 55% of students live in detached houses while the remaining 45% reside in apartment blocks. The educational level of the parents is high, since the parents of 40% of the students surveyed are higher education graduates, with the parents of 27.4% and 12% of the students surveyed being high school and secondary high school graduates, respectively. Concerning the professional occupation of the father, the predominant profession is "freelancer" with 27%, followed by "private employee" with 22.6%. Moreover, 21.9% are civil servants and 21.5% are farmers. Regarding the mother's professional situation, the highest percentage of 27.4% are unemployed/households followed by 25.2% and 18.5% who are civil servants and private employees, respectively.

The overall responses of students to the three environmental-driven sections of the survey are outlined in Tables 2–4. In Table 2, Section A focuses on students' self-reported knowledge on general environmental subjects and the various energy sources (see Appendix A).

Table 2. Questionnaire Section A: Answers Synopsis.

Positive Answers on Dichotomous Type Questions	Responses	Percent of Cases
Are you familiar with Renewable Energy Sources?	260	96.30%
Are you familiar with non-Renewable Energy Sources?	246	91.11%
Do you know the meaning of the word "Ecology"?	219	81.11%
Do you know about the ecological problem?	216	80.00%
Are you familiar with the term "Energy Crisis"?	160	59.26%
Does your school use Renewable Energy Sources?	35	12.96%
Have you joint environmental programs in your school?	146	54.07%

Table 2. *Cont.*

Positive Answers on Dichotomous Type Questions	Responses	Percent of Cases
Are you taking part in an environmental program now?	63	23.33%
Is there currently an environmental program running?	148	54.81%
Have you used your PC for accessing environmental information?	78	28.89%
Do you use the School Library for environmental information?	52	19.26%
Do you have home internet connection?	256	94.81%
Do you know if there is a Public Library in your area?	222	82.22%
Do you turn off classroom lights during breaks?	229	84.81%
Do you use public transportation for school?	69	25.56%
Do you go to school on foot?	180	66.67%
Do you use a bike for moving to and from school?	85	31.48%
Is there a recycle bin in your school?	231	85.56%
Do you use the recycle bin in your school?	155	57.41%
Do you use school supplies derived from recycled materials?	129	47.78%
Are you familiar with low energy consumption appliances?	188	69.63%
Would you buy a low energy consumption electric device with higher price?	164	60.74%

Table 3. Questionnaire Section B: Answers Synopsis.

(Questionnaire Section B—Variables QB1 to QB10)	Var Name	Mean	Std. Deviation
How important is the existence of a Recycle Bin in your school?	QB1	3.98	0.987
How important do you consider school's contribution to shaping environmental conscience?	QB2	3.88	0.976
Do you think your school should participate in recycling programs?	QB3	4.11	0.845
Do you think that your school operates in an energy efficient way?	QB4	3.04	1.016
Do you think that using a bicycle reduces energy consumption?	QB5	4.10	0.936
Do you believe that Renewable Energy Sources can help solve the ecological problems of the planet?	QB6	3.99	0.925
Do you think that participation in ecologic activities helps maximizing environmental awareness?	QB7	3.95	0.966
Do you agree with the view that harmonious coexistence of "human-nature" is a prerequisite for the survival of us all?	QB8	4.12	0.881
To what extent do you believe co-participation of children—parents will help in developing energy-saving behavior?	QB9	3.81	0.931
To what extent do you think your parents are sensitive to energy saving?	QB10	3.42	1.009

Most students report that they are informed about diverse types of energy sources as well as energy use and consumption. According to Table 2, the descriptive statistics revealed that around 96% of the students are familiar with the term "renewable energy sources" and around 80% are familiar with the word "ecology". On the contrary, only 60% are familiar with the word "energy crisis" and only 13% of the students answered positively regarding the use of renewable energy technologies at school. Around half of the students responded positively concerning their current or previous participation in environmental programs/actions. A lower percentage of the students of about 30% responded positively concerning Internet use for retrieving environmental information and around 20% answered positively about using the school library for environmental information.

In Table 3, Section B includes questions on students' attitudes concerning school and the role of the family towards environmentalism, as well as general ecological sensitivity questions (see Appendix A).

Concerning schools' role, results reveal that around 70% students agree or strongly agree on schools' contribution to shaping environmental conscience. Around 80% of the students believe that their school should participate in recycling programs and 75% of the students believe that participation in ecological activities motivates environmental awareness. On the contrary, only 35% of the students agree that their school operates in an energy efficient way while 30% of the students disagree or totally disagree. By examining students attitude concerning the role of the family, around 45% of the students agree or totally agree with the content of the question concerning their parents' environmental sensitivity while 70% of the students agree that the co-operation of children and parents leads to energy-saving behavior. There were also three general ecological sensitivity questions concerning the role of RES towards solving the ecological problem, participation in ecological activities and a question about the harmonious coexistence of humankind with nature. Students agreed or strongly agreed with these broad questions, with rates of above 75%.

In Table 4, Section C refers to student environmental knowledge, environmental education and willingness to participate in environmental actions.

Table 4. Questionnaire Section C: Answers Synopsis.

Questionnaire Section C	Var Name	Mean	Std. Deviation
How would you rate your environmental knowledge level?	QC1	3.01	0.871
How would you rate the adequacy of environmental education you receive?	QC5	3.06	0.935
In what extend do you believe that environmental participation would activate energy behavior?	QC6	3.81	0.882
Rate the degree in which the economic crisis has contributed to energy saving behavior?	QC10	3.71	1
Do you believe that your family economic status plays a part in practicing energy saving behavior?	QC11	3.5	0.955

Environmental self-reported knowledge of students is moderate. Only 5% of the students believe that they have very strong environmental knowledge and 19% believe that they have good knowledge. On the other hand, around 25% of the students evaluate their environmental knowledge level as low or very low while 50% of the students have moderate knowledge. Concerning students' opinion on environmental education in school, around 30% of the sample yielded a positive answer. Around 45% believe they receive moderate environmental education at schools and 6% believe they receive no environmental education at all. Concerning the question on students' beliefs whether participation in ecological activities leads to environmental awareness, around 68% agree or totally agree while 8% disagree or totally disagree. A graphical representation of students' answers can be found in the Appendix (students' attitude towards energy and environment).

Reliability analysis is performed using Cronbach's alpha coefficient. The coefficient equals 0.757 and is considered acceptable, according to the empirical scale provided by Darren and Mallery [98]. However, by looking at the last column of Table 5 (Cronbach's Alpha if Item Deleted), it was noticed that questions marked in italics seem to reduce overall reliability, i.e., a higher overall alpha index was reported if these questions were not used at all. Therefore, for reasons of analysis simplification and increased reliability, a decision to drop variables QB4, QB10 and QC10 was taken.

After removing the three variables marked in bold in Table 5, running a corrected trial revealed that Cronbach's Alpha index is higher (0.771).

By calculating the correlation matrix, we noticed that statistically significant correlation exists between most of the variables, thus several tests to check the appropriateness of data for factor analysis were deployed. KMO measure of sampling adequacy and Bartlett's test of sphericity were used to test data appropriateness for factor analysis. KMO measure for our data equals 0.822, indicative that the correlations between the variables are satisfactory for factor analysis to be performed. Ideally, the KMO index should take values >0.8, although values >0.55 are considered acceptable. Another test of appropriateness for factor analysis is Bartlett's test of sphericity. It tests the null hypothesis that the sample comes from a normal multivariate population, by using the x^2 distribution. The value of the x^2 test function is 602.912 with 66 degrees of freedom, indicating statistical significance at the 0.001 level, which satisfies the assumption that the data are suitable for factor analysis.

Table 5. Cronbach's Alpha reliability analysis.

Var Name	Scale Mean If Item Deleted	Scale Variance If Item Deleted	Corrected Item-Total Correlation	Cronbach's Alpha If Item Deleted
QB1	51.51	40.519	0.314	0.749
QB2	51.60	38.017	0.537	0.727
QB3	51.38	39.827	0.459	0.736
QB5	51.39	41.376	0.264	0.753
QB6	51.49	39.909	0.400	0.741
QB7	51.54	38.525	0.498	0.731
QB8	51.36	40.039	0.414	0.740
QB9	51.67	39.306	0.450	0.736
QC1	52.48	40.057	0.419	0.739
QC5	52.43	40.625	0.330	0.747
QC6	51.67	39.864	0.430	0.738
QC11	51.99	40.093	0.366	0.744
QB4	52.45	42.166	0.169	0.763
QB10	52.06	41.963	0.188	0.761
QC10	51.77	41.665	0.214	0.758

The number of factors to be used for the analysis was estimated using the scree plot and confirmed with the Kaiser criterion (Figure 2).

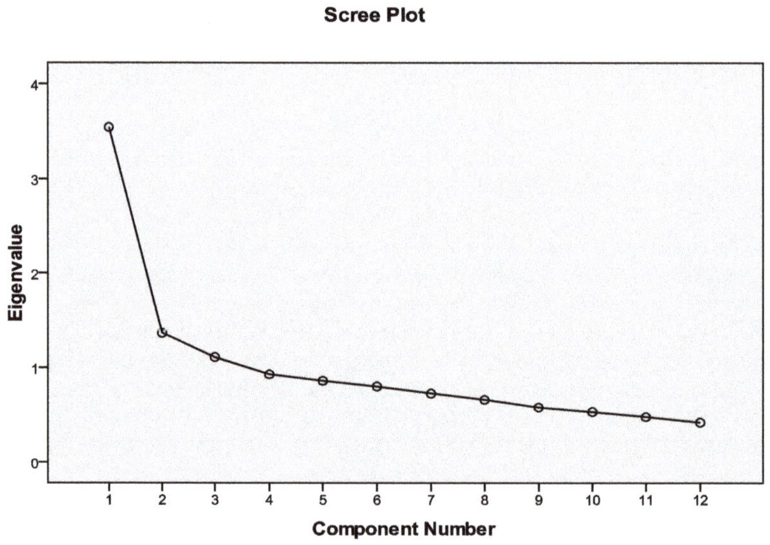

Scree Plot

Figure 2. Components with eigenvalue greater than 1.

By looking at the scree plot and using Kaiser's empirical criterion, which suggests setting the number of components equal to the components having eigenvalue > 1, we concluded that three factors are appropriate for this analysis.

By using Principal Component Method (PCA), loadings were calculated for the three factors (components), as presented in Table 6.

Table 6. Initial component matrix.

Initial Component Matrix	Component		
	1	**2**	**3**
How important do you consider school's contribution to shaping environmental conscience?	0.722	−0.311	−0.247
Do you think that participation in ecologic activities helps minimizing environmental awareness?	0.680	−0.264	0.210
Do you think your school should participate in recycling programs?	0.634	−0.339	0.132
To what extent do you believe co-participation of children—parents will help in developing energy-saving behavior?	0.586	−0.156	−0.296
In what extend do you believe that environmental participation would activate energy behavior?	0.586		0.312
Do you agree with the view that harmonious coexistence of "human-nature" is a prerequisite for the survival of us all?	0.549	0.330	−0.159
Do you believe that Renewable Energy Sources can help solve the ecological problems of the planet?	0.545		0.448
How important is the existence of a Recycle Bin in your school?	0.480	−0.322	−0.396
Do you believe that your family economic status plays a part in practicing energy saving behavior?	0.445	0.140	0.136
How would you rate the adequacy of environmental education you receive?	0.371	0.626	−0.226
Extraction Method: Principal Component Analysis.			

In Table 6, high factor loadings do not support the identification of the factors, so an orthogonal rotation of the initial matrix is required. The total variance explained by the factors is presented in Table 7 under "Extraction of Sums of Squared Loadings".

Table 7. Total variance explained.

Total Variance Explained						
Component	Extraction Sums of Squared Loadings			Rotation Sums of Squared Loadings		
	Total	**% of Variance**	**Cumulative %**	**Total**	**% of Variance**	**Cumulative %**
1	3.541	29.509	29.509	2.289	19.073	19.073
2	1.363	11.358	40.868	2.020	16.835	35.908
3	1.109	9.241	50.109	1.704	14.201	50.109
Extraction Method: Principal Component Analysis.						

According to Table 7, the first component explains 29.5% of the total variance, and in total all three components explain 50.11% of the variance. This outcome remarked that all three factors seem to explain a relative low percentage of the total variance, but it has been reported that in social sciences information collected usually by questionnaires includes less precision, so a solution that accounts for 60% of the total variance (and in some instances even less) is commonly considered satisfactory [99].

Since the questionnaires were filled by secondary education students and the questions included their opinion on vague topics, it was decided to keep the proposed factor number according to Kaiser's criterion of an eigenvalue greater than 1.

By using the Varimax method and removing scores <0.5, the Rotated Component Matrix and factor loadings were calculated, and the corresponding outcomes are presented in Table 8.

Table 8. Rotated Component Matrix: Varimax method.

Rotated Component Matrix	Component		
	1	**2**	**3**
How important do you consider school's contribution to shaping environmental conscience?	0.779		
How important is the existence of a Recycle Bin in your school?	0.692		
To what extent do you believe co-participation of children—parents will help in developing energy-saving behavior?	0.622		
Do you think your school should participate in recycling programs?	0.559		
Do you believe that Renewable Energy Sources can help solve the ecological problems of the planet?		0.689	
In what extend do you believe that environmental participation would activate energy behavior?		0.605	
Do you think that using a bicycle reduces energy consumption?		0.598	
Do you think that participation in ecologic activities helps minimizing environmental awareness?	0.510	0.561	
Do you believe that your family economic status plays a part in practicing energy saving behavior?			
How would you rate your environmental knowledge level?			0.773
How would you rate the adequacy of environmental education you receive?			0.756
Do you agree with the view that harmonious coexistence of "human-nature" is a prerequisite for the survival of us all?			0.563

Extraction Method: Principal Component Analysis.
Rotation Method: Varimax with Kaiser Normalization.

The questions assigned to each component to reflect students' perceptions fall into three distinct categories. In the first component of the rotated component matrix, there are questions emphasizing student perception concerning school and family role towards environmental conscience. Variables included in the second component refer to "Student's Environmental conscience". Variables in the third component represent "Student's perception on Environmental Education". The first component of the rotated matrix explains 19.1% of the variance while the second component explains 16.8% and the third component explains 14.2% (see Table 7). The three components are presented on Table 9, and are referred to as such in the next sections of the analysis.

Table 9. Factors (Components) identification.

Factor Interpretation
1. School and family role towards environment
2. Student's environmental awareness
3. Student's environmental education

Our main aim was to divide student population into groups with common characteristics according to their environmental behavior. To identify the appropriate number of clusters, K-Means method in SPSS was applied, by inputting all three components that were saved as standardized scale variables during the factor analysis stage. According to the algorithm, the user must input the initial number of cluster centers, and observations are assigned to each center with the criterion of the closer distance so that a cluster is formed. The algorithm continues locating new data centers and stops if there is no noticeable difference between two consecutive iterations. For our data, the initial number of cluster centers was set to four.

K-Means method provides a reliability test via the produced ANOVA. Table 10 depicts the initial cluster centers for the three factors, in standardized z values (mean = 0, St. dev. = 1).

Table 10. Initial cluster centers: k-means method.

	Initial Cluster Centers			
	1	2	3	4
School and family role towards environment	0.060	−1.045	−3.290	2.131
Environmental awareness	1.225	1.367	−2.757	−3.133
Environmental Education	−3.620	2.451	−2.325	0.380

After 10 iterations, the algorithm stopped by locating no further difference between iterations. In Table 11, final cluster centers, which were used for the analysis, are located.

Table 11. Final cluster centers, k-means method (highest score is in bold).

	Final Cluster Centers			
	1	2	3	4
School and family role towards environment	−0.091	−0.053	−1.859	*0.769*
Environmental awareness	*0.764*	0.305	−1.316	−0.717
Environmental Education	−0.748	*0.962*	−0.192	−0.320

By using the ANOVA method, it is observed that the differences in the mean between the clusters are statistically significant at the 99.9% confidence level (sig. < 0.001) (Table 12).

Table 12. Statistical significance of identified clusters, ANOVA method.

ANOVA						
	Cluster		Error		F	Sig.
	Mean Square	df	Mean Square	df		
School and family role towards environment	44.092	3	0.514	266	85.782	0.000
Environmental awareness	45.463	3	0.499	266	91.193	0.000
Environmental Education	45.514	3	0.498	266	91.401	0.000

Concerning the 1st factor, "School and family role towards environment", Cluster 4 yielded the highest positive normalized score, as can be seen in Table 10, which presents the final cluster centers. Clusters 1–3, on the contrary, yielded a below average score. Cluster 3, which includes 25 students, has the lowest score.

Concerning the second factor, "Students' Environmental Awareness", Clusters 1 and 2 have an above average normalized score. Clusters 3 and 4 sustain scores lower than average. Furthermore, around 2/3 of the students (Clusters 3 and 4 include 62% of the students) scored above average in the factor "environmental awareness", thus it seems most secondary level students are currently environmentally active.

In Table 13, the total number of students, as calculated by the K-means, is presented.

Table 13. Cluster size.

Number of Students in Each Cluster		
Cluster	1	78
	2	91
	3	25
	4	76
Total		270

Concerning the third factor, "Students' Environmental Education", Cluster 2 revealed an above average normalized score (see Table 11). Clusters 1, 3 and 4 yielded a negative normalized average score and Cluster 1 sustained the lowest score on this factor. These important findings revealed that most students considered their environmental knowledge and education to be inadequate.

By looking at the four identified student groups from cluster analysis, Cluster 1 seems to experience low environmental education, is environmentally active, and has a low expectation for school and family contribution towards environmental awareness. This student group seems to disregard school and family role towards the environment, exhibits environmental awareness and yielded a low score in evaluating their environmental education level. They can be characterized as "the ecologists, need education". Cluster 2 is the biggest student cluster, representing around 1/3 of the sample. Secondary education students belonging to Cluster 2 seem to have an appropriate level of environmental education, are environmentally sensitive and take the role of school and family into account towards environmental activation. These students are characterized as the "environmentally activated". Cluster 3 represents the minority of the student sample (9.2%), disregards school and family role towards the environment, exhibits low environmental awareness and yielded a low score in evaluating their environmental education level. They can be characterized as "environmentally indifferent". Cluster 4 represents 28.1% of the sample, considers school and family role towards environmental motivation to be very significant and sustained low scores on environmental awareness and environmental education. Cluster 4 is the most challenging student cluster, as it seems that those students depend on school role and are willing to become environmentally active in the future within the "green" school environment. This student minority seems to be interested in forming an environmentally focused behavior and can be characterized as the "school motivated and potentially active".

5. Discussion and Conclusions

This study investigated the accumulated knowledge, the prevailing attitudes, and the energy habits of students at secondary education in the prefecture of Larissa. The study focused on determining the students' environmental behavior and their attitudes towards environmental sustainability. Specifically, the key-areas examined are: (a) the familiarity of students with the concept of energy; (b) their accumulated knowledge and their assimilation of the topics of renewable and conventional energy; (c) students' attitudes towards energy saving and integrated/multifaceted environmental protection; (d) their habits concerning efficient energy use; and (e) the applicability of environment-driven knowledge to students' attitudes and habits.

Concerning students' wider knowledge on various environmental issues and energy sources, the results of the study were noteworthy, since 96.30% of students self-reported to be extensively knowledgeable about the diverse types of energy sources while, on the other hand, only 60% reported to be aware of the term of "energy crisis". Students' awareness towards environment was evident in their daily habits. A proportion of 85% turns off the lights during breaks, 54% have joined voluntary environmental programs organized by their schools, 78% believe that it is important for the school to have recycle bins and around 60% would buy a low energy consumption electric device with higher price. Concerning school role, around 80% of the students believe their schools should participate in recycling, 2/3 of the students consider that the school environment can contribute and support their environmentally-based initiative and 75% of them find participation in ecological activities to be equally important in improving awareness. In parallel, the role of family is of utmost importance since 70% of the students believe that the family environment in collaboration with their teachers' creativity can encourage them in environmental-oriented educational activities, even though 45% of students believe that they have received modest environmental education, while only 30% state that they have received adequate environmental education. In a methodological overview, for several Likert scale questions, factor analysis under the PCA approach was used. This method extracted three factors, interpreting students' behavior. Consequently, according to the identified factors, priority should be given to three dimensions: "School and family role towards environmental conscience", "Student's Degree of Environmental conscience and "Student's Degree of Environmental Education".

Concerning the factor of "School and family role", it has been reported that individual-oriented energy behavior is an extremely complex issue that is determined by many parameters, such as the intrinsic socio-economic status [100]. Thus, adult family members should be aware that the energy behavior within the household greatly influences the energy behavior of their children. Furthermore, the purpose of a "sustainable school" is not only to help children acquire environmental literacy and understand the dimensions of sustainability. The main goal of such an educational reform is to change the school itself in promoting sustainability and helping today students to follow sustainable practices in their everyday life. Such schools have managed to raise students' environmental consciousness [101]. Concerning the factor named as "Student's Degree of Environmental conscience", emphasis should be placed even at primary education level, for the availability of resource material and information—such as renewable energy educational software or laboratory applications—which should be available to students, in the form of educational games and activities. Concerning student's environmental education, it is reported that the extent of curricular materials on environmental and energy subjects could contribute to developing skills, values, and attitudes aligned with sustainable development perspectives [81].

Grouping the sample into clusters by using the K-means method enabled the identification of four different student clusters. These clusters reflected differences of environmental attitudes that were reported among students. More specifically, the clusters segmented the student sample under the following sub-groups: (1) the ecologists, need education; (2) the environmentally activated, which is the largest group; (3) the environmentally indifferent, which is the smallest group; and (4) the school motivated and potentially active. Except for Cluster 2 (environmentally activated), all other clusters require more environmental education. Furthermore Clusters 1 and 2 need more support from family and school towards environmental issues. Custer 4, representing 28% of the sample, has adequate support from family but needs more motivation and education to become active. Finally, there is a minority of students (9.2%) that seem completely indifferent to the environmental issues.

It is noteworthy that students require a more solid environmental education than current taught courses offer. By looking at students' responses, it becomes evident that school and family background are expected to play a key role in assisting students overcome the future challenges in the energy sector. In this respect, younger and senior family members must be soundly educated and collaborate with each other. Moreover, initiatives taken by the formal educational system must be encouraged, enabling students to become aware of energy and inspire changes from conservative perspectives of energy behavior. Energy education taught courses can be further incorporated into humanitarian, social, and natural sciences, respectively. Education is an interdisciplinary area of a wide consortium of sciences where decisions about taught content, resources' allocation, and learning deliverables/outcomes are made at regional and national level. Cultural and national aspects are also directly affecting the learning environment and the abiding public-driven policies. However, energy issues are prevailing across Europe and their inclusion in the school curricula should be a European-level priority, too. Furthermore, the core of learning process is still focusing on local action and it must be adaptable to each student cognitive background. To this end, focusing environmental topics on energy education, the abiding energy-based policies should bring together energy, environment, and economics, giving a rational basis for decision-making. Many educational courses on environmental issues also incorporate energy studies—but usually only towards the viewpoint of sustainable development.

However, there is still an imperative need of specific energy education programs to be developed that could formulate the basis of cultivating the behavioral consciousness of current and future energy consumers. These programs should not only focus on environmental deterioration but also on the advantages abided to RES, since the expansion of RES usage can positively contribute to life quality [102]. The main policies concerned must increase all students' awareness by educating them on the capabilities, pricing, and multifaceted impact of the various energy sources (both renewable and conventional). Such policies should further consider local energy availability/backup technologies and requirements, together with localized climatic and cultural characteristics. In parallel, the content of educational curricula should remain consistent with national and international priorities, reflecting the values of "thinking globally, acting locally". By understanding the measures established through currently applied energy policies, students should become capable of being creatively involved in shaping a viable RES-driven future.

Author Contributions: V.P. designed the experimental framework and collected the data. V.P., S.N. and M.C. carried out the implementation, performed the calculations and the computer programming. G.L.K. and G.A. gathered and implemented all the theoretical background of the paper, having the input from the experimental development. G.A., G.L.K., and S.N. reviewed and discussed the results of the study.

Conflicts of Interest: The authors declare no conflict of interest.

Appendix A

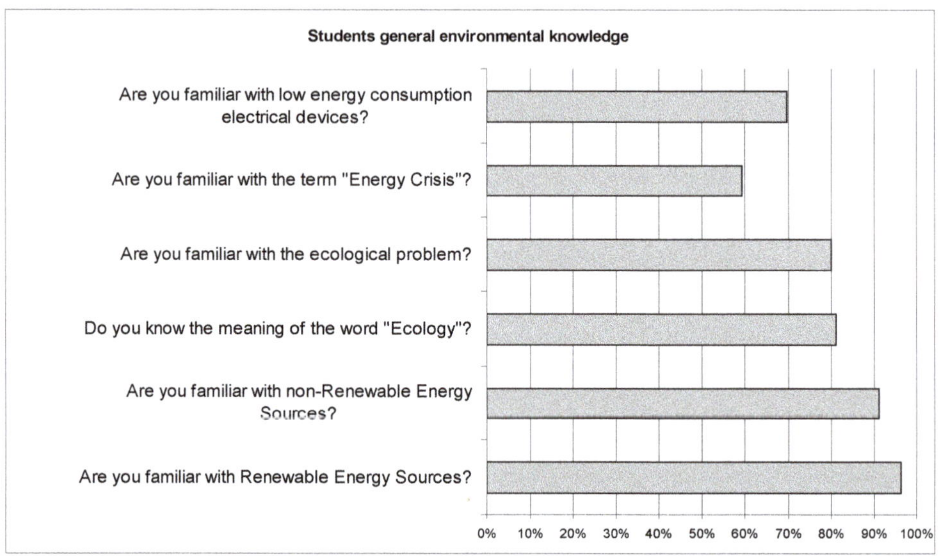

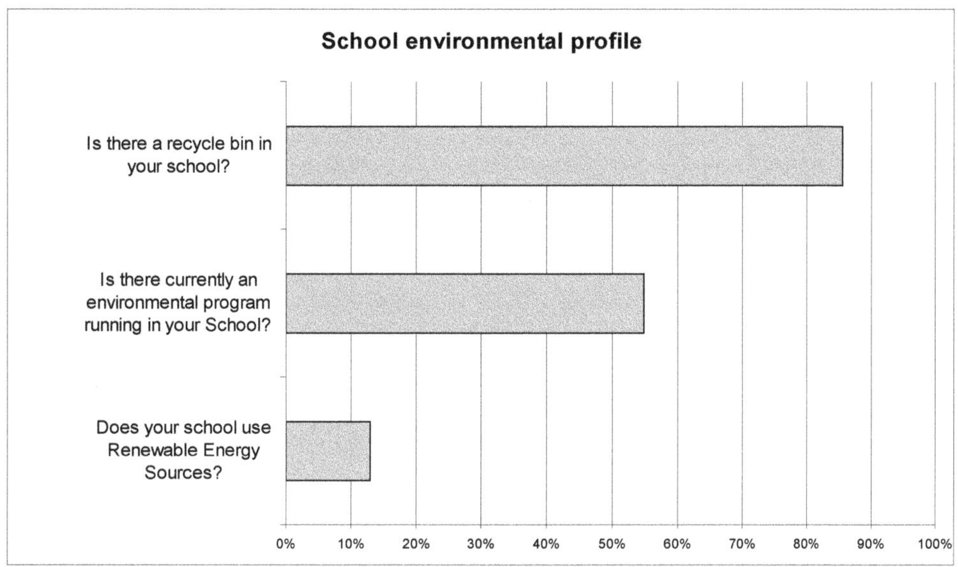

School environmental profile
Is there a recycle bin in your school?
Is there currently an environmental program running in your School?
Does your school use Renewable Energy Sources?
0%   10%   20%   30%   40%   50%   60%   70%   80%   90%   100%

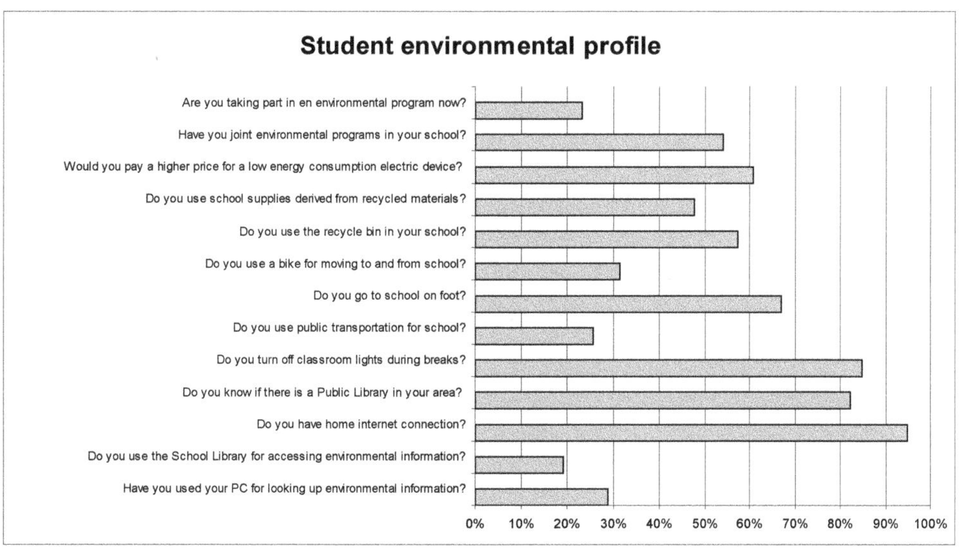

Student environmental profile
Are you taking part in en environmental program now?
Have you joint environmental programs in your school?
Would you pay a higher price for a low energy consumption electric device?
Do you use school supplies derived from recycled materials?
Do you use the recycle bin in your school?
Do you use a bike for moving to and from school?
Do you go to school on foot?
Do you use public transportation for school?
Do you turn off classroom lights during breaks?
Do you know if there is a Public Library in your area?
Do you have home internet connection?
Do you use the School Library for accessing environmental information?
Have you used your PC for looking up environmental information?
0%   10%   20%   30%   40%   50%   60%   70%   80%   90%   100%

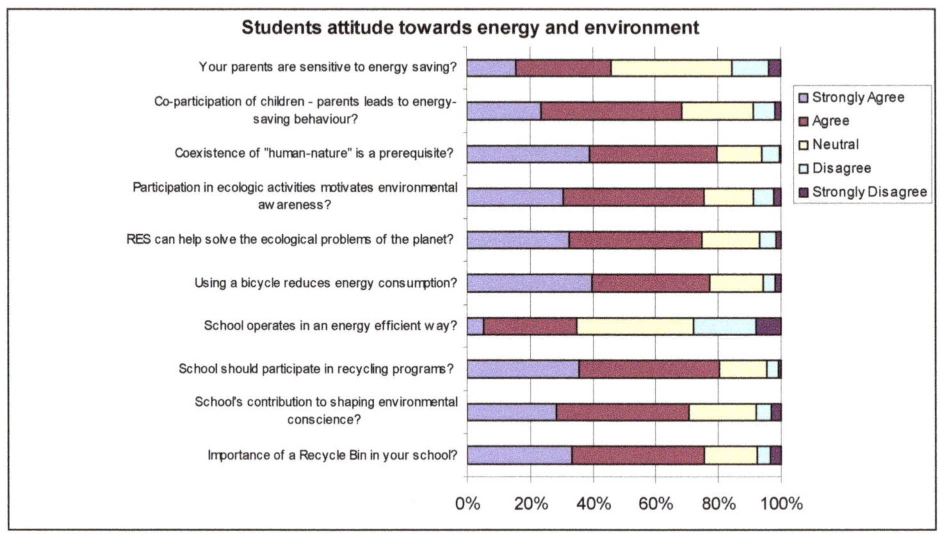

References

1. Van Gent, H.A.; Rietveld, P. Road transport and the environment in Europe. *Sci. Total Environ.* **1993**, *129*, 205–218. [CrossRef]

2. Robinson, J.A.; Srinivasan, T.N. Long-term consequences of population growth: Technological change, natural resources, and the environment. In *Handbook of Population and Family Economics*; Gulf Professional Publishing: Houston, TX, USA, 1997; Volume 1, pp. 1175–1298.

3. Lam, C.W.; Lim, S.; Schoenung, J.M. Environmental and risk screening for prioritizing pollution prevention opportunities in the U.S. printed wiring board manufacturing industry. *J. Hazard. Mater.* **2011**, *189*, 315–322. [CrossRef] [PubMed]

4. Ockenden, M.C.; Deasy, C.; Quinton, J.N.; Bailey, A.P.; Surridge, B.; Stoate, C. Evaluation of field wetlands for mitigation of diffuse pollution from agriculture: Sediment retention, cost and effectiveness. *Environ. Sci. Policy* **2012**, *24*, 110–119. [CrossRef]

5. Rehman, M.U.; Rashid, M. Energy consumption to environmental degradation, the growth appetite in SAARC nations. *Renew. Energy* **2017**, *111*, 284–294. [CrossRef]

6. Azam, M. Does environmental degradation shackle economic growth? A panel data investigation on 11 Asian countries. *Renew. Sustain. Energy Rev.* **2016**, *65*, 175–182. [CrossRef]

7. Zeb, R.; Salar, L.; Awan, U.; Zaman, K.; Shahbaz, M. Causal links between renewable energy, environmental degradation and economic growth in selected SAARC countries: Progress towards green economy. *Renew. Energy* **2014**, *71*, 123–132. [CrossRef]

8. Brundtland, G.H. *Report of the World Commission on Environment and Development: Our Common Future*; United Nations: New York, NY, USA, 1987.

9. Mantzava, G. Climate Change. Master's Thesis, National Technical University of Athens, Athens, Greece, 2003. (In Greek)

10. Tatsidjodoung, P.; Dabat, M.H.; Blin, J. Insights into biofuel development in Burkina Faso: Potential and strategies for sustainable energy policies. *Renew. Sustain. Energy Rev.* **2012**, *16*, 5319–5330. [CrossRef]

11. Deng, Y.; Xu, J.; Liu, Y.; Mancl, K. Biogas as a sustainable energy source in China: Regional development strategy application and decision making. *Renew. Sustain. Energy Rev.* **2014**, *35*, 294–303. [CrossRef]

12. European Commission. Directive 2001/77/EC of the European Parliament and of the Council of 27 September 2001 on the Promotion of Electricity Produced from Renewable Energy Sources in the Internal Electricity Market. European Commission: Luxembourg, 2001. Available online: http://eur-lex.europa.eu/legal-content/EN/TXT/HTML/?uri=CELEX:32001L0077&from=EN (accessed on 10 May 2018).

13. NDDS Greece. National Strategy for Sustainable Development Greece—Executive Summary, Hellenic Republic Ministry for the Environment, Physical Planning and Public Works. 2002. Available online: http://www.minenv.gr/4/41/000/nssd-english-final.pdf (accessed on 10 December 2015).
14. Xu, L.J.; Fan, X.C.; Wang, W.Q.; Xu, L.; Duan, Y.L.; Shi, R.J. Renewable and sustainable energy of Xinjiang and development strategy of node areas in the "Silk Road Economic Belt". *Renew. Sustain. Energy Rev.* **2017**, *79*, 274–285. [CrossRef]
15. Gaigalis, V.; Skema, R. Analysis of the fuel and energy transition in Lithuanian industry and its sustainable development in 2005–2013 in compliance with the EU policy and strategy. *Renew. Sustain. Energy Rev.* **2015**, *52*, 265–279. [CrossRef]
16. Gaigalis, V.; Markevicius, A.; Skema, R.; Savickas, J. Sustainable energy strategy of Lithuanian Ignalina Nuclear Power Plant region for 2012–2035 as a chance for regional development. *Renew. Sustain. Energy Rev.* **2015**, *51*, 1680–1696. [CrossRef]
17. Kyriakopoulos, G.; Chalikias, M. The Investigation of Woodfuels' Involvement in Green Energy Supply Schemes at Northern Greece: The Model Case of the Thrace Prefecture. *Procedia Technol.* **2013**, *8*, 445–452. [CrossRef]
18. Kolovos, K.; Kyriakopoulos, G.; Chalikias, M. Co-evaluation of basic woodfuel types used as alternative heating sources to existing energy network. *J. Environ. Prot. Ecol.* **2011**, *12*, 733–742.
19. Chalikias, M.; Christopoulou, O. Factors affecting the forest plantations establishment in the frame of the common Agricultural policy. *J. Environ. Prot. Ecol.* **2011**, *12*, 305–316.
20. Kyriakopoulos, G.; Kolovos, K.; Chalikias, M. Woodfuels Prosperity towards a More Sustainable Energy Production. In *Track I: Social & Humanistic Computing for the Knowledge Society: Emerging Technologies and Systems for the Society and the Humanity, Proceedings of the 3rd World Summit on the Knowledge Society (WSKS 2010), Corfu, Greece, 22–24 September 2010*; Springer: Berlin/Heidelberg, Germany, 2010; pp. 19–25.
21. Abrahamse, W.; Steg, L.; Vlek, C.; Rothengatter, T. The effect of tailored information: Goal setting, and tailored feedback on household energy use, energy-related behaviors and behavioral antecedents. *J. Environ. Psychol.* **2007**, *27*, 265–276. [CrossRef]
22. Abrahamse, W.; Steg, L. How do socio-demographic and psychological factors relate to households' direct and indirect energy use and savings? *J. Econ. Psychol.* **2009**, *30*, 711–720. [CrossRef]
23. Hargreaves, T.; Nye, M.; Burgess, J. Making energy visible: A qualitative field study of how householders interact with feedback from smart energy monitors. *Energy Policy* **2010**, *38*, 6111–6119. [CrossRef]
24. Martiskainen, M.; Coburn, J. The role of information and communication technologies (ICTs) in household energy consumption—Prospects for the UK. *Energy Effic.* **2011**, *4*, 209–221. [CrossRef]
25. McCalley, L.T.; Midden, C.J.H. Energy conservation through product-integrated feedback: The roles of goal-setting and social orientation. *J. Econ. Psychol.* **2002**, *23*, 589–603. [CrossRef]
26. Ueno, T.; Sano, F.; Saeki, O.; Tsuji, K. Effectiveness of an energy-consumption information system on energy savings in residential houses based on monitored data. *Appl. Energy* **2006**, *83*, 166–183. [CrossRef]
27. Van Dam, S.S.; Bakker, C.A.; van Hal, J.D.M. Home energy monitors: Impact over the medium-term. *Build. Res. Inf.* **2010**, *38*, 458–469. [CrossRef]
28. Willis, R.; Stewart, R.; Panuwatwanich, K.; Jones, S.; Kyriakides, A. Alarming visual display monitors affecting shower end use water and energy conservation. Australian residential households. *Resour. Conserv. Recycl.* **2010**, *54*, 1117–1127. [CrossRef]
29. Galis, V.; Gyberg, P. Energy behaviour as a collective. *Energy Effic.* **2011**, *4*, 303–319. [CrossRef]
30. Leighty, W.; Meier, A. Accelerated electricity conservation in Juneau, Alaska: A study of household activities that reduced demand 25%. *Energy Policy* **2011**, *39*, 2299–2309. [CrossRef]
31. Crosbie, T. Household energy consumption and consumer electronics: The case of television. *Energy Policy* **2008**, *36*, 2191–2199. [CrossRef]
32. Wall, R.; Crosbie, T. Potential for reducing electricity demand for lighting in households: An exploratory socio-technical study. *Energy Policy* **2009**, *37*, 1021–1031. [CrossRef]
33. Barr, S.; Gilg, A.W.; Ford, N. The household energy gap: Examining the divide between habitual- and purchase-related conservation behaviours. *Energy Policy* **2005**, *33*, 1425–1444. [CrossRef]
34. Ek, K.; Sfderholm, P. The devil is in the details: Household electricity saving behavior and the role of information. *Energy Policy* **2010**, *38*, 1578–1587. [CrossRef]

35. Maréchal, K. Not irrational but habitual: The importance of behavioural lock-in in energy consumption. *Ecol. Econ.* **2010**, *69*, 1104–1114. [CrossRef]
36. Martinsson, J.; Lundqvist, L.J.; Sundström, A. Energy saving in Swedish households. The (relative) importance of environmental attitudes. *Energy Policy* **2011**, *39*, 5182–5191. [CrossRef]
37. Nair, G.; Gustavsson, L.; Mahapatra, K. Factors influencing energy efficiency investments in existing Swedish residential buildings. *Energy Policy* **2010**, *38*, 2956–2963. [CrossRef]
38. Naassen, J.; Holmberg, J. Quantifying the rebound effects of energy efficiency improvements and energy conserving behaviour in Sweden. *Energy Effic.* **2009**, *2*, 221–231. [CrossRef]
39. Wang, Z.; Zhang, B.; Yin, J.; Zhang, Y. Determinants and policy implications for household electricity-saving behaviour: Evidence from Beijing, China. *Energy Policy* **2011**, *39*, 3550–3557. [CrossRef]
40. Ntanos, S.; Kyriakopoulos, G.; Chalikias, M.; Arabatzis, G.; Skordoulis, M. Public Perceptions and Willingness to Pay for Renewable Energy: A Case Study from Greece. *Sustainability* **2018**, *10*, 687. [CrossRef]
41. Cayla, J.M.; Maizi, N.; Marchand, C. The role of income in energy consumption behaviour: Evidence from French households data. *Energy Policy* **2011**, *39*, 7874–7883. [CrossRef]
42. Moussaoui, I. De la société de consommation à la société de modération. Ce que les Français disent, pensent et font en matière de maîtrise de l'énergie. *Ann. Rech. Urbaine* **2007**, *103*, 112–119. [CrossRef]
43. Arabatzis, G.; Malesios, C. Pro-Environmental attitudes of users and not users of fuelwood in a rural area of Greece. *Renew. Sustain. Energy Rev.* **2013**, *22*, 621–630. [CrossRef]
44. Dillman, D.A.; Rosa, E.A.; Dillman, J.J. Lifestyle and home energy conservation in the United States: The poor accept lifestyle cutbacks while the wealthy invest in conservation. *J. Econ. Psychol.* **1983**, *3*, 299–315. [CrossRef]
45. Black, J.S.; Stern, P.C.; Elsworth, J.T. Personal and contextual influences on household energy adaptations. *J. Appl. Soc. Psychol.* **1985**, *70*, 3–21. [CrossRef]
46. Winter, C. Education for sustainable development and the secondary curriculum in English schools: Rhetoric or reality? *Camb. J. Educ.* **2007**, *37*, 337–354. [CrossRef]
47. Winter, C.; Firth, R. Knowledge about Education for Sustainable Development: Four case studies of student teachers in English secondary schools. *J. Educ. Teach.* **2007**, *33*, 341–358. [CrossRef]
48. Tsekos, C.A.; Christoforidou, E.I.; Tsekos, E.A. Planning an environmental education project for kindergarten under the theme of "The Forest". *Rev. Eur. Stud.* **2012**, *4*, 111–117. [CrossRef]
49. Zakaria, S.Z.S. Science education in primary school towards environmental sustainability. *Res. J. Appl. Sci.* **2011**, *6*, 330–334.
50. Ballantyne, R.; Fien, J.; Packer, J. Program effectiveness in facilitating intergenerational influence in environmental education: Lessons from the field. *J. Environ. Educ.* **2001**, *32*, 8–15. [CrossRef]
51. Lillemo, S.C. Measuring the effect of procrastination and environmental awareness on households' energy-saving behaviours: An empirical approach. *Energy Policy* **2014**, *66*, 249–256. [CrossRef]
52. Chen, L. Professional development of energy education teachers in rural areas from the perspective of environmental factors. *Energy Educ. Sci. Technol. Part A Energy Sci. Res.* **2014**, *32*, 2657–2664.
53. Ku, C.K.; Chen, Y.W.; Kao, T.S.; Chien, S.C. The environmental education strategy of integration of universities, NGOs and elementary schools to develop Taiwan's energy education program. *WIT Trans. Ecol. Environ.* **2012**, *167*, 165–175.
54. Nabalegwa, M.; Buyinza, M. Prospects of petroleum exploration and local community environmental education in the Albertine Graben, Western Uganda. *Res. J. Appl. Sci.* **2012**, *7*, 409–412.
55. Flogaiti, E. *Environmental Education*; Pedio Publications: Athens, Greece, 2011. (In Greek)
56. Khan, A. A vision of a 21st-century community learning centre. In *Education for Sustainability*; Huckle, J., Sterling, S., Eds.; Earthscan: London, UK, 1996; pp. 222–227.
57. Posch, P. The ecologisation of schools and its implications for educational policy. *Camb. J. Educ.* **1999**, *29*, 341–348. [CrossRef]
58. Braun, T.; Cottrell, R.; Dierkes, P. Fostering changes in attitude, knowledge and behavior: Demographic variation in environmental education effects. *Environ. Educ. Res.* **2017**, 1–22. [CrossRef]
59. Cruz, A.R.; Selby, S.T.; Durham, W.H. Place-based education for environmental behavior: A 'funds of knowledge' and social capital approach. *Environ. Educ. Res.* **2017**, *24*, 627–647. [CrossRef]
60. Chankrajang, T.; Muttarak, R. Green Returns to Education: Does Schooling Contribute to Pro-Environmental Behaviours? Evidence from Thailand. *Ecol. Econ.* **2017**, *131*, 434–448. [CrossRef]

61. Mullenbach, L.E.; Green, G.T. Can environmental education increase student-athletes' environmental behaviors? *Environ. Educ. Res.* **2018**, *24*, 427–444. [CrossRef]
62. Barata, R.; Castro, P.; Martins-Loução, M.A. How to promote conservation behaviours: The combined role of environmental education and commitment. *Environ. Educ. Res.* **2017**, *23*, 1322–1334. [CrossRef]
63. Meyer, A. Does education increase pro-environmental behavior? Evidence from Europe. *Ecol. Econ.* **2015**, *116*, 108–121. [CrossRef]
64. Fernández-Manzanal, R.; Serra, L.M.; Morales, M.J.; Carrasquer, J.; Rodríguez-Barreiro, L.M.; Del Valle, J.; Murillo, M.B. Environmental behaviours in initial professional development and their relationship with university education. *J. Clean. Prod.* **2015**, *108*, 830–840. [CrossRef]
65. Zsóka, Á.; Szerényi, Z.M.; Széchy, A.; Kocsis, T. Greening due to environmental education? Environmental knowledge, attitudes, consumer behavior and everyday pro-environmental activities of Hungarian high school and university students. *J. Clean. Prod.* **2013**, *48*, 128–138. [CrossRef]
66. Damerell, P.; Howe, C.; Milner-Gulland, E.J. Child-orientated environmental education influences adult knowledge and household behaviour. *Environ. Res. Lett.* **2013**, *8*, 015016. [CrossRef]
67. Repka, P.; Švecová, M. Environmental education in conditions of National Parks of Slovak Republic. *Procedia-Soc. Behav. Sci.* **2012**, *55*, 628–634. [CrossRef]
68. Hatzigeorgiou, E.; Haralambopoulos, H. CO_2 emissions, GDP and energy intensity: A multivariate cointegration and causality analysis for Greece, 1977–2007. *Appl. Energy* **2011**, *88*, 1377–1385. [CrossRef]
69. Kandpal, T.; Broman, L. Renewable energy education: A global status review. *Renew. Sustain. Energy Rev.* **2014**, *34*, 300–324. [CrossRef]
70. Alvarez Suárez, P.; Vega Marcote, P. Developing sustainable environmental behavior in secondary education students (12–16): Analysis of a didactic strategy. *Procedia-Soc. Behav. Sci.* **2010**, *2*, 3568–3574.
71. Marcinkowski, T.J.; Volk, T.J.; Hungerford, H.R. *An Environmental Education Approach to the Training of Middle Level Teachers a Teacher Education Program Specialization*; UNESCO/UNEP: Paris, France, 1990. Available online: http://www.unesco.org/education/information/pdf/333_52.pdf (accessed on 27 November 2013).
72. Monroe, M.C.; Oxarart, A.; Plate, R.R. A Role for Environmental Education in Climate Change for Secondary Science Educators. *Appl. Environ. Educ. Commun.* **2013**, *12*, 4–18. [CrossRef]
73. Acikgoz, C. Renewable energy education in Turkey. *Renew. Energy* **2011**, *36*, 608–611. [CrossRef]
74. Çelikler, D.; Aksan, Z. The opinions of secondary school students in Turkey regarding renewable energy. *Renew. Energy* **2015**, *75*, 649–653. [CrossRef]
75. Evangelinos, K.; Jones, N.; Panoriou, E. Challenges and opportunities for sustainability in regional universities: A case study in Mytilene, Greece. *J. Clean. Prod.* **2009**, *17*, 1154–1161. [CrossRef]
76. Lefkeli, S.; Manolas, E.; Ioannou, K.; Tsantopoulos, G. Socio-cultural impact of energy saving: Studying the behaviour of elementary school students in Greece. *Sustainability* **2018**, *10*, 737. [CrossRef]
77. Daskolia, M.; Dimos, A.; Kampylis, P.G. Secondary teachers' conceptions of creative thinking within the context of environmental education. *Int. J. Environ. Sci. Educ.* **2012**, *7*, 269–290.
78. Macko, J.; Blahútová, D.; Stollárová, N. Space for the environmental education in the system of secondary education in Slovakia. *Informatologia* **2013**, *46*, 256–260.
79. Bokhoree, C.; Jheengut, A.; Bholah, R. Environmental clubs as vehicles for promoting education for environmental sustainability in Mauritian secondary schools. *Int. J. Environ. Cult. Econ. Soc. Sustain.* **2012**, *7*, 178–190. [CrossRef]
80. Torkar, G. Secondary school students' environmental concerns and attitudes toward forest ecosystem services: Implications for biodiversity education. *Int. J. Environ. Sci. Educ.* **2016**, *11*, 11019–11031.
81. Capelo, A. Integration of education for sustainable development into formal secondary curricula of East Timor. In *Handbook of Research on Pedagogical Innovations for Sustainable Development*; IGI Global: Hershey, PA, USA, 2014; pp. 54–66.
82. Sorgo, A.; Kamensek, A. Implementation of a curriculum for environmental education as education for sustainable development in Slovenian upper secondary schools. *Energy Educ. Sci. Technol. Part B Soc. Educ. Stud.* **2012**, *4*, 1067–1076.
83. Lamas, J.G.; Cuadrado, V.A. Argumentative skills in the design of webquests in environmental education for secondary students. *Int. J. Learn.* **2012**, *18*, 63–72. [CrossRef]

84. Van Ongevalle, J.; van Petegem, P.; Deprezc, S.; Chimbodza, I.J.M. Participatory planning for project sustainability of environmental education projects: A case study of the secondary teacher training environmental education project (St^{2}eep) in Zimbabwe. *Environ. Educ. Res.* **2011**, *17*, 433–449. [CrossRef]

85. Siardos, G. *Multivariate Statistical Analysis Methods. Part One*; Ziti Publications: Thessaloniki, Greece, 1999. (In Greek)

86. SPSS. *The SPSS Two-Step Cluster Component—A Scalable Component Enabling More Efficient Customer Segmentation*; White Paper-Technical Report; SPSS Inc.: Chicago, IL, USA, 2001.

87. Statistics Solutions. Conduct and Interpret a Cluster Analysis. 2015. Available online: http://www.statisticssolutions.com/cluster-analysis-2/ (accessed on 20 November 2017).

88. Ntona, E.; Arabatzis, G.; Kyriakopoulos, G. Energy saving: Views and attitudes of students in secondary education. *Renew. Sustain. Energy Rev.* **2015**, *46*, 1–15. [CrossRef]

89. Local Union of Municipalities and Communities of Prefecture of Larissa. *Prefecture of Larissa, Nature-History-Development*; Local Union of Municipalities and Communities of Prefecture of Larissa: Larissa, Greece, 2002. (In Greek)

90. Locator Map of Larisa Prefecture in Greece. Greece, 2017. Available online: http://www.mykosmos.gr/loc_mk/images/maps/map_greece_larissa.gif (accessed on 21 June 2017).

91. Larissa Prefecture Depicting Main Cities. 2017. Available online: https://pirgetoslarisas.files.wordpress.com/2011/02/map.png?w=300&h=300 (accessed on 20 June 2017).

92. Directorate of Secondary Education of Larissa. *Statistical Data of Schools of Prefecture of Larissa*; Directorate of Secondary Education of Larissa: Larissa, Greece, 2014. (In Greek)

93. CheckMarket. Sample Size Calculator. 2018. Available online: https://www.checkmarket.com/sample-size-calculator/ (accessed on 20 April 2018).

94. Seale, C.; Filmer, P. doing social research. In *Researching Society and Culture*; Seale, C., Ed.; Sage Publications: London, UK, 2000.

95. The United Nations Convention on the Rights of the Child, 1990. Available online: https://www.unicef.org.uk/what-we-do/un-convention-child-rights/ (accessed on 10 April 2018).

96. Cohen, L.; Lawrence, M.; Morrison, K. *Research Methods in Education*; Routledge: London, UK, 2007.

97. Denscombe, M. *The Good Research Guide: For Small-Scale Social Research Projects*, 4th ed.; Volume: Open UP Study Skills; McGraw-Hill Open University Press: Maidenhead, UK, 2010.

98. Darren, G.; Mallery, P. *SPSS for Windows Step by Step: A Simple Guide and Reference. 11.0 Update*, 4th ed.; Allyn & Bacon: Boston, MA, USA, 2003.

99. Hair, J.F.; Black, W.C.; Babin, B.J.; Anderson, R.E. Explanatory Factor Analysis. Chapter 3. In *Multivariate Data Analysis*, 7th ed.; Prentice Hall International, Inc.: Englewood Cliffs, NJ, USA, 2010; pp. 89–150.

100. Stephenson, J.; Barton, B.; Carrington, G.; Gnoth Lawson, R.; Thorsnes, P. Energy cultures: A framework for understanding energy behaviours. *Energy Policy* **2010**, *38*, 6120–6129. [CrossRef]

101. Berglund, T.; Gericke, N.; Chang Rundgren, S.N. The implementation of education for sustainable development in Sweden: Investigating the sustainability consciousness among upper secondary students. *Res. Sci. Technol. Educ.* **2014**, *32*, 318–339. [CrossRef]

102. Ntanos, S.; Kyriakopoulos, G.; Chalikias, M.; Arabatzis, G.; Skordoulis, M.; Galatsidas, S.; Drosos, D. A Social Assessment of the Usage of Renewable Energy Sources and Its Contribution to Life Quality: The Case of an Attica Urban Area in Greece. *Sustainability* **2018**, *10*, 1414. [CrossRef]

Article

A Social Assessment of the Usage of Renewable Energy Sources and Its Contribution to Life Quality: The Case of an Attica Urban Area in Greece

Stamatios Ntanos [1,*], Grigorios L. Kyriakopoulos [2], Miltiadis Chalikias [3], Garyfallos Arabatzis [1], Michalis Skordoulis [1], Spyros Galatsidas [1] and Dimitrios Drosos [4]

[1] Department of Forestry and Management of the Environment and Natural Resources, School of Agricultural and Forestry Sciences, Democritus University of Thrace, 68200 Orestiada, Greece; garamp@fmenr.duth.gr (G.A.); mskordoulis@gmail.com (M.S.); sgalatsi@fmenr.duth.gr (S.G.)

[2] School of Electrical and Computer Engineering, National Technical University of Athens, 15780 Zografou, Greece; gregkyr@chemeng.ntua.gr

[3] Department of Tourism Management, School of Business, Economics and Social Sciences, University of West Attica, 12244 Egaleo, Greece; mchalikias@hotmail.com

[4] Department of Business Administration, School of Business, Economics and Social Sciences, University of West Attica, 12244 Egaleo, Greece; drososd@puas.gr

* Correspondence: sdanos@ath.forthnet.gr or sdanos@puas.gr

Received: 28 March 2018; Accepted: 28 April 2018; Published: 3 May 2018

Abstract: The aim of this paper is to analyze and evaluate the use of Renewable Energy Sources (RES) and their contribution to citizens' life quality. For this purpose, a survey was conducted using a sample of 400 residents in an urban area of the Attica region in Greece. The methods of Principal Components Analysis and Logit Regression were used on a dataset containing the respondents' views on various aspects of RES. Two statistical models were constructed for the identification of the main variables that are associated with the RES' usage and respondents' opinion on their contribution to life quality. The conclusions that can be drawn show that the respondents are adequately informed about some of the RES' types while most of them use at least one of the examined types of RES. The benefits that RES offer, were the most crucial variable in determining both respondents' perceptions on their usage and on their contribution to life quality.

Keywords: renewable energy sources; life quality; RES public acceptance; logit regression

---

1. Introduction

Nowadays the key-determinants of public attitudes towards green energy schemes are the accelerated pace of energy demand—based on limited resources in conventional energy sources—and the understanding for a greater penetration of "greener" energy due to devastating climate changes on the planet [1]. The link between energy, economic development, and carbon release is a critical research topic [2,3]. The ongoing regional adaptability of Renewable Energy Sources (RES) to national energy mixes attracted global interest, including that of countries such as Greece [4–7], Turkey [8], Spain [9,10], Ukraine [11], Western Europe [12–15], Japan [16], and China [17,18].

Social perceptions vary according to the type of RES investment. Concerning wind investments, social perceptions show that there exist largely approved benefits such as competitiveness, sustainability, lower energy costs, energy independence and local development. On the other hand, local communities often tend to contrast the development of RES due to the relevant costs burdened by the society. Such critical aspects of consideration are the relative aesthetic and acoustic impacts as well as impacts on the territory, in alignment with the spatial localization of wind farms that can undermine the viability of the relevant projects [19]. Local citizens could endanger the objectivity of the outcomes, since they could be prejudiced and concerned about the project consequences [20]. Besides, co-ownership is effectively manipulating the financial constraints of large RES-based projects, which fall beyond the financial possibilities of most communities, leaving the co-ownership perspective as a viable option of large-scale development of RES technologies [21].

Small hydropower (SHP) stations are beneficial for electricity production. The development of SHP sustains a wide spectrum of opportunities to the rural and suburban areas, including the installation of hydraulic works made for other purposes, such as irrigation canals, and dams for water supply purposes. Additionally, these investments have low maintenance costs and extended useful life. Nevertheless, social disproval and opposition can be possibly expressed against hydroelectricity, especially in areas where large dams are built. In this respect, the construction and operation of hydropower stations apparently affect the environmental, social, economic, and political aspects. The social adaptation of SHP, especially in Greece, should be in alignment with a long-term energy policy plan [22]. It is also noteworthy that—based on the qualitative and empirical evidence on hydropower research—the participation and involvement of local communities in hydropower projects are positively associated with their acceptance [23].

Electricity produced by photovoltaic (PV) stations is another type of RES. In many countries, the public communities overwhelmingly support the development of large-scale solar installations [24]. However, when these investments are near residential areas, social opposition and communal objections arise from various stakeholders, thus, the direct benefits to residents should be offered. In a behavioral-based survey, the variables of perceived costs, maintenance requirements, and environmental concerns were evaluated, showing significant differences between RES users and non-users [25]. Marketable cost and operational performance of PVs vary, from place to place. If no subsidy is given, there should be a significant drop in the installation cost of PVs while governmental policies can be drawn under the specifications of solar radiation levels and the maximum income tax rates per installation area [26]. Efficiency is a parameter of utmost importance for the diffusion of PVs while for site space adequacy, the built-in PVs as roof-PV mounting or as wall PVs were suggested [27]. Photovoltaic installations can be ideally applied in Greece, due to county's abundant sunlight, while governments must lift the prohibition on issuing new photovoltaic licenses and take all the measures needed for market expansion [28].

As we may conclude from the above analysis, public acceptance is an important issue for RES policy implementations and its targets achievement. Thus, many researchers have dealt with the social acceptance of RES. Devine-Wright [29] in a review article, has classified a range of potential factors explaining social perceptions on RES. These factors are, namely, personal (age, gender, class, income), social-psychological (knowledge and direct experience, environmental and political beliefs, place attachment), and contextual (technology type and scale, institutional structure, and spatial context) [29]. Furthermore, there is clear evidence that RES positively contributes to citizens' life quality [30].

Previous research results show that citizens in Greece are sufficiently informed and willing to invest in RES [31]. Thus, it is a fact that nowadays, most of the citizens are demanding more incentives to use RES than in the past, as they are not only willing to invest in RES, but also believe that those investments can improve their lives' quality [30].

Attica is studied as a case that bears particular significance for Greece and the broader region, given both the lack of research on its citizens' views about RES and the fact that it is a highly populous metropolitan area. It is easy to realize that the majority of the contemporary studies about social acceptance of RES in Greece, concern provincial regions such as these of Lesvos [6], Pella [22], Andros [32], Crete [30], Larissa [33,34], and Ioannina [35]. In fact, such regions are in the spotlight as their climate supports energy production based on RES [36]. However, it is important to analyze citizens' views on RES in metropolitan areas where energy needs are significantly higher [37]. Since half of the Greek population resides in Attica where there is a huge problem in energy allocation, the understanding of citizens' views on RES is of vital importance in order to motivate them to pay for energy produced by RES or even invest in them [31,37]. This is because citizens' perceptions of the environment and RES can significantly influence public policies [38]. Thus, by measuring and understanding Attica's residents' views in order to form a proper policy to motivate them, the metropolitan area of Athens would become a "greener" constant consumer of energy produced by RES [37]. This "greener" character is needed to be achieved, as Attica is an environmentally compromised region because of its metropolitan character. An effective allocation of the energy sources could allow the development of an energy plan for the rest of the country without the constraints of Attica; this would significantly contribute to the citizens' life quality improvement both in Attica and in the rest of the country [37,39,40].

The above facts are the main drivers of this study's development. Thus, the aim here is to analyze the social acceptance of RES by examining the variables which are correlated with citizens' perceptions of them. More specifically, the variables underlying the differences between RES users and non-users and, the variables encouraging citizens' positive views towards RES' contribution to their life quality will mainly be analyzed. The contribution of this work consists in examining RES in relation to their contribution to life quality since there is no other research to make this correlation. In this sense, understanding the citizens' perception on RES contribution to their lives' quality is very important as it will be easier to point out the incentives that will drive them to use RES.

2. Materials and Methods

The survey took place in a representative urban area of Attica, with a population of 69,946 residents. Previous Greek surveys on the public perceptions on RES were evaluated to form the questionnaire [22,29–31,33]. Questionnaires were filled-out during the period of September 2016 to October 2016. The delivered questionnaire included 16 composite questions which led to the creation of 73 variables, covered various aspects of renewable energy sources such as familiarization, utility, knowledge of technologies, and social acceptance.

Concerning sample size, by retrieving the relevant questionnaire surveys on the social assessment of green investments in Greece, we noticed that in most of those studies, sample size varied between 300–400 cases [6,22,32,34,35,41,42]. The estimation of the final sample size of our research was done by using the equation of simple random sampling with substitution [43,44]. For the calculations, we set the confidence level at 95%; thus, we accept an error of 5%. A confidence interval of 95% indicates a range that would account for 95% of the results of a study that was theoretically repeated countless times. The confidence interval when the population dispersion is available, is calculated by using Equation (1) [44]; there will be no correction of the finite population, as the sample represents less than 5% of the total population [45]:

$$\overline{x} - Z_{1-\frac{a}{2}}\frac{\sigma}{\sqrt{n}}, \ \overline{x} + Z_{1-\frac{a}{2}}\frac{\sigma}{\sqrt{n}} \tag{1}$$

When the population variability is unknown and for a large sample, the appropriate function is the following [43]:

$$n = \frac{4s^2\left(Z_{1-\frac{a}{2}}\right)}{D^2} \tag{2}$$

where n is the estimated sample size, s is the calculated standard deviation derived from the control sample, the $Z_{1-\frac{a}{2}}$ value is that derived from the confidence level chosen by the investigator based on the normal distribution table, and D is the total width of the desired confidence level, as determined by the researcher or as given by similar studies.

Subsequently, when the variables are expressed in percentages (proportions), the equation for sample size takes the form below [43]:

$$n = \frac{4(Zcrit)^2 p(1-p)}{D^2} \tag{3}$$

In our sample, the variable with the higher standard deviation is "age" (mean = 40.5, s = 14.24). By using Equation (2), the sample size is estimated as follows:

$$n = \frac{4 \times 203 \times 1.96}{2^2} = 397.88$$

The appropriate sample size was rounded up to be set at 400 persons since all other variables led to smaller estimates. The final sample size of 400 is compatible with the mean sample size of the studies reviewed [6,22,30,34]. Regarding the response rate of the reviews studies, we noticed that it was averaged at 48.8% while in our study is equal to 45.7%.

Concerning the analysis methods, the initially Principal Components Analysis is applied to all Likert scale questions. To validate the sampling adequacy, the Kaiser Meyer Olkin index and Bartlett test were used. To locate the factors associated with variable "RES usage", we applied binary logit regression. Furthermore, we created an ordinal logistic regression model for discovering the factors that shape respondents' agreement on a 5-point Likert statement about "RES contribution to life quality". For the purposes of the analysis, the SPSS v.17 and STATA MP/13 statistical packages were used.

3. Results and Discussion

3.1. Reliability Analysis

To assess the questionnaire's reliability, the Alpha-Cronbach's test was used. The Alpha-Cronbach's value equaled to 0.884 which indicates high internal consistency and valid questions; by performing an Alpha Cronbach analysis for each individual item, we did not notice reliability issues in any of the questions used, hence, we concluded that the applied questionnaire is properly designed, and the recorded data can be statistically analyzed.

3.2. Sample Demographics

In this section, we include the socio-demographic characteristics of the people that took part in the survey. According to Table 1, most of the respondents are males (52.3%), while the majority belongs to the age group of 41–44 years old (35.5%). Besides this, the high school educational level is at 38.0%, followed by university graduates (35.0%). Most of the sample population holds an annual family income of up to 20,000 €, while it should be noted that around 30% of the sample population stated that their annual income does not exceed 10,000 €. Concerning the occupational status, 34.3% and 22.3% of the sample population are employees at the private and at the public sector, respectively, 14.3% are self-employed, while around 25% of the sample's population are students, unemployed, or homemakers.

Table 1. The sample demographics.

Variable	Categories	%
Gender	Male	52.3
	Female	47.8
Age	18–30	28.3
	31–40	26.5
	41–55	35.5
	56–65	8.5
	>65	1.3

Table 1. *Cont.*

Variable	Categories	%
Education	Primary education	2.3
	Secondary education	2.0
	High school	38.0
	Vocational education	8.3
	Higher education	35.0
	MSc/PhD	14.5
Household annual income	<10,000 Euro	33.6
	10,001–20,000 Euro	31.74
	20,001–30,000 Euro	21.45
	>30,000 Euro	13.21
Occupation	Private employee	36.8
	Public employee	22.3
	Self-employed	15.8
	Student	15.0
	Unemployed	10.3

3.3. Citizens' Perceptions of RES

Respondents' perceptions on RES are examined in this section. Figure 1, depicts the respondents' knowledge about RES types.

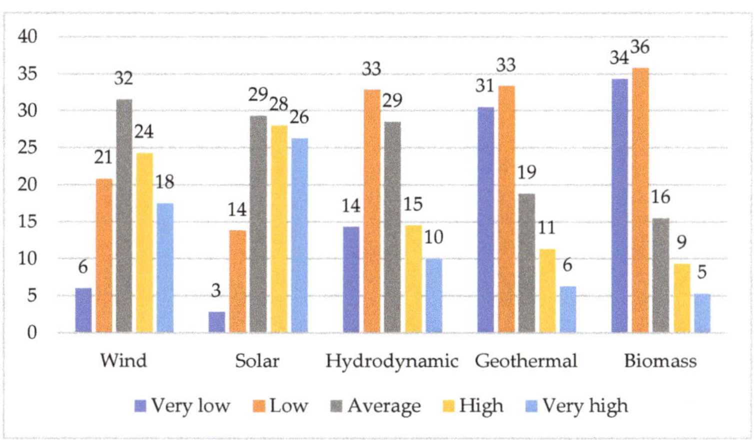

Figure 1. The knowledge about RES types (% percent).

According to Figure 1, the respondents seem to have a low level of knowledge concerning hydrodynamic, geothermal, and biomass-based sources of energy. On the contrary, they have a fair level of knowledge concerning wind and solar power sources.

As shown in Figure 2, most of the sample (59%) uses at least one type of RES. Remarkably, out of the RES users, most of them (95%) use solar water heaters while 11% have installed solar PVs; on the contrary, just 0.85% of them use geothermal sources of power. The above results are compatible with the respondents' knowledge level about RES types since solar power is the most familiar and, at the same time, the most commonly used renewable energy source.

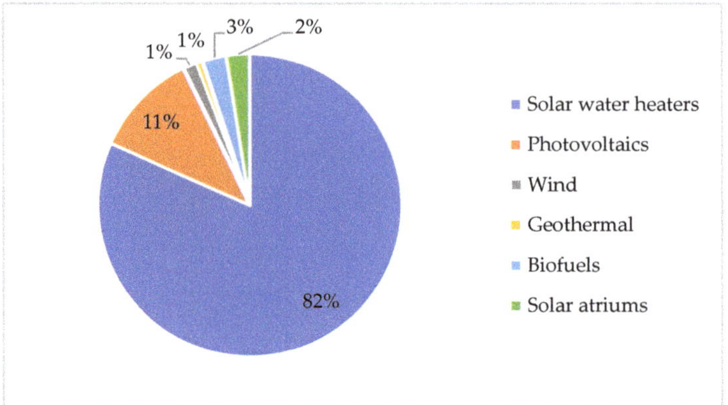

Figure 2. The RES usage by type (%).

Next, the motives to use energy produced by RES are analyzed. According to the data in Figure 3, we may conclude that the most important measure to be taken in the context of an effective adoption of RES by citizens is installation subsidies as 87.2% of the respondents have positive perceptions of it. On the other hand, the least important incentive is credit provision as 34.5% of the respondents express positive views on it. The above analysis shows not only how citizens would be motivated to buying energy produced by RES, but also how to invest in energy production using RES. Thus, an effective public policy should focus on providing incentives for both the purchase of energy produced by RES and the production of it.

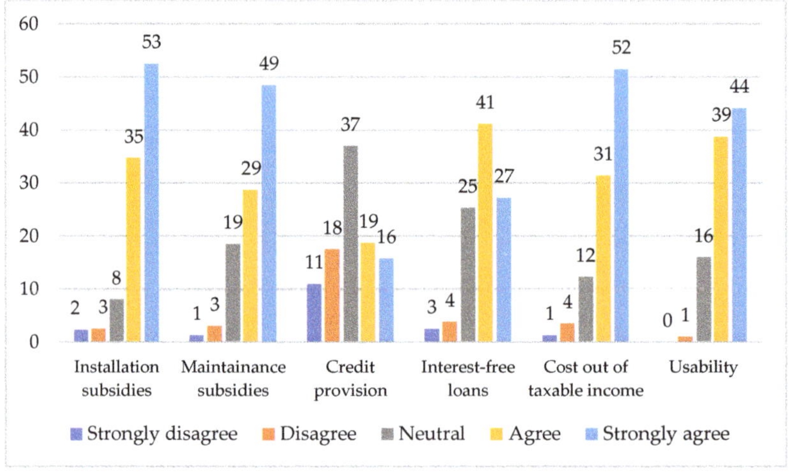

Figure 3. The motives to use energy produced by RES (%).

In Figure 4, the respondents' perceptions of RES contribution towards increased life quality is analyzed. Most of the respondents reported that RES improve life quality (85%) since environmental degradation due to fuel consumption is minimized.

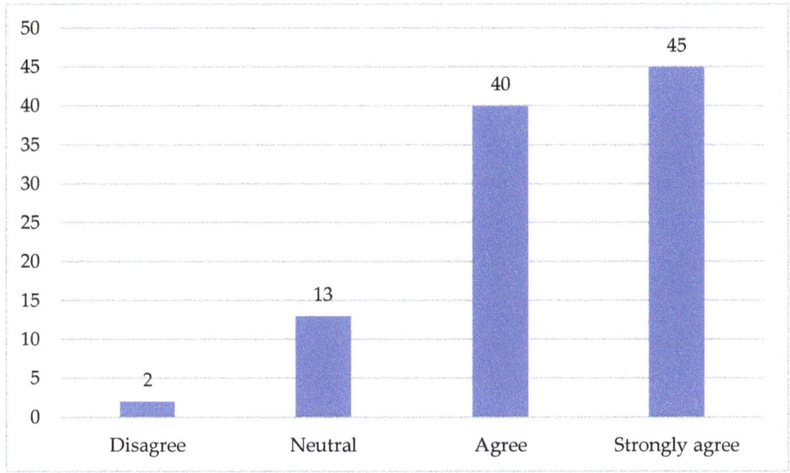

Figure 4. The public perceptions of the RES contribution to life quality (%).

In response to the other perceived advantages of RES, according to Table 2, the respondents (88.7%) see environmental protection as the most important parameter followed by the reduced oil dependence. By looking at the "agree" category about RES contribution to reduced oil dependence, it was concluded that this parameter received a portion of 40%. In all the cases, disagreement levels are extremely low which confirms a positive public perspective about RES and their positive effects.

Table 2. The RES' perceived advantages (%).

	Strongly Disagree	Disagree	Neutral	Agree	Strongly Agree
Environmental protection	0.3	1.5	11.5	35.3	51.4
Economic development	0.3	1.3	19.3	41.6	37.5
"Green" development	0.5	2.8	13.4	39.3	44.0
New labor positions	0.5	2.3	20.3	38.4	38.5
Reduced oil dependence	0.0	1.3	13.3	40.0	45.4
Energy independence	0.0	1.5	16.8	35.0	46.7

3.4. Citizens' Perceptions Analysis of RES Usage and Their Contribution to Life Quality

The Principal Component Analysis (PCA) method is used to facilitate the logit models on questions concerning respondents' opinion on RES. In this method, each identified component interprets a rate of variance that has not been interpreted by previous components. A proportion of 60% of the variance is needed to be interpreted by the factors that arise in social sciences [46]. The criterion for the selection of factors is for the eigenvalue to be greater than 1, known as the Kaiser criterion. The Kaiser–Meyer–Olkin sample measure equals to 0.86; thus, it is proven that factor analysis is acceptable. This is also validated by Bartlett's test of Sphericity, where sig. = 0. The final number of factors was determined by applying the Principal Components method based on varimax rotation. Nine factors that have eigenvalues greater than 1 have emerged, explaining a total of 68% of the observed variance. An internal affinity test was performed by using Cronbach's alpha coefficient for the 40 questions used in the factorial analysis, returning a value of 0.884 which is considered to be high [46].

Regarding the nature of the questions that have been assigned to the factors, the following profile of factor interpretation was concluded, as presented in Table 3.

Table 3. The interpretation of the factors.

Factor (Component)	Interpretation
F1	RES perceived benefits
F2	RES perceived disadvantages
F3	RES economic incentives
F4	RES actions for expansion
F5	RES social promotion barriers
F6	RES economical promotion barriers
F7	RES price compared with fossil fuels
F8	Influence of social-legal framework
F9	RES purchase with interest-free installments

As it can be seen in Table 3, a new set of 9 variables—out of the initial 40 Likert scale questions of the questionnaire—was formulated. The interpretation of each component separately is carried out by commenting on the social assessment variables that they represent.

The first component (F1) is identified as "RES perceived benefits". It explains 13.7% of the total variance of the variables that are included in the analysis and it is considered as the most important factor. The questions/variables that are associated with the highest loadings in this factor are: "RES promote green growth" (84.4) and "RES promote environmental protection" (83.7).

The second component (F2) explains 11.4% of the total fluctuation and is identified as "RES perceived disadvantages". This component is mainly determined by the questions/variables: "RES have a low rate of return" (86.1) and "are not profitable throughout the year" (83.4).

The third component (F3) refers to investment incentives for RES and explains 8.7% of the total variance. It is mainly formed by questions/variables such as "subsidized system maintenance" (78.8), "deduction of installation costs from taxable income" (77.1) and others.

The fourth component (F4) explains 7.6% of the total variance and is mainly composed of the following questions: "Public information from the local authorities" (75.8), "Public information from the state" (71.2), "well defined legal framework" (63.7). This component is identified as "RES actions for expansion".

The fifth component (F5) explains 7.4% of total variance and is identified as "Social Barriers to RES Promotion" since the variables representing the highest load on this factor are "Lack of Knowledge" (83.0) and "Lack of Information" (79.9).

The sixth component (F6) explains 5.8% of the total variance and is identified as "Economic barriers to the promotion of RES" since the variable representing the highest load on this component is "High installation costs" (84.6).

The seventh component (F7) explains 5.1% of the total variance and is identified as "Fossil fuel price relative to RES" as the variable representing the highest load on this factor is "If the cost of oil is appreciably expensive" (90.7).

The eighth component (F8) explains 4.9% of the total variance and is identified as "Effect of a social-legal framework on RES use" since the variables that represent the highest load on this component are "I would use RES if it were also used by fellow citizens" (83.0) and "Lack of complete legal framework" (70.7).

Last, the ninth component (F9) explains 3.1% of the total fluctuation and is identified as "Purchase of RES system with interest-free installments" with the factor load being 71.8.

In the first stage of our analysis, we focused on exploring the variables that are associated with whether a respondent is a RES user or not. For this purpose, we applied a binary logit model where the variable "use of RES (yes/no)" was determined as the dependent. The previously identified factors were used as explanatory variables based on a relevant study [47]. The selection of the most appropriate model was based on the applicability of the backward method. Hosmer-Lemeshow's test (sig. = 0.001) further indicated that the dependent variable values did not sustain a statistically significant difference from the values provided by the model, thus, the model is considered applicable [48]. Nagelkerke's pseudo-R Square statistic showed that the final iteration (step 6) explained a percentage of 15% of the dependent variable [49]. Out of the 9 initial independent variables (F1 to F9), the stepwise binary logistic model retained 4 variables at the 90% confidence level. Those statistically significant variables are F1 (RES perceived benefits), F5 (Institutional promotion barriers for RES), F6 (Economic barriers for RES), and F7 (RES price compared with conventional fuels). The final model for the estimation of RES users is presented in Table 4.

Table 4. The variables included in the final model for assessing RES usage (yes/no).

	Variable	B	S.E.	Wald	Df	Sig.	Exp(B)
	F1	0.618	0.113	29.742	1	0.000	0.539
	F5	−0.257	0.110	5.470	1	0.019	1.292
Step 6	F6	−0.193	0.110	3.062	1	0.080	1.213
	F7	0.263	0.108	5.981	1	0.014	0.769
	Constant	−0.389	0.108	12.897	1	0.000	0.678

The final model based on the above table data is the following one:

$$log\left(\frac{p}{1-p}\right) = -0.389 + 0.618F1 - 0.257F5 - 0.193F6 + 0.263F7 \tag{4}$$

By estimating Exp(B), the odds ratio was calculated. For example, the odds ratio coefficient, under column Exp(B) of F1 means that by keeping all the other explanatory variables at a fixed value, we will see 0.54% increase in the odds of a respondent belonging to the category of "RES user", for a one unit increase in F1 (RES perceived benefits), since Exp(0.618) = 0.539. The same explanation applies to variable F7. On the other hand, the negative coefficient of variables F5 (RES social promotion barriers) and F6 (RES economical promotion barriers) mean that they are negatively associated with RES use. This means that non-RES users consider those barriers (high cost and social barriers as information lack, lack of confidence, the role of the state) to be determining and, at the same time, they seem to overlook the RES advantages.

To validate the proposed model of estimation of RES users, we tested the relationship between each of the independent variables with the dependent variable "RES use (yes/no)", by applying the Mann–Whitney U method, as presented in Table 5. By looking at the statistical significance index (sig. < 0.05) in Table 5, all four independent variables were found to be related to the dependent variable.

Table 5. The Mann–Whitney U between RES use and factors 1, 5, 6, and 7.

	Factor 1	Factor 5	Factor 6	Factor 7
Mann-Whitney U	13.579.500	16.853.500	17.406.500	16.353.500
Wilcoxon W	27.109.500	44.583.500	45.136.500	29.883.500
Z	−5.021	−2.132	−1.644	−2.573
Asymp. Sig. (2-tailed)	0.000	0.033	0.100	0.010

The binary logistic model correctly identified 70.2% of all cases. The success rate for "RES users" is 87.7%, as it correctly identifies 206/235 of the respondents, whereas the success rate range for the "non-RES users" category is narrowed down to just 45.1%, as it correctly identifies 74/164 of the respondents.

In the second stage of our analysis, we focused on examining the factors that shape respondents' opinion about RES' contribution to life quality improvement. All nine factors generated by the above factor analysis procedure were used. Carrying an ordinal regression with the stepwise method in STATA, it was noticed that the final model retained only four factors as independent variables, as the others were removed due to the pr (0.10) criterion. The reference category was that of "strongly agree" as shown in Table 6.

Table 6. The ordinal logistic regression with the stepwise method for the variable "Life quality".

Life Quality	B	Std. Err.	Z	$P > z$	95% conf.	Interval
F1	2.799	0.204	13.740	0.000	2.400	3.198
F2	−0.415	0.135	3.070	0.002	0.150	0.679
F3	0.502	0.125	4.000	0.000	0.256	0.748
F4	0.742	0.128	5.800	0.000	0.491	0.993
/cut1	−8.098	0.647			−9.366	−6.830
/cut2	−3.715	0.291			−4.286	−3.144
/cut3	0.763	0.162			0.445	1.080

The final model here, based on the above table data is the following one:

$$log\left(\frac{P(Y_i \le j)}{P(Y_i \le j)}\right) = a^j(2.799\text{F1} - 0.415\text{F2} + 0.502\text{F3} + 0.742\text{F4}) \tag{5}$$

In the above model, $j = 1, 2, 3$ are the categories of the dependent variable $(4 - 1 = 3)$. The *p*-value (sig. = 0) indicated that the model was statistically significant compared to the null model without any explanatory variables. The pseudo-R^2 coefficient equaled to 0.4665 suggesting a strong model in accordance with a relevant statistical table [49]. By estimating Exp(B), the odds ratio was calculated and noted to be higher than 1 for the four independent variables (F1, F2, F3, and F4), suggesting, in most of the cases, a positive correlation between the independent variables and the dependent variable. More specifically, for a one-unit increase in variable F1 while keeping the other variables constant, the likelihood of the category "strongly agree" increases at $1 - \text{Exp}(2.799) = 1542\%$. Respectively, for an increase of one unit in variables F3 and F4, the probability of the category "fully agree" is increased by 65%, and 110%, respectively. Lastly, for an increase of one unit in variable F2, the probability of the category "fully agree" is decreased by 34%.

To validate the proposed ordinal model, we verified the condition of proportionality with the combined utilization of the Brant test in conjunction with the parallel lines in STATA. Finally, three stepwise binary logistic regression models are presented in Table 7, by using life quality as the dependent variable (whether respondents agree that the use of renewable energy improves life quality) and setting as independent variables the four factors (F1, F2, F3, and F4) that were statistically significant in the ordinal logistic regression. A filter was used for the data selection to compare two categories at a time, for the four-category variable life quality (disagree, neutral, agree, and strongly agree). Thus, by taking the "strongly agree" statement as a reference category, three logit models were formulated, all meeting the acceptance criterion of Hosmer and Lemeshow [48].

Moreover, by checking the goodness of fit for the three models with the Nagelkerke pseudo-R Square index, the model between "strongly agree" and "neutral" sustained the highest level of adaptation to the data with $R^2 = 0.805$ as presented in Table 8.

Concerning the predictability of the three binary logistic models, they can determine in which category a respondent belongs concerning his views about RES contribution to life quality, as captured by F1 to F4. Regarding the Exp(B) column of Table 8, we concluded that in all three models, variable F1 "RES perceived benefits" is the main determinant of "strongly agree". Model 1 includes F1–F4 as significant between the categories of "agree" and "strongly agree". Model 2 retained F1 and F4, the "RES actions for expansion", as statistically significant. This model distinguishes between the neutral position towards RES and the strong positive position. Model 3 determines between the categories of "strongly agree" and "disagree" while the stepwise method retained only variable F1 as statistically significant.

Table 7. The variables and coefficients on the regression models for "Life quality".

Logit Models	Variables in Model	B	S.E.	Wald	df	Sig.	Exp(B)
Model 1: odds between "strongly agree and agree"	F1	2.912	0.296	96.442	1	0.000	18.386
	F2	0.562	0.186	9.140	1	0.003	1.754
	F3	0.718	0.168	18.242	1	0.000	2.051
	F4	0.918	0.182	25.484	1	0.000	2.504
	Constant	−0.763	0.187	16.693	1	0.000	0.466
Model 2: odds between "strongly agree and neutral"	F1	2.901	0.415	48.792	1	0.000	18.199
	F4	0.879	0.308	8.133	1	0.004	2.410
	Constant	1.759	0.339	26.957	1	0.000	5.806
Model 3: odds between "strongly agree and disagree"	F1	2.545	0.731	12.134	1	0.000	12.741
	Constant	4.422	1.009	19.193	1	0.000	83.301

Table 8. The R^2 tests for regression models on "Life quality".

Logit Models	−2 Log Likelihood	Cox & Snell R Square	Nagelkerke R Square
Model 1: odds between "strongly agree and agree"	252.228	0.472	0.630
Model 2: odds between "strongly agree and neutral"	75.192	0.532	0.805
Model 3: odds between "strongly agree and disagree"	15.244	0.184	0.741

By looking at Table 9, we notice that out of the three proposed models, the second one has the highest predictability of 94.4%.

Table 9. The binary logit models—the percentage of the correct interpretation of the variable "Life quality".

		Predicted Values		
		Agree	Totally Agree	Percentage Correct
Model 1	Agree	132	28	82.5
	Totally Agree	24	155	86.6
Overall Percentage				84.7
		Neutral	Totally Agree	Percentage Correct
Model 2	Neutral	46	8	85.2
	Totally Agree	5	174	97.2
Overall Percentage				94.4
		Disagree	Totally Agree	Percentage Correct
Model 3	Disagree	4	2	66.7
	Totally Agree	1	178	99.4
Overall Percentage				98.4

By examining the logit models, we noticed that if a person has a completely negative attitude towards RES contribution to life quality and is found on the "disagree" category of the 5-point Likert Scale, it is possible to move to the "agree" category by a minor increase in the perceived benefits from RES. Furthermore, if a person is already found in the "agree" category, an increase in all the four variables is needed to move to the "strongly agree" point of the scale. Finally, if a person has a neutral position towards RES contribution to life quality, an increase is needed to the variables concerning RES perceived benefits and RES actions for expansion to move to the "strongly agree" category.

4. Conclusions

The aim of this study is to analyze the social acceptance of RES by examining the variables which are correlated with citizens' perceptions towards them and specifically "RES usage" and "citizens' perceptions on RES contribution to their lives' quality".

The research results show that respondents are adequately informed about some of the RES types, while 59% of them use at least one RES investment, mainly solar heaters and solar PVs. Furthermore, the respondents have a good amount of knowledge on solar and wind investments.

RES' acceptance is directly affected by the respondents' perception on the benefits abiding their use. This variable of the perceived RES benefits is the most crucial in determining whether a person is a RES user or not. In parallel, economics and social issues, as well as the government's role, are negatively related to the respondents' attitudes towards RES in the case of Greece. Those issues may also include high installation and maintenance cost, lack of confidence, lack of knowledge, and insufficient support of the RES investments by the state. It is noteworthy that the benefits arising from RES' usage and actions for RES expansion incited the perception that RES can be proven highly beneficial to end-users, since they can actively contribute to improving their life quality. According to the research results, most of the respondents are convinced that RES expansion can significantly contribute to their lives' quality improvement.

Strategies that can strengthen RES' acceptance are possible to be developed. Based on the research results, it can be drawn that RES' acceptance is not difficult to be increased, as the binary logit analysis shows that if a person has a completely negative attitude towards RES contribution to life quality, it is possible to move that person to a positive category by a minor increase in the perceived benefits. Thus, RES' benefits must be highlighted. Social support and information provision on the potential benefits from technological advances in renewable energy can promote the interaction and participation of local communities to RES' acceptance. An increase in the role of local authorities would result in an effective policy solution to renewable energy projects [19]. The challenge for project developers is to identify salient stakeholders who understand what it is that they really care about and prioritize.

Moreover, all stakeholders should remember that their effect of participation in energy decisions clearly exists and—as many delayed or canceled projects suggest—failing to take participatory decision-making into account can be costly. Besides, the psychographic factors such as the level of information, membership in environmental organizations, emotional and value components, along with political views, can shape public opinions about RES-based projects more than physical proximity [21,23]. Indeed, the installation and operation of any RES technology require social acceptance and social-driven contradictions resolving—even before the establishment and the consultation with the local community—to persuade those skeptical citizens and reconcile all competitive interests [22]. Last but not least, the research results point out that the authorities should limit the economic promotion obstacles of RES'.

Regarding the future studies' orientation concerning RES, it can be noted that Greece has shown an enduring reliance on fossil-based fuels, mainly charcoal. Nevertheless, due to its geographical configuration, Greece has an abundance of renewable energy sources, mainly solar and wind. Based on this observation there should be a focus on energy production by solar and wind sources. For the case of solar power, such investments can be easily installed in urban areas. This finding bears significance for Greece, as Attica hosts almost half of the country's population [50]. As a result, Attica's residents should be motivated to purchase energy produced by renewable sources or even to produce it on their own in order to meet their specific energy needs. Citizens' motivation would be relatively easy, as the binary—logit models show that a minor increase in the perceived benefits of RES can move a citizens' attitude from a negative to a positive category. In this way, RES usage would be significantly increased in Attica, permitting a better allocation of the available energy resources for the whole country and, at the same time, improving citizens' life quality. It should be noted that state funding programs are already underway in this direction.

The recent European legislation on gas emissions, sustainable energy production, and the ongoing participative role of RES, has gained the interest in accepting energy autonomy schemes based on RES [51]. Thus, the study of the European legislation adaptation to the national legislative framework offers numerous opportunities for the wider development of renewables—wind power, solar energy, biomass and energy crops, geothermal sources, tidal and hydropower potentials—in supporting the Greek energy demand in both the mainland and offshore areas.

Lastly, an extension of the current research would be on the correlation of a region's specific energy needs and its citizens' perceptions on RES and their contribution to life quality. In this way, the energy needs would be in the spotlight, aiming to explain citizens' perceptions on RES.

Author Contributions: D.D. provided the questionnaire and performed data collection. S.N., M.C., and M.S. designed the experimental framework and analyzed the data. S.N. and M.C. carried out the implementation, performed the calculations and the computer programming. G.K., G.A. and S.G. gathered and implemented all the theoretical background of the paper, having the input from the experimental development. G.A., G.K., and S.N. reviewed and discussed the results of the study.

Conflicts of Interest: The authors declare no conflict of interest.

References

1. Coburn, T.C.; Farhar, B. Public Reaction to Renewable Energy Sources and Systems. In *Encyclopedia of Energy*; Cutler, J., Ed.; Elsevier: New York, NY, USA, 2004.
2. Kaya, Y.; Yokobori, K. *Environment, Energy, and Economy: Strategies for Sustainability*; United Nations University Press: New York, NY, USA, 1997.
3. Chalikias, M.S.; Ntanos, S. Countries clustering with respect to carbon dioxide emissions by using the IEA database. In Proceedings of the 8th International Conference on Information and Communication Technologies in Agriculture, Food and Environment, Kavala, Greece, 17–20 September 2015; pp. 347–351.
4. Kyriakopoulos, G.L.; Chalikias, M.S. The Investigation of Woodfuels' Involvement in Green Energy Supply Schemes at Northern Greece: The Model Case of the Thrace Prefecture. *Procedia Technol.* **2013**, *8*, 445–452. [CrossRef]
5. Tsantopoulos, G.; Tampakis, S.; Andrea, V.; Komitsa, H. Views on Environmental Problems and Assessment of Solutions. The Case of Larisa, Greece. *J. Environ. Prot. Ecol.* **2013**, *14*, 646–654.
6. Kontogianni, A.; Tourkolias, C.; Skourtos, M. Renewables portfolio, individual preferences and social values towards RES technologies. *Energy Policy* **2013**, *55*, 467–476. [CrossRef]
7. Voumvoulakis, E.; Asimakopoulou, G.; Danchev, S.; Maniatis, G.; Tsakanikas, A. Large scale integration of intermittent renewable energy sources in the Greek power sector. *Energy Policy* **2012**, *50*, 161–173. [CrossRef]
8. Baris, K.; Kucukali, S. Availability of renewable energy sources in Turkey: Current situation, potential, government policies and the EU perspective. *Energy Policy* **2012**, *42*, 377–391. [CrossRef]
9. López-Peña, Á.; Pérez-Arriaga, I.; Linares, P. Renewables vs. energy efficiency: The cost of carbon emissions reduction in Spain. *Energy Policy* **2012**, *50*, 659–668. [CrossRef]
10. Lueken, C.; Carvalho, P.S.; Apt, J. Distribution grid reconfiguration reduces power losses and helps integrate renewables. *Energy Policy* **2012**, *48*, 260–273. [CrossRef]
11. Trypolska, G. Feed-in tariff in Ukraine: The only driver of renewables' industry growth? *Energy Policy* **2012**, *45*, 645–653. [CrossRef]
12. Wassermann, S.; Reeg, M.; Nienhaus, K. Current challenges of Germany's energy transition project and competing strategies of challengers and incumbents: The case of direct marketing of electricity from renewable energy sources. *Energy Policy* **2015**, *76*, 66–75. [CrossRef]
13. Fuss, S.; Szolgayová, J.; Khabarov, N.; Obersteiner, M. Renewables and climate change mitigation: Irreversible energy investment under uncertainty and portfolio effects. *Energy Policy* **2012**, *40*, 59–68. [CrossRef]

14. Kirsten, S. Renewable Energy Sources Act and Trading of Emission Certificates: A national and a supranational tool direct energy turnover to renewable electricity-supply in Germany. *Energy Policy* **2014**, *64*, 302–312. [CrossRef]
15. Kosenius, A.K.; Ollikainen, M. Valuation of environmental and societal trade-offs of renewable energy sources. *Energy Policy* **2013**, *62*, 1148–1156. [CrossRef]
16. Moe, E. Vested interests, energy efficiency and renewables in Japan. *Energy Policy* **2012**, *40*, 260–273. [CrossRef]
17. Shyu, C.W. Rural electrification program with renewable energy sources: An analysis of China's Township Electrification Program. *Energy Policy* **2012**, *51*, 842–853. [CrossRef]
18. Zhou, H. Impacts of renewables obligation with recycling of the buy-out fund. *Energy Policy* **2012**, *46*, 284–291. [CrossRef]
19. Caporale, D.; Delucia, C. Social acceptance of on-shore wind energy in Apulia Region (Southern Italy). *Renew. Sustain. Energy Rev.* **2015**, *52*, 1378–1390. [CrossRef]
20. Enevoldsen, P.; Sovacool, B. Examining the social acceptance of wind energy: Practical guidelines for onshore wind project development in France. *Renew. Sustain. Energy Rev.* **2016**, *53*, 178–184. [CrossRef]
21. Musall, F.D.; Kuik, O. Local acceptance of renewable energy—A case study from southeast Germany. *Energy Policy* **2011**, *39*, 3252–3260. [CrossRef]
22. Arabatzis, G.; Myronidis, D. Contribution of SHP Stations to the development of an area and their social acceptance. *Renew. Sustain. Energy Rev.* **2011**, *15*, 3909–3917. [CrossRef]
23. Tabi, A.; Wüstenhagen, R. Keep it local and fish-friendly: Social acceptance of hydropower projects in Switzerland. *Renew. Sustain. Energy Rev.* **2017**, *68*, 763–773. [CrossRef]
24. Carlisle, J.; Kane, S.L.; Solan, D.; Bowman, M.; Joe, J.C. Public attitudes regarding large-scale solar energy development in the U.S. *Renew. Sustain. Energy Rev.* **2015**, *48*, 835–847. [CrossRef]
25. Zhai, P.; Williams, E. Analyzing consumer acceptance of photovoltaics (PV) using fuzzy logic model. *Renew. Energy* **2012**, *41*, 350–357. [CrossRef]
26. Swift, K.D. A comparison of the cost and financial returns for solar photovoltaic systems installed by businesses in different locations across the United States. *Renew. Energy* **2013**, *57*, 137–143. [CrossRef]
27. Tyagi, V.V.; Rahim, N.A.; Rahim, N.A.; Jeyraj, A.; Selvaraj, L. Progress in solar PV technology: Research and achievement. *Renew. Sustain. Energy Rev.* **2013**, *20*, 443–461. [CrossRef]
28. Tsantopoulos, G.; Arabatzis, G.; Tampakis, S. Public attitudes towards photovoltaic developments: Case study from Greece. *Energy Policy* **2014**, *71*, 94–106. [CrossRef]
29. Devine-Wright, P. *Reconsidering Public Attitudes and Public Acceptance of Renewable Energy Technologies: A Critical Review*; School of Environment and Development, University of Manchester: Manchester, UK, 2007.
30. Zografakis, N.; Sifaki, E.; Pagalou, M.; Nikitaki, G.; Psarakis, V.; Tsagarakis, K.P. Assessment of public acceptance and willingness to pay for renewable energy sources in Crete. *Renew. Sustain. Energy Rev.* **2010**, *14*, 1088–1095. [CrossRef]
31. Ntanos, S.; Kyriakopoulos, G.; Chalikias, M.; Arabatzis, G.; Skordoulis, M. Public Perceptions and Willingness to Pay for Renewable Energy: A Case Study from Greece. *Sustainability* **2018**, *10*, 687. [CrossRef]
32. Tampakis, S.; Tsantopoulos, G.; Arabatzis, G.; Rerras, I. Citizens' views on various forms of energy and their contribution to the environment. *Renew. Sustain. Energy Rev.* **2013**, *20*, 473–482. [CrossRef]
33. Kyriakopoulos, G.; Kolovos, K.; Chalikias, M. Environmental sustainability and financial feasibility evaluation of woodfuel biomass used for a potential replacement of conventional space heating sources. Part II: A Combined Greek and the nearby Balkan Countries Case Study. *Oper. Res.* **2010**, *10*, 57–69. [CrossRef]
34. Arabatzis, G.; Malesios, C. An econometric analysis of residential consumption of fuelwood in a mountainous prefecture of Northern Greece. *Energy Policy* **2011**, *39*, 8088–8097. [CrossRef]
35. Malesios, C.; Arabatzis, G. Small hydropower stations in Greece: The local people's attitudes in a mountainous prefecture. *Renew. Sustain. Energy Rev.* **2010**, *14*, 2492–2510. [CrossRef]
36. Kaldellis, J.K. Social attitude towards wind energy applications in Greece. *Energy Policy* **2005**, *33*, 595–602. [CrossRef]
37. Xydis, G. Development of an integrated methodology for the energy needs of a major urban city: The case study of Athens, Greece. *Renew. Sustain. Energy Rev.* **2012**, *16*, 6705–6716. [CrossRef]

38. Arabatzis, G.; Malesios, C. Pro-Environmental attitudes of users and not users of fuelwood in a rural area of Greece. *Renew. Sustain. Energy Rev.* **2013**, *22*, 621–630. [CrossRef]

39. Lambert, J.G.; Hall, C.A.S.; Balogh, S.; Gupta, A.; Arnold, M. Energy, EROI and quality of life. *Energy Policy* **2014**, *64*, 153–167. [CrossRef]

40. Akella, A.K.; Saini, R.P.; Sharma, M.P. Social, economical and environmental impacts of renewable energy systems. *Renew. Energy* **2009**, *34*, 390–396. [CrossRef]

41. Kaldellis, J.K.; Kapsali, M.; Kaldelli, E.; Katsanou, E. Comparing recent views of public attitude on wind energy, photovoltaic and small hydro applications. *Renew. Energy* **2013**, *52*, 197–208. [CrossRef]

42. Savvanidou, E.; Zervas, E.; Tsagarakis, K. Public acceptance of biofuels. *Energy Policy* **2010**, *38*, 3482–3488. [CrossRef]

43. Eng, J. Sample Size Estimation: How Many Individuals Should Be Studied? *Radiology* **2003**, *227*, 309–313. [CrossRef] [PubMed]

44. Chalikias, M.S.; Manolesou, A.; Lalou, P. *Research Methodology and Introduction to Statistical Data Analysis with IBM SPSS STATISTICS*; Hellenic Academic Literature: Athens, Greek, 2015. (In Greek)

45. Rose, S.; Spinks, N.; Canhoto, A. *Management Research: Applying the Principles*; Routledge: New York, NY, USA, 2015.

46. Hair, J.F.; Black, W.C.; Babin, B.J.; Anderson, R.E. Explanatory Factor analysis. In *Multivariate Data Analysis*, 7th ed.; Prentice Hall International: Englewood Cliffs, NJ, USA, 2010.

47. Kolovos, K.G.; Kyriakopoulos, G.; Chalikias, M.S. Co-evaluation of basic woodfuel types used as alternative heating sources to existing energy network. *J. Environ. Prot. Ecol.* **2011**, *12*, 733–742.

48. Hosmer, D.W.; Lemeshow, S. *Applied Logistic Regression*, 2nd ed.; John Wiley & Sons: Hoboken, NY, USA, 2000.

49. Hu, B.; Shao, J.; Palta, M. Pseudo-R2 in Logistic Regression Model. *Stat. Sin.* **2006**, *16*, 847–860.

50. Diakakis, M.; Priskos, G.; Skordoulis, M. Public perception of flood risk in flash flood prone areas of Eastern Mediterranean: The case of Attica Region in Greece. *Int. J. Disaster Risk Reduct.* **2018**, *28*, 404–413. [CrossRef]

51. Papageorgiou, A.; Skordoulis, M.; Trichias, C.; Georgakellos, D.; Koniordos, M. Emissions trading scheme: Evidence from the European Union countries. In *Communications in Computer and Information Science*; Kravets, A., Shcherbakov, M., Kultsova, M., Shabalina, O., Eds.; Springer International Publishing: Cham, Switzerland, 2015; pp. 222–233.

Article

Agricultural Commodities and Crude Oil Prices: An Empirical Investigation of Their Relationship

Eleni Zafeiriou [1,*], **Garyfallos Arabatzis** [2], **Paraskevi Karanikola** [2], **Stilianos Tampakis** [2] and **Stavros Tsiantikoudis** [2]

[1] Department of Agricultural Development, Democritus University of Thrace, Xanthi 671 00, Greece
[2] Department of Forestry and Management of the Environment and Natural Resources, Democritus University of Thrace, Xanthi 671 00, Greece; garamp@fmenr.duth.gr (G.A.); pkaranik@fmenr.gr (P.K.); stampaki@fmenr.duth.gr (S.T.)
* Correspondence: ezafeir@agro.duth.gr

Received: 11 March 2018; Accepted: 11 April 2018; Published: 16 April 2018

Abstract: Within the last few decades, the extended use of biodiesel and bioethanol has established interlinkages between energy markets and agricultural commodity markets. The present work examines the bivariate relationships of crude oil–corn and crude oil–soybean futures prices with the assistance of the ARDL cointegration approach. Our findings confirm that crude oil prices affect the prices of agricultural products used in the production of biodiesel, as well as of ethanol, validating the interaction of energy and agricultural commodity markets. The practical value of the present work is that the findings provide policy makers with insight into the interlinkages between agricultural and energy markets to promote biodiesel or bioethanol by affecting crude oil prices. The novelty of the present work stands on the use of futures prices that incorporate all available information and thus are more appropriate to identify supply and demand shocks and price spillovers than real prices. Finally, the period of study includes extremely low, as well as extremely high, crude oil prices and the results illustrate that biofuels cannot be substituted for crude oil and protect economies from energy volatility.

Keywords: soybean; corn; crude oil; energy markets; sustainable development

---

1. Introduction

Climate change, the volatility of crude oil prices, the issue of food security, and the global economic crisis have motivated decision-makers, industries, and economic agents involved in the biofuel sector to expand the use and production of biofuels [1]. However, persistent low crude oil prices since 2014 seem to be an impediment to this effort.

The expansion of biofuels has established interlinkages between energy markets and agricultural markets given that corn is used to produce biofuels and soybean is used to produce biodiesel [2].

The major objectives of a national policy for a healthy economy are food security, limitations on GHG emissions, economic growth, and compliance with objectives set by the Kyoto protocol. The biofuel market is an artificial market, the function of which is regulated by state [3,4]. The role of this market has become crucial in economic growth, satisfying the priority of sustainability. On the other hand, the alternative use of agricultural products such as corn and soybean (either for food or to produce biofuels) is leading to indirect land use change and to a deterioration in the problem of climate change.

Therefore, given the use of the two agricultural products in the production of biofuels, as well as their use for food, the existing mutual interdependencies reflect the mutual market interdependencies. Increasing prices of corn may well be attributed to the dominant conditions in the food market that are also providing an indication for the interdependency of corn price with biofuels [5]. The specific relationship is related to many factors including the country, the model, and the dimension of time [6].

On the other hand, given that corn and soybean evolved into a major motor-fuel energy source, a close relationship between corn (the major feedstock for ethanol) and soybean production (major input in biodiesel) with crude oil, the main feedstock for gasoline production, may be an expected result. The data used extends from late 2004 through mid-2008, when rapid expansion of the ethanol industry was occurring, which is the reason for which the relationship has been mainly established [2,7].

Most of issues mentioned above are reflected in the futures prices formatted on the NY stock exchange. A study on the co–movement of futures prices of agricultural products used especially in the production of bioenergy would be of great interest, while the role of crude oil prices seems to have had a great impact on those prices. Natanelov et al. [8] have made a similar effort, and, according to their results, co-movement is a dynamic concept, and economic and policy development may affect the relationship between commodities, while they have also shown that biofuel policy has a strong impact in the co-movement of crude oil and corn futures until the crude oil prices surpass a certain threshold.

The concept of excess co-movement among futures prices has been subject to extended study with the assistance of different methodologies. The most widely used methodology is cointegration implemented for different time periods and providing conflict results (Pindyck and Rodenberg [9], Natanelov et al. [8]). Confirmation of co-movement is based on the herd behavior of financial prices [9]. Ali et al. [7], with the assistance of the partial equilibrium model, attributed the behavior of the commodity prices to specific demand and supply conditions. On the other hand, Natanelov et al. [8] focused on price movements between crude oil futures and a series of agricultural commodities and gold futures by employing a comparative framework to identify changes in relationships over time. According to their findings, co-movement is validated as a dynamic concept, while policy development may reform the relationship between commodities and their determinants that are studied, that is, crude oil prices.

The present study is timely, given the particularities in the evolution of crude oil prices within the last few years. To be more specific, the global economic crisis led to extremely high values of crude oil prices in 2010, while since 2014 the crude oil prices have tended to be persistently low, a fact that may well lead to a high level of carbon emissions and limited substitution with renewable energy prices. The volatility of crude oil prices has a different impact in the short term rather than in the long term. In addition, a strand of literature developed on this specific issue focuses on the fact that the crude oil market has significant volatility spillover effects on non-energy commodity markets, which demonstrates its core position among commodity markets [10]. This impact and its implications for the substitution of crude oil with other renewable or non–renewable resources seems to affect the level of carbon emissions and vice versa [11].

Furthermore, the futures prices for each commodity reflect the market condition, as well as the agricultural policies implemented concerning their production, market conditions, assistant aid, or even energy policy that promotes their use to produce biomass.

The innovation of our study stands on the fact that we employ ARDL bounds cointegration to test the crude oil–corn prices and soybean–crude oil prices. The structure of the present work is organized as follows; Section 2 describes material and methods, Section 3 provides a brief literature review, Section 4 provides the results and a short discussion, and Section 5 concludes.

2. Literature Review

The energy–agricultural market interlinkages have become a subject of extended study within the last decade. The global food crises in 2007/08 and 2010/11, as well as negative environmental and social impacts of promoting biofuels, have given governments second thoughts regarding the promotion of biofuels.

Since the outbreaks of biofuels industry, an abundance of manuscripts have focused on dependency between fossil fuel, biofuel, and feedstock prices [12–15]. In a few papers and due to data availability issues, the biofuel prices are ignored [2]. In those studies, ignoring biofuel prices generally relies on the hypothesis that a change in the food–fuel price relationship post the outbreak of the biofuels industry is related to the impact of biofuels.

In most of these studies, the analysis used is cointegration and/or estimation of a VECM, or one of its generalized nonlinear versions. For the case of US, cointegration is validated between agricultural prices and energy prices, while energy prices are considered to drive the feedstock prices [12,16,17] (Saghaian 2010; Serra et al. 2011a; Wixson and Katchova 2012). The works conclude that the interlinkages are attributed to the existence of biofuel markets. Only Zhang et al. [14] confirm no evidence of cointegration between energy and agricultural commodity prices. Furthermore, only Serra et al. (2011) [12] and Rajcaniova and Pokrivcak (2011) [18] confirm that biofuel markets may shape fossil fuel prices.

In general, in most cases it is concluded that energy prices drive long-run world agricultural price levels [2]. In global terms, Nazlioglu and Soytas [19] appraise the link between world crude oil prices, USD exchange rates, and a long list of world agricultural commodity prices. In addition, Nazlioglu [20] finds evidence of cointegration between corn, soybean, and wheat with oil prices, especially in recent years. In another work, Yu et al. [21], focusing on the relationship between crude oil and edible oil prices, find no evidence of energy prices driving food prices.

The methodology of autoregressive distributed lag models (ARDL models) has been employed in many works [22–24].

To be more specific, Cooke and Robles [22] confirm that recent increases in world corn, wheat, rice, and soybean prices are strongly affected by oil prices. Chen et al. [23] find evidence of positive short-run links between crude oil and grain prices, while Esmaeili and Shokoohi [24] investigate co-movement of world food prices, oil prices, and different macroeconomic variables. According to their findings, crude oil prices affect food prices only indirectly through the food production index.

Global data, with a time period including peaks of crude oil prices and persistent low prices with the assistance of ARDL cointegration approach, and the futures prices that are considered more informative than real prices, constitute the innovation of the present manuscript.

3. Material and Methods

The data employed in the present work are monthly futures prices of crude oil (CME NYMEX WTI Crude Oil Futures in US dollars per ton), corn, and soybeans (CME CBOT grain futures), derived by Bloomberg [25]. The study period is from July 1987 until February 2015. To account for the problem of comparing disparate price units, the data is indexed based on the price of August 1999 for each commodity.

Trend analysis of the data used for empirical investigation on the selected variables can be seen in Figures 1 and 2. To be more specific, Figure 1 illustrates the evolution of futures prices of crude oil, Figure 2 illustrates the evolution of corn futures prices, and, finally, Figure 3 illustrates the evolution of soybean futures prices.

Evidently, an increasing trend is validated for the case of crude oil prices with two structural breaks observed in 1999 and 2008 that will be further investigated with the assistance of a break unit root test.

On the other hand, for the case of the corn futures prices, Figure 2 presents a series of peaks that document the necessity of a break unit root test to examine the stationarity. Similar behavior can be seen for the case of soybean, since similar peaks are observed for the same time periods, as can be seen in Figure 3.

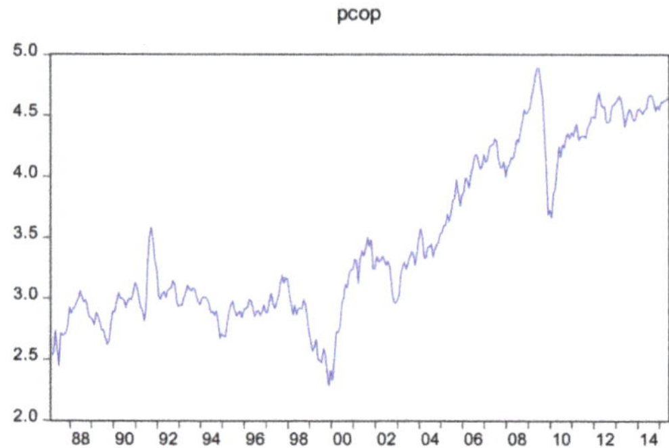

Figure 1. The evolution of futures crude oil prices in US dollars per barrel (in logarithmic form). Source: Bloomberg, [25].

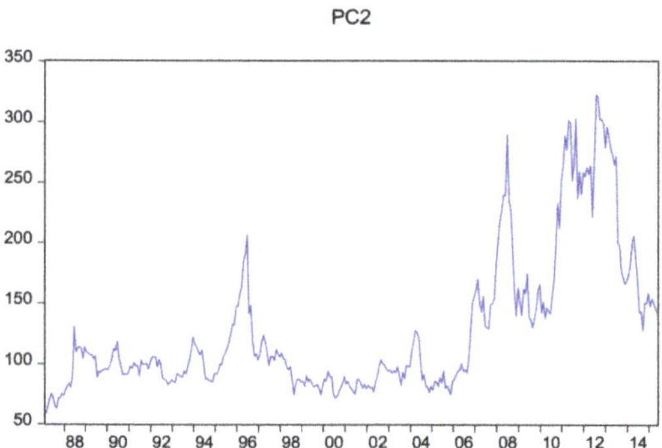

Figure 2. The evolution of futures corn prices in USD per ton (raw data). Source: Bloomberg, 2015 [25].

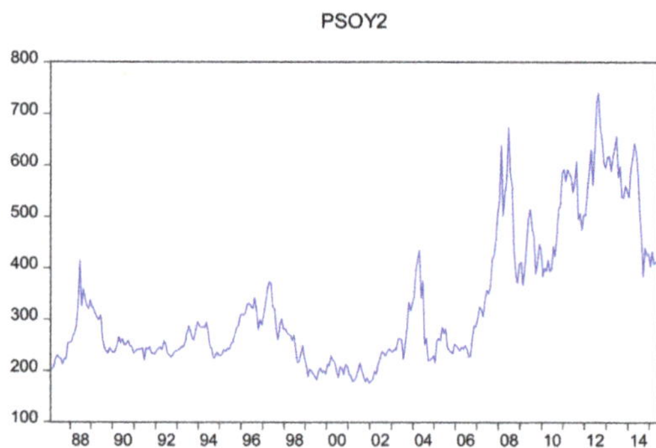

Figure 3. The evolution of futures soybean prices USD/ton (raw data). Source: Bloomberg [25].

3.1. Model Specification

We employed ARDL bounds cointegration process to estimate the relationship between the prices of the two agricultural commodities and crude oil prices. This relationship involves futures prices for soybean and corn, with crude oil prices as formatted in NY stock exchange. The advantage for the use of futures prices is that they represent not only the supply and demand conditions in the agricultural market but also a trend in the global economy (including economic crises, changes in exchange rates, etc.). Therefore, significant events that have an impact either on crude oil prices or on the agricultural commodities or changes in the major lines of energy policy (biofuels, issue of food security, and others) should be reflected in the estimated model in order for useful conclusions to come out.

The relationship to be examined for the case of corn is specified as follows:

$$Z_{1t} = f(P_{co}, \ P_{tc})$$

(1)

In addition, in the case of soybean, the bivariate relationship to be surveyed is the following:

$$Z_{3t} = f(P_{soy}, \ P_c)$$

(2)

3.2. Theoretical Econometric Framework

The methodology implemented is the ARDL bounds cointegration, given the limited data available [26]. Prior to the empirical investigation of co-movement among the variables under review with the assistance of ARDL cointegration approach, in order to check stationarity in the data, it is necessary to employ a unit root test. The unit root test used is the Augmented Dickey Fuller (ADF) break unit root test, given the structural breaks observed in the variables studied. The specific test involves a two-step procedure. In the first step the intercept, trend, and breaking variables are used to detrend the series with the assistance of ordinary least squares (OLS), and in the second step the detrended series are employed to test the existence of a unit root with the assistance of a modified Dickey-Fuller regression [27]. The first model may involve non-trending data with intercept break, trending data with intercept break, trending data with intercept and trend break, and finally trending data with trend break. The results refer either the trend specification (trend, intercept, or both) or to the break specification (trend, intercept, or both).

The problem of selecting an appropriate method of unit root test may well lead to misspecifications. For that reason, in the present manuscript, we have adopted the sequential procedure proposed by Shrestha and Chowdhury [28] in selecting the optimal method and model of the unit root test. The Shrestha-Chowdhury general-to-specific model selection procedure has been presented subtly in the work of Shrestha and Chowdhury [28].

The main objective of the particular test is to ensure that the time series employed are not I(2), a condition that implies robustness for the results derived by the ARDL bounds cointegration test.

Having validated the existence of cointegration, we estimate the Unrestricted Error Correction Model (UECM). The lag selection through which the data generating process is captured is based on the general-to-specific framework [29].

The ARDL model employed is, however, slightly modified in an appropriate way in order residual serial correlation and endogeneity problems to be corrected simultaneously [30]. The following UECM is used for our purpose.

The ARDL model employed is modified in an appropriate way in order for residual serial correlation and endogeneity problems to be corrected simultaneously [30]. The following UECM is used for our purpose.

$$lnY_t = c_{1+} + c_2 T + \beta_1 \ln(X_t) + \sum_{i=1}^{p} \beta_2 \Delta \ln(X_c) + \sum_{i=1}^{p} \beta_3 \Delta \ln(Y_{t-1}) + u_t$$

(3)

in which Y_t demotes the dependent variable and X_t denotes the independent variable in every individual case.

In Equation (3), we may identify the long term as well as the short-term parameters. To be more specific, β_1 represents the long-term parameter while rejecting the null hypothesis, and $\beta_1 = 0$ (equivalent to no cointegration) against the alternate, according to which $\beta_1 \neq 0$ (which implies that the variables are cointegrated), The test for cointegration is based on the computed F-statistic compared to the values of the tabulated critical bounds. The upper critical bound (UCB) is employed under the condition that the regressors are I(1) or I(0), while the lower critical bound is used only under the condition that they are I(0). The potential results of the test are the following:

(1)   If the F statistic exceeds the Upper Critical Bound, cointegration is confirmed;
(2)   In case the F statistic is less than the lower Critical Value, the null hypothesis of no cointegration is confirmed;
(3)   An area of uncertainty is determined within the two critical bounds, a case in which our decision relies on the lagged error correction term for a long run relationship.

The validity of the existence of a long run relationship allows the estimation of the error correction model (short run dynamics are surveyed), for which also sensitivity test, parameter stability, and goodness of fit using cumulative sum of squares of recursive residuals (CUSUM) is implemented.

The error correction model estimated for each bivariate relationship examined for the variables under review is provided by Equation (4).

$$(1-L)\begin{bmatrix} lnY_t \\ \ln(X)_t \end{bmatrix} = \begin{bmatrix} \varphi_1 \\ \varphi_2 \end{bmatrix} + \sum_{i=1}^{p}(1-L)\begin{bmatrix} a_{11i} & a_{12i} & a_{13i} \\ b_{21i} & b_{22i} & b_{23i} \\ \delta_{31i} & \delta_{32i} & \delta_{33i} \end{bmatrix} + \begin{bmatrix} \beta \\ \gamma \\ \vartheta \end{bmatrix} ECM_{t-1} + \begin{bmatrix} \eta_{1i} \\ \eta_{2i} \\ \eta_{3i} \end{bmatrix} \quad (4)$$

L denotes the lag operator, and ECM denotes the error correction term generated by the cointegrating equation, while the η terms are serially independent random error terms. The F statistic for the parameters of first differences validates the short run causality, while on the other hand the statistical significance of φ_i coefficients with the assistance of t statistics provides evidence for the long run causality.

4. Results–Discussion

According to our findings provided in Table 1, with the assistance of a breakpoint ADF unit root test, all the variables employed are I(1), a result that allows us to bounds test.

Table 1. Results of breakpoint ADF unit root test.

Variables	Trend/Break Specification	t-Statistic	Critical Values (5%)	Break Date
pc	Both/trend	−4.001	−4.859	November 1995
pco	Both/both	−4.082	−5.175	January 2003
psoy	Both/intercept	−4.129	−4.859	March 2005
Δpc	Both/Intercept	−19.75 ***	−4.859	November 1987
Δpco	Both/Intercept	−16.079 ***	−4.859	April 2008
Δpsoy	Both/Intercept	−21.006 ***	−4.859	January 2007

*** denotes reject of unit root hypothesis in 1% level of significance.

The second stage in our methodology involves the bounds testing to confirm or reject cointegration. The results are provided in the following Table 2.

Table 2. The results of bounds testing to cointegration.

Estimated Models	F-Statistics
F_c (pco/pc)	4.229 ***
F_c (psoy/pc)	5.659 ***

Critical Value Bounds for 10%, 5%, 2.5%, and 1% are for I0 Bound 2.44, 3.15, 3.88, and 4.81, and for I1 Bound 3.28, 4.11, 4.92, and 6.03, respectively. *** reject of no cointegration in 1% level of significance.

The next Table 3 provides the estimated results on the long run relationships.

Table 3. Results of long run relationships.

Dependent Variable Pco		
Corn–crude oil	Variable	Coefficient
	pc	1.449 ***
Dependent Variable Psoy		
Soybean–crude–oil	Variable	Coefficient
	pc	1.640312 ***

*** denotes reject of null hypothesis of no statistic significance of the independent variable in 1% level of significance.

According to our findings, the results derived indicate the existence of interactions among agricultural commodities and crude oil prices. This is an expected result, and it validates the interactions of agricultural markets and crude oil markets in global terms. The specific results are interpreted as follows: given that corn is the major input for ethanol, a substitute for crude oil prices, the behavior of crude oil prices may well reflect the behavior of corn prices. To be more specific, high crude oil prices lead to high ethanol prices and therefore higher corn prices. Conversely, decreases in crude oil price will lead to a lower gasoline price, a lower ethanol price, and a lower breakeven corn price.

Furthermore, the results derived should be free from misspecification and autocorrelation or heteroscedasticity of the residuals. For that reason, all the tests concerning specification, autocorrelation, and heteroscedasticity tests are presented in Table 4. According to our findings, whiteness of the residuals for all the models, as well as no misspecification problems, were confirmed.

Table 4. Results of the diagnostic tests for the estimated models.

Corn–Crude Oil Prices	BG autocorrelation test	0.334573
	BGP heter. test	1.650531
	RESET test	1.2318
Soybean–Crude Oil Prices	BG autocorrelation test	0.913924
	BGP heter. test	1.338892
	RESET test	0.45172

The next step in our analysis involves the presentation of the short run dynamics of the error correction model estimated in order to examine the interactions among the corn prices and the crude oil prices in the short run and to highlight the differences with the results derived in the long run. These results are provided in Tables 5 and 6. According to our findings, the results are not different from those derived in the long run. To be more specific, in the short term crude oil prices do not seem to have a statistically significant impact on the formation of a future's price of an agricultural product, i.e., corn. The return to the steady state is statistically significant, with a speed of 6% per month.

Table 5. Short run dynamics of the corn–crude oil prices relationship.

Regressor	Coefficient	St. Error	*t*-Statistic	Probability
D(PC1(-1))	−0.0125	0.0536	−0.234	0.815
D(PC1(-2))	0.181	0.0532	3.406	0.0007
ECM(-1) *	−0.060	0.017	−3.572	0.0004

Diagnostic Test Statistics; $R_{squared}$ = 0.05; Cointeq = PC1 − (1.4494 * PCOP1 − 0.6718).

Table 6. Short run dynamics of the soybean–crude oil prices relationship.

Regressor	Coefficient	St. Error	*t*-Statistic	Probability
D(PSOY1(-1))	−0.001270	0.053900	−0.023555	0.9812
D(PSOY1(-2))	0.138412	0.053465	2.588835	0.0101
D(PSOY1(-3))	−0.098150	0.053946	−1.819404	0.0698
ECM(-1) *	−0.056283	0.016705	−3.369328	0.0008

Diagnostic Test Statistics; Rsquared = 0.05; Cointeq = PSOY1 − (1.6403 * PCOP).

Regarding the last step in our analysis, the parameter stability of the estimated model is conducted with the assistance of the CUSUM of squared residual test.

Evidently, based on Figure 4, the graph of the corn–crude oil prices remains partially between the lines, a result indicating limited stability of the estimated coefficients for the period studied.

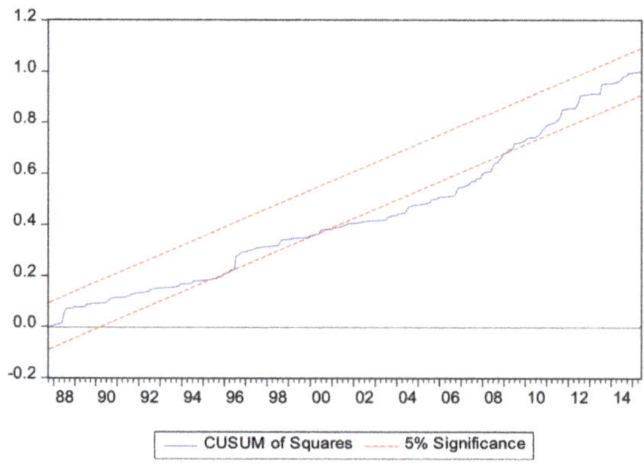

Figure 4. Results of CUSUM of squared residual test for crude oil–corn price relationship.

On the other hand, for the case of soybean crude oil prices relationship as illustrated in Figure 5, the CUSUM test of squared residuals confirms parameter stability.

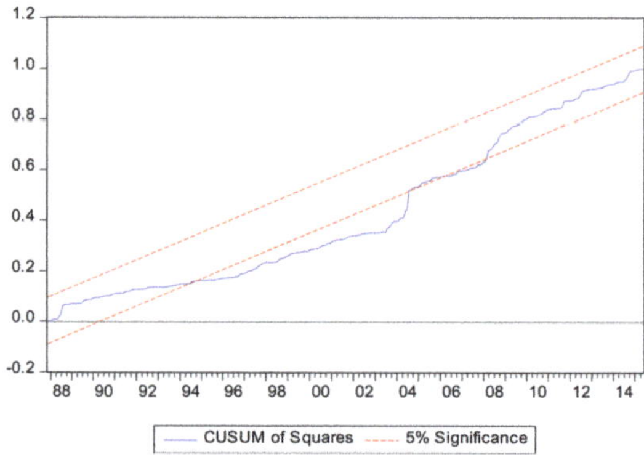

Figure 5. Results of CUSUM of squared residual test for crude oil–soybean price relationship.

Limited parameter stability is validated for the estimated model according to Figure 5. The findings based on our data indicate a long run relationship for both bivariate relationships, though in the short run no relationship is validated. This is an expected result given that the interaction between energy and agricultural market has evolved in the long run and not in the short run. This result is in line with that of Natanelov et al. [8], while the policies under the condition of the assumption of the long-run presence of a certain degree of linkage between energy markets and the market of food products that substitutes for other food products used as energy feedstock, and between food products used as energy feedstock and their substitutes for food exhibiting strong seasonality in production, would need to be flexible to be effective. The periodical shocks in the supply and demand of these agricultural commodities should be taken into consideration given that they may have a severe impact on the market interlinkages between different production periods.

The study of this relationship is important due also to the changes in biofuels industry. To be more specific, global biofuel production has been growing for the decade 2005–2015 with result that in the year 2014 ethanol production reached about 114 billion liters and biodiesel production reached about 30 billion liters [31]. The implementation of biofuel policies in global terms can expand the use of biofuels leading to a reduction in greenhouse gas emissions, along with limitation in the dependency on fossil fuel.

Though the interlinkages with the agricultural markets have led to high food prices, the simultaneous increase in feedstock prices has been harmful for biofuel competitiveness in the liquid fuels market, necessitating the need for subsidization and other protectionist policies. The solution to this problem is the promotion of second generation biofuels for which no food crops are used, since it will limit the competition for agricultural land and crops, reducing in turn the impact on agricultural prices. A final implication of our findings (the impact of energy prices on feedstock prices) can provide the food policy makers with a tool to forecast food prices and implement food policies with regard to the evolution of energy prices.

5. Conclusions

The present work provides an insight into the interaction between the crude oil futures market and the soybean and corn futures markets. The analysis of this relationship (co-movement) is based on the ARDL cointegration approach. The prices used were the futures and not the real market prices. The reason that we used the future prices is that they incorporate, by definition, all available information and thus are more efficient at identifying supply and demand shocks.

The selection of the specific agricultural commodities is related to their role in the production biofuels. An issue that is also worth mentioning is the great volatility observed within the last decade in the crude oil market. The confirmation of interlinkages among feed prices and energy prices may well interpret the transfer of instability from energy markets to food markets, a fact that has been particularly intensive since the 2010 due to the global biofuels industry boom. Therefore, it is necessary for the policies implemented to aim at limiting the linkages in order for the food crisis to be limited. The design of these policies can become plausible if it is studied adequately at what extent biofuels can limit the impact of extreme crude oil price changes relative to fossil fuels such as diesel or gasoline, and as a sequence to the prices of agricultural products used for the production of biodiesel. Copula modeling provides a plausible methodology for the study of this relationship, providing a solution to the global problem of interlinkages between energy and agricultural markets. [2].

Author Contributions: Eleni Zafeiriou conceived and designed the experiments. Eleni Zafeiriou and Garyfallos Arambatzis performed the experiments. All the authors analyzed the data. All the authors contributed reagents/materials/analysis tools; Eleni Zafeiriou wrote the paper.

Conflicts of Interest: The authors declare no conflict of interest.

References

1. De Gorter, H.; Drabik, D.; Just, D.R. *The Economics of Biofuel Policies: Impacts on Price Volatility in Grain and Oilseed Markets*; Palgrave Studies in Agricultural Economics and Food Policy; Palgrave Macmillan: New York, NY, USA, 2015; p. 282, ISBN 9781137414847.
2. Serra, T.; Zilberman, D. Biofuel–related price transmission literature: A review. *Energy Econ.* **2013**, *37*, 141–151. [CrossRef]
3. Zafeiriou, E.; Arabatzis, G.; Tampakis, S.; Soutsas, K. The impact of energy prices on the volatility of ethanol prices and the role of gasoline emissions. *Renew. Sustain. Energy Rev.* **2014**, *33*, 87–95. [CrossRef]
4. Zafeiriou, E.; Karelakis, C. Income volatility of energy crops: The case of rapeseed. *J. Clean. Prod.* **2016**, *122*, 113–120. [CrossRef]
5. Zhang, Q.; Reed, M.R. Examining the Impact of the World Crude Oil Price on China's Agricultural Commodity Prices: The Case of Corn, Soybean, and Pork. In Proceedings of the 2008 Annual Meeting, Dallas, TX, USA, 2–6 February 2008.
6. Zilberman, D.; Kim, E.; Kirschner, S.; Kaplan, S.; Reeves, J. Technology and the future bioeconomy. *Agric. Econ.* **2013**, *44*, 95–102. [CrossRef]
7. Ai, C.R.; Chatrath, A.; Song, F. On the comovement of commodity prices. *Am. J. Agric. Econ.* **2006**, *88*, 574–588. [CrossRef]
8. Natanelov, V.; Alam, M.J.; McKenzie, A.M.; Van Huylenbroeck, G. Is there co-movement of agricultural commodities futures prices and crude oil? *Energy Policy* **2011**, *39*, 4971–4984. [CrossRef]
9. Pindyck, R.S.; Rotemberg, J.J. The excess co-movement of commodity prices. *Econ. J.* **1990**, *100*, 1173–1189. [CrossRef]
10. Ji, Q.; Fan, Y. How does oil price volatility affect non-energy commodity markets? *Appl. Energy* **2012**, *89*, 273–280. [CrossRef]
11. Apergis, N.; Payne, J.E. Renewable energy, output, CO_2 emissions, and fossil fuel prices in Central America: Evidence from a nonlinear panel smooth transition vector error correction model. *Energy Econ.* **2014**, *42*, 226–232. [CrossRef]
12. Serra, T.; Zilberman, D.; Gil, J.M.; Goodwin, B.K. Nonlinearities in the U.S. corn-ethanolgasoline price system. *Agric. Econ.* **2011**, *42*, 35–45. [CrossRef]
13. Serra, T.; Gil, J.M. Biodiesel as a motor fuel price stabilization mechanism. *Energy Policy* **2012**, *50*, 689–698. [CrossRef]
14. Zhang, Z.; Lohr, L.; Escalante, C.; Wetzstein, M. Food versus fuel: What do prices tell us? *Energy Policy* **2010**, *38*, 445–451. [CrossRef]
15. Francisco, P.; Augusto, R.M. *The Emergence of Biofuels and the Co-Movement between Crude Oil and Agricultural Prices*; Department of Economics and Business, Universitat PompeuFabra: Barcelona, Spain, 2009.
16. Saghaian, S.H. The impact of the oil sector on commodity prices: Correlation or causation? *J. Agric. Appl. Econ.* **2010**, *42*, 477–485. [CrossRef]
17. Wixson, S.E.; Katchova, A.E. Price asymmetric relationships in commodity and energy markets. In Proceedings of the 123rd EAAE Seminar in Dublin, Ireland, 23–24 February 2012.
18. Pokrivcak, J.; Rajcaniova, M. Crude oil price variability and its impact on ethanol prices. *Agric. Econ. Czech* **2011**, *57*, 394–403. [CrossRef]
19. Nazlioglu, S.; Soytas, U. Oil price, agricultural commodity prices, and the dollar: A panel cointegration and causality analysis. *Energy Econ.* **2012**, *34*, 1098–1104. [CrossRef]
20. Nazlioglu, S. World oil and agricultural commodity prices: Evidence from nonlinear causality. *Energy Policy* **2011**, *39*, 2935–2943. [CrossRef]
21. Yu, L.; Wang, S.; Lai, K.K. Forecasting crude oil price with an EMD–based neural network ensemble learning paradigm. *Energy Econ.* **2008**, *30*, 2623–2635. [CrossRef]
22. Cooke, B.; Robles, M. *Recent Food Prices Movements: A Time Series Analysis*; IFPRI Discussion Papers 942; International Food Policy Research Institute (IFPRI): Washington, DC, USA, 2009.
23. Chen, S.T.; Kuo, H.I.; Chen, C.C. Modeling the relationship between the oil price and global food prices. *Appl. Energy* **2010**, *87*, 2517–2525. [CrossRef]
24. Esmaeili, A.; Shokoohi, Z. Assessing the effect of oil price on world food prices: Application of principal component analysis. *Energy Policy* **2011**, *39*, 1022–1025. [CrossRef]

25. Bloomberg. Impact Report Update 2015. Available online: https://data.bloomberglp.com/sustainability/sites/6/2016/04/16_0404_Impact_Report.Q12 (accessed on 10 April 2018).
26. Haug, A.A. Temporal aggregation and power of cointegration tests: A Monte Carlo study. *Oxf. Bull. Econ. Stat.* **2002**, *64*, 399–412. [CrossRef]
27. Vogelsang, T.J.; Perron, P. Additional tests for a unit root allowing for a break in the trend function at an unknown time. *Int. Econ. Rev.* **1998**, *39*, 1073–1100. [CrossRef]
28. Shrestha, M.B.; Chowdhury, K. *ARDL Modelling Approach to Testing the Financial Liberalisation Hypothesis*; Working Paper 05-15; Department of Economics, University of Wollongong: Wollongong, Australia, 2005.
29. Laurenceson, J.; Chai, C.H. *Financial Reform and Economic Development in China*; University of Queensland, Australia, Edward Elgar Publishing, Inc.: Oxford, UK, 2003, ISBN 1-84064-988-7.
30. Pesaran, M.H.; Shin, Y. An autoregressive distributed lag modelling approach to cointegration analysis. In *Econometrics and Economic Theory in the 20th Century: The Ragnar Frisch Centennial Symposium*; Strom, S., Ed.; Cambridge University Press: Cambridge, UK, 1999.
31. OECD–FAO. *OECD-FAO Agricultural Outlook 2015*; Organisation for Economic Cooperation and Development, and Food and Agricultural Organization of the United Nations: Rome, Italy, 2015.

sustainability

Article

First Approach to a Holistic Tool for Assessing RES Investment Feasibility

José María Flores-Arias [1,*], Lucio Ciabattoni [2], Andrea Monteriù [2], Francisco José Bellido-Outeiriño [1], Antonio Escribano [1] and Emilio José Palacios-Garcia [3]

[1] R&D Group 'Instrumentation and Industrial Electronics, TIC-240', Universidad de Córdoba, E-14071 Córdoba, Spain; fjbellido@uco.es (F.J.B.-O.); z52esesa@uco.es (A.E.)

[2] Department of Information Engineering, Università Politecnica delle Marche, 60131 Ancona, Italy; l.ciabattoni@univpm.it (L.C.); a.monteriu@univpm.it (A.M.)

[3] Department of Energy Technology, Aalborg University, 9220 Aalborg East, Denmark; epg@et.aau.dk

* Correspondence: jmflores@uco.es; Tel.: +34-957-212-223

Received: 28 February 2018; Accepted: 9 April 2018; Published: 11 April 2018

Abstract: Combining availability, viability, sustainability, technical options, and environmental impact in an energy-planning project is a difficult job itself for the today's engineers. This becomes harder if the potential investors also need to be persuaded. Moreover, the problem increases even more if various consumptions are considered, as their patterns depend to a large extent on the type of facility and the activity. It is therefore essential to develop tools to assess the balance between generation and demand in a given installation. In this paper, a valuable tool is developed for the seamless calculation of the integration possibilities of renewable energies and the assessment of derived technical, financial and environmental impacts. Furthermore, it also considers their interaction with the power grid or other networks, raising awareness of the polluting emissions responsible for global warming. Through a series of Structured Query Language databases and a dynamic data parameterization, the software is provided with sufficient information to encode, calculate, simulate and graphically display information on the generation and demand of electric, thermal and transport energy, all in a user-friendly environment, finally providing an evaluation and feasibility report.

Keywords: renewable energy sources; energy mix; smart grid integration; energy balance

---

1. Introduction

As renewable energy sources (RES) have become an important part of the power generation mix, the availability of appropriate management tools is an important issue to be achieved. Management involves the integration of RES into the grid and, subsequently, a complete balance of both available energy resources and consumer profiles [1,2].

Likewise, detailed technical and economic studies must be performed in parallel to ensure the feasibility of the project. All these aspects must consider the environmental impact derived from related industrial activities [3]. The importance of this facet is evident since as early as 1995, the U.S. Department of Energy published a report with the manual for the economic evaluation of energy efficiency and renewable energy technologies [4].

Technical and economic issues are widely covered by several management tools, whereas environmental audit techniques have their appropriate set of tools. Nevertheless, it is beyond the scope of this paper to delve deeper into these fields.

Therefore, how to combine energy availability with economic viability, technical options, and the resulting carbon footprint, and how to issue a holistic report to persuade potential investors is a coming challenge for technicians. The tool proposed in this paper seeks to lay the foundations of such a holistic approach, being able to provide several important aspects to support the decision to install an RES facility in a certain location.

To build the aforementioned software tool, the programming language Python 3.6 and the integrated development environment (IDE) PYQT Designer 5 are used. The system also made use of nine SQL (Structured Query Language) relational databases, developed for that purpose and which can be deployed on a local server or on a Data as a Service (DaaS) cloud system.

The structure of the paper is as follows. First, Section 2 provides a brief overview of the common tools used to estimate or calculate both energy production and consumption. Secondly, Section 3 addresses the process of modeling different energy sources and characterizing the systems, including the generation of demand profiles, feasibility evaluations, modeling of consumptions and balance among the elements. An illustrative example of the application of the proposed tool is developed in Section 4. Finally, Section 5 presents some conclusions.

2. State-of-the-Art of Energy Estimation Tools

It can easily be seen that designing a tool to estimate the total production of energy from a specific renewable source is not a novel idea [5]. However, it can also be verified that the vast majority of these current tools are focused on a single type of RES at a specific location [6–8] and only a few of them are on a pair [9]. Furthermore, financial and economic project analysis tools are widely used.

Only a few works manage both the estimation of energy production and the economic analysis, as in [10]. These implementations addressed how to deal with an immediate consumption scenario, determining which loads must be shed to fit and match the demand profile with the short-term production forecast and assessing the economic feasibility of a household installation. In addition, some approaches have made use of holistic assessment tools for a micro-turbine combined thermal RES [11], for building thermal insulation solutions [12] or for the evaluation of investments in different energy market scenarios [13].

On the contrary, the definition and construction of consumption profiles for complex and various consumer choices remain a hot topic for researchers [14,15]. There are many research groups working on establishing the best methodology to determine the most accurate consuming profile to model and predict the energy demand of a particular residential area, service or facility [16–18].

For instance, stochastic data analysis, temporal series measurement, aggregation or disaggregation of the electric consumption data are several commonly used modeling techniques [19]. Otherwise, local authorities such as the Spanish Government [20] might provide standard profiling methods in absence of data.

3. Profiling Methodology

The idea proposed by Ciabattoni et al. of sizing a particular RES installation ensuring its economic convenience in varying consumption patterns [21] has been taken as a starting point. From there, the present proposal has been completely constructed in its entirety, extending it both technically and geographically so that any of the renewable energy sources available in a given geographical demarcation and the conventional sources available in it can be considered, as well as the complex set of consumption needs.

In general, the production profiles, which conform the SQL databases, can be obtained from two main sources. On the one hand, public repositories such as the solar radiation data taken from the NASA (National Aeronautics and Space Administration) website can be used with a top-down approach [22]. On the other hand, time series monitored in different installations can be employed with a bottom-up philosophy, by aggregating the individual records or applying time series modeling techniques [10].

The same procedure is applied for demand profiles [2,15–20]. Thus, whatever the profile processing used to obtain an operative model, in most cases, top-down methods are used to define resources availability, deterministic and behavioral ones to determine generators capacity and mainly bottom-up methods for shaping individuals and facilities consumption.

The analysis application includes several calculations and estimation methods both to determine the amount of energy available and to describe the consumption profile of the appliances.

Each energy source or load is modeled according to its behavioral equations and recorded data, publically available from various repositories. From all these data, a relational SQL database system is built which also provide a seamless maintenance, flexibility and periodic updates.

Furthermore, the developed graphical user interface (GUI) allows for an intuitive interaction with the user while hiding all the algorithmic complexity. To structure the system, the resources have been organized into a two-level tab hierarchy. On the top level, tabs for the energy resources, the energy generators, the consumer demand and the balance among all of them can be found. Following this hierarchy, every time a resource is selected, a group of various subtabs is shown where detailed definitions can be performed. This is addressed in detail in the following subsections.

3.1. Profiling Energy Resources

As stated above, a top-bottom profiling method was mainly used to obtain the availability of each energy resource under consideration. Four main types of resources were considered in this proposal, i.e., radiated energy, flow energy, potential energy and fuel energy.

3.1.1. Solar Radiation Estimation and Profiling

Two scenarios were considered when estimating the irradiation model of a given location.

On the one hand, the global solar radiation (*Ho*) in a specific period (*t1–t2*), i.e., the total amount of energy received during a considered period, was determined by:

$$Ho = \int_{t1}^{t2} Isc.Eo(cos\theta t.cos\delta.cosL + sen\delta.senL)dt \tag{1}$$

where θ is the solar incidence angle, $Isc \cdot Eo \cdot cos\theta$ is the extraterrestrial irradiance *Isc* corrected with the equation of time *Eo*, δ is the declination angle and *L* is the latitude, all of them at mid-month for all the months within the period considered, according to Duffie and Beckman [23].

In this first approximation, the data taken from the repository published by NASA are combined with the geographical data of the chosen location and applied to Equation (1). This results in the estimation of the monthly mean value of the horizontal extraterrestrial radiation value on the surface.

Combining these estimated data in the form of a monthly averaged surface solar radiation database, several algorithms may be used for determining the direct radiation, the diffuse radiation and the total radiation in any inclination, orientation, altitude or orientation applied to one- or two-axis solar trackers.

On the left side of the tab shown in Figure 1 is the selector that allows including the option of parameterization regarding the types of the supportive infrastructure for the solar collectors and its orientation related geometrical data. Below these, the user can find the dials that allows incorporating the corrective factors depending on both the clarity and transparency of the atmosphere and the albedo of the ground into the profile calculation algorithm [23]. No other on-execution-time environmental or weather modifying parameters had been considered in this proposal.

On the other hand, some laboratories such as the National Renewable Energy Laboratory [9]. which has large series of measured local data, provide prediction algorithms that can be directly used to estimate the solar irradiation of a specific place over a determined time horizon by means of the learning based on these recorded data. This set of values in the form of SQL database is directly usable as a source in this tab. As most SQL code is not completely portable between different database systems without adjustments, a convenient formatting is needed.

Since this method would always be available for any kind of resources if registered data were provided, it will not be referred again in the following sections of this document.

Figure 1. Window for estimating the average monthly solar radiation at a specified location and installation conditions.

An example of the interface window from the solar radiation profiling tab is shown in Figure 1, where the annual profile of generated energy per square meter for different configurations can be seen. On the right side of the tab, dials and sliding bars were included to allow for a basic variation of location and/or atmospheric conditions. For the sake of simplicity, data profiles have been smoothed with the monthly average.

3.1.2. Kinetic Energy Estimation and Profiling

The calculation of this value set the first limit for establishing the power rate of a wind turbine (WT) or a hydro generator (HG). The kinetic energy (Ec) of a fluid of density δ flowing across blades of a flat area (A) at a specified speed (v) per time unit (t) was determined by:

$$Ec = \int \frac{1}{2}\delta\ Av^3 dt. \tag{2}$$

Nevertheless, not all of this energy could be converted to torque/speed in the WT because of Betz Law. In addition, mechanical (Rm) and electric (Rg) transformations reduce the total efficiency, so the usable generator's energy was

$$Ec(usable) = \int 4\left(a^2 - a^3\right) Pv \cdot Rh \cdot Rm \cdot Rg \cdot Rt \cdot dt. \tag{3}$$

It can easily be understood that being provided with the most accurate historic weather records of the wind speed or river flow rate regimes at a specific location, produces the most accurate estimation profiles to feed our SQL databases.

As mentioned in the previous section, the data were provided per surface unit. An example of the wind and water flow power availability at a specific location is shown in Figure 2.

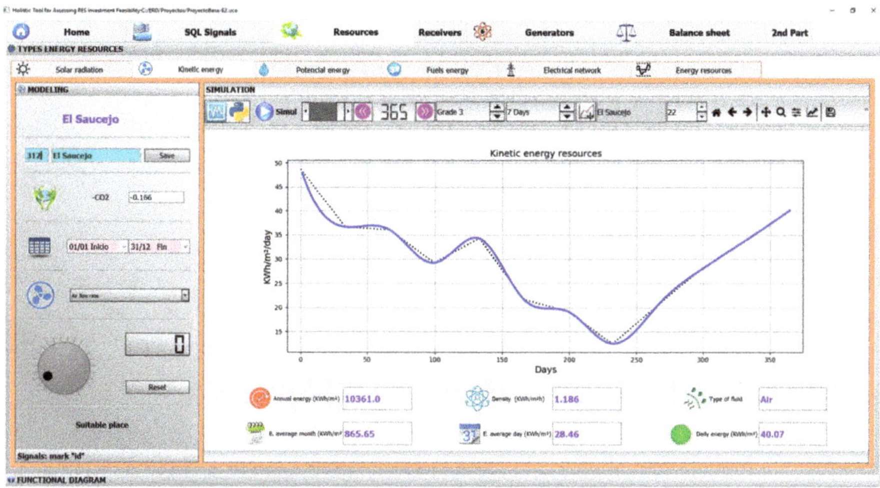

Figure 2. Window of average monthly wind energy estimations at a specified location.

3.1.3. Water Potential Energy Estimation and Profiling Tab

Water accumulated in a dam by rainfall in a river basin can be used to produce energy by means of turbines. The water flow energy was determined by:

$$Et = \int_0^t \delta \cdot Q \cdot H \cdot dt, \tag{4}$$

where δ is the density of water, Q is the flow rate and H is the elevation of the water column, according to Agüera [24].

Hence, the energy per square meter available in a basin due to runoff waters from expected rainfalls stored in a dam could be estimated within a specified period, as shown in Figure 3.

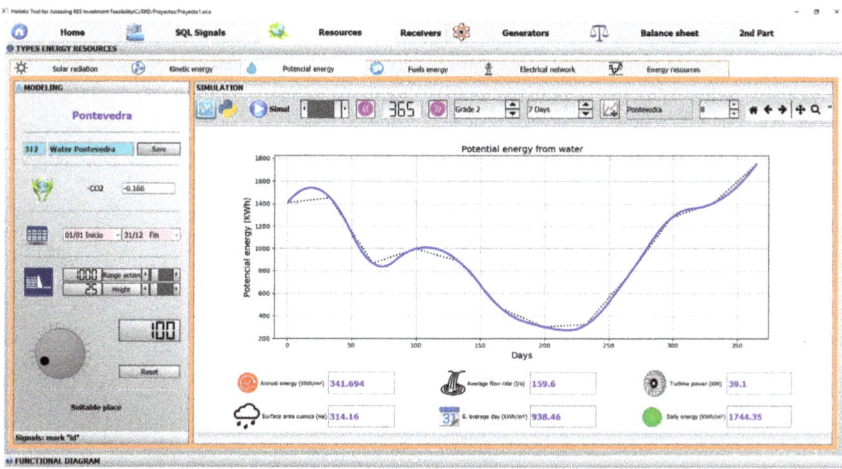

Figure 3. Window for estimating the average and smoothed monthly-dammed rainfall power estimation at a specified area.

As stated before, valid geographical and meteorological data are essential for a valid estimation. Particularly, the reservoir of potential energy that a river basin can provide is difficult to estimate so the use of averaged models is even more justifiable than for other RESs.

3.1.4. Fuel Energy Estimation and Modeling, Additional Data and Comparison Features Tab

As a rule, there is no particular regime regarding the supply of any fuel type. Subsequently, the energy produced in a boiler due to fuel combustion was determined by the relationship between its lower calorific value and the weight of the burnt fuel.

Note that this model also included an additional aspect not considered before, i.e., the weight of carbon dioxide released into the atmosphere during the combustion process, expressed in kg.

Figure 4 displays the estimated energy production estimation using coal as the fuel.

This program tab also offers the possibility to consider existing line grids, whose tariffs and emissions data can easily be entered into the program.

Figure 4. Window for estimating the average monthly coal energy and associated CO_2 emissions.

Finally, this tab allows for the selection within several resources from the ones described above and either to perform a comparison between them or to compose a so-called "energy mix", i.e., the total availability of modeled and predicted energy at a particular area or location.

Figure 5 displays the data input interface for existing grids and an example of the graphic comparing feature is shown in Figure 6, respectively.

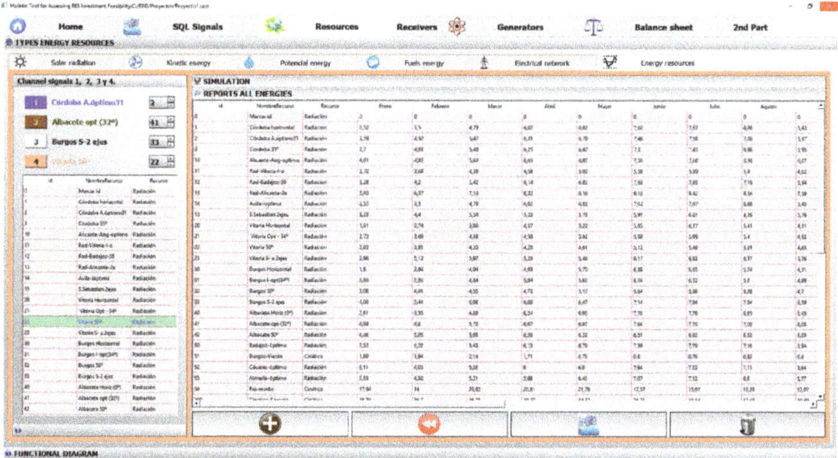

Figure 5. Window with the data from existing grids.

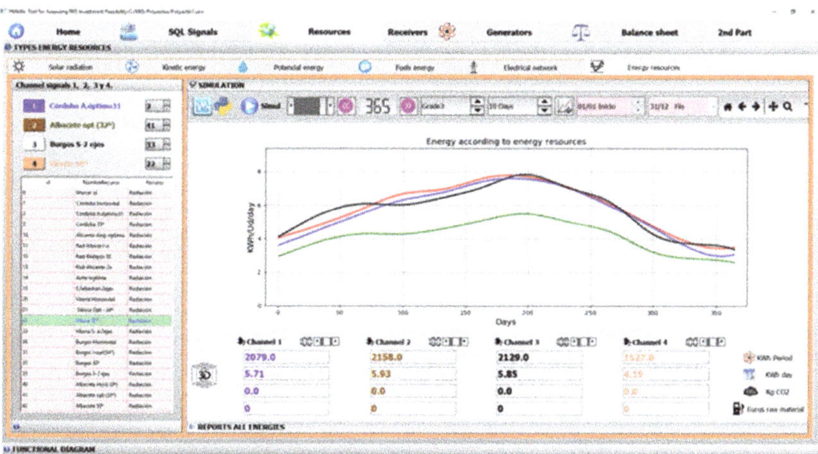

Figure 6. Window for comparing various existing energy resources.

3.2. Generation Profiling Tab

As stated at the beginning of Section 3, in the profiling tab, both the description and the characteristics of each listed generation system were incorporated based on the information provided by manufacturers and suppliers. Efficiency data, energy losses in HV and LV distribution grid and the thermal effects were also recorded in the database for each source. Thus, the final behavior and capacity of each listed generator were defined by applying the necessary correction parameters to the predefined equations. Consequently, the entire supplying system was considered from the generator itself to the point of common coupling (PCC).

The left side of Figure 7 shows the interface for entering the description and parameters of each selected generator type. These generators are shown in the spreadsheet on the right side after they were registered. Then, the generator is assigned a unique ID number, which can be chosen by the user to aggregate this unit to the project.

To ensure a complete analysis of any generation system, the model includes its evaluation and studies from different points of view, such as technical feasibility and economic and environmental viability. Therefore, the purchase price and other costs are included in this definition tab, as well as information regarding CO_2 emissions.

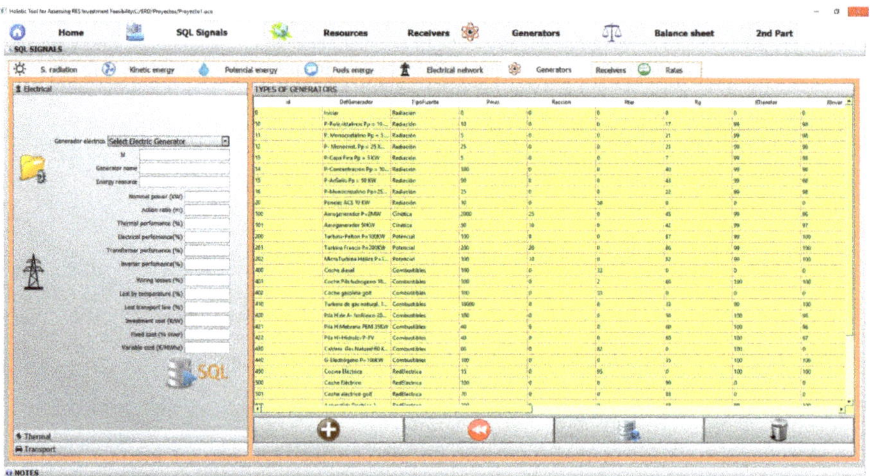

Figure 7. Window with the SQL database that includes the generators details.

3.2.1. Technical Feasibility

To estimate the technical feasibility of a previously defined generation system, the data corresponding to the associated energy source at a defined location were applied. In this process, it is necessary to choose the right relationship between the selected generator and its primary source of energy. Failing to do this will lead to erroneous calculations and eventually to wrong decisions.

Combining the data from an ES with that of a selected generator with a recorded efficiency and associated losses provides the total amount of energy that the installation can produce in a defined period. Otherwise, a forecast estimation of a given installation for a determined time horizon, as exposed in [10], can also give its scheduled power profile to be used in this evaluating tab.

Figures 8 and 9 show two examples of generation systems at different locations, one for a photovoltaic plant and the other for a hydropower system.

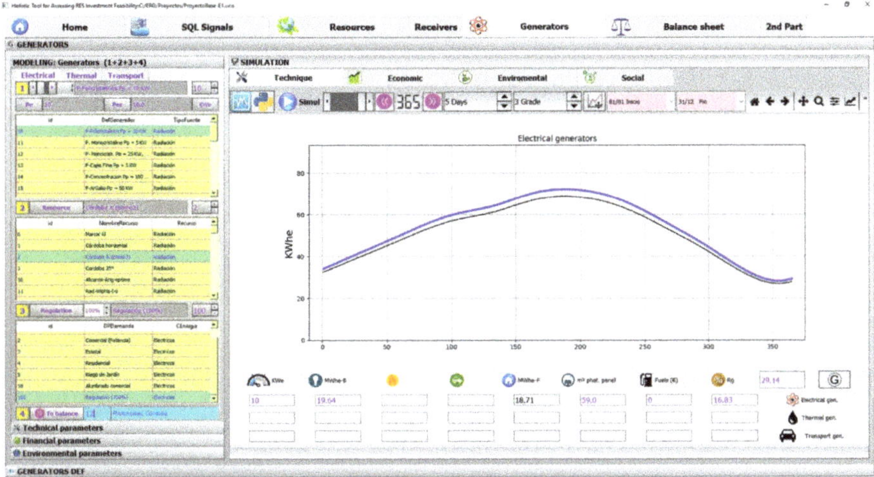

Figure 8. Window with the annual energy production profile of a PV plant in Southern Spain.

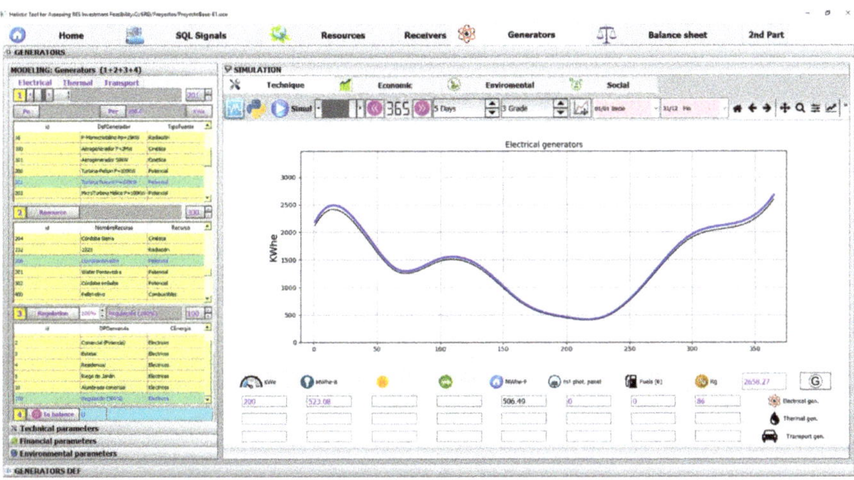

Figure 9. Window with the annual energy production profile of a hydro turbine in Northern Spain.

3.2.2. Economic Viability

Several parameters were considered to estimate the economic viability of a determined installation. Factors such as cash flow, net present value (NPV) or cost–profit ratio were included in this assessment. Likewise, the production day-period and the daily tariffs with variable hourly prices were considered and can be tuned as well.

The return on investment (ROI) based on the retained cash flow is displayed in a graph. The parameterization area in the tab allows for the selection among various demand characteristics (Figure 10), the prices which can be modified by some sliders in the upper area of the tab (Figure 11), or the introduction of financial evaluation settings in the right side (Figure 12). All these parameters concern either the cost–profit relationship or the cash flow and, therefore, the economic viability.

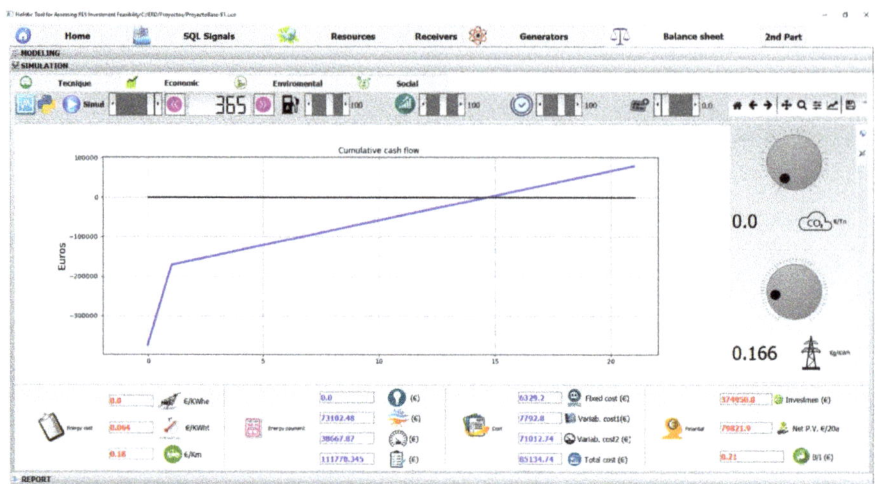

Figure 10. Window with the accumulated cash flow chart.

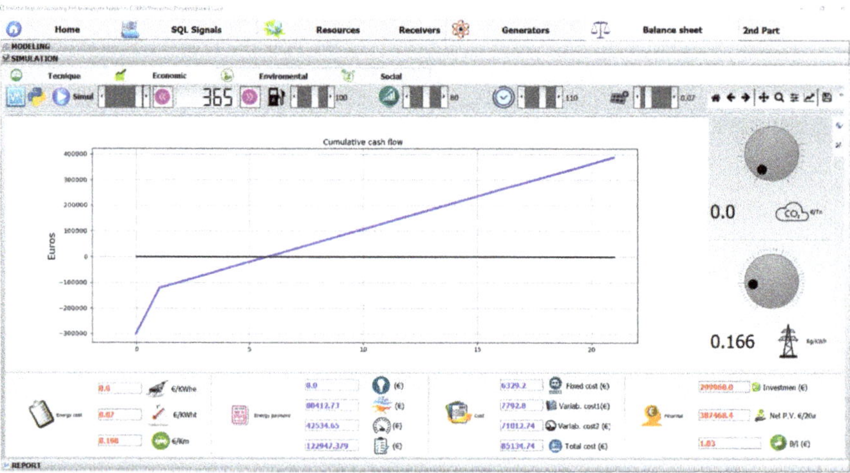

Figure 11. Sensitivity analysis (upper bar): Investment factor = 0.80; Rate electrical = 1.1; and Price Sale surplus energy = 0.07 €/Kwh (note these values to the right of the sliders).

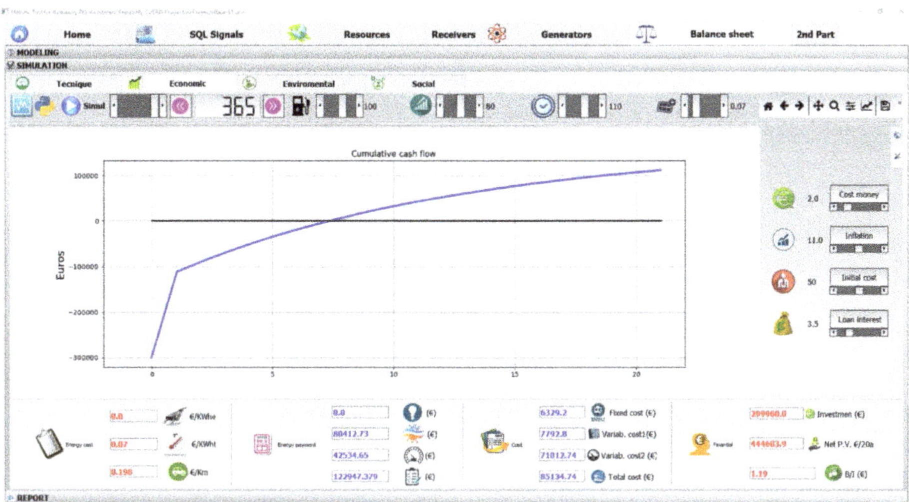

Figure 12. Financial result with inflationary scenario.

3.2.3. Environmental Viability

Considering the stored data concerning the variety of energy sources and their generation characteristics, this tab provides the environmental feasibility in a graphical report in which the amount of carbon dioxide emitted or saved in a year is displayed (Figure 13).

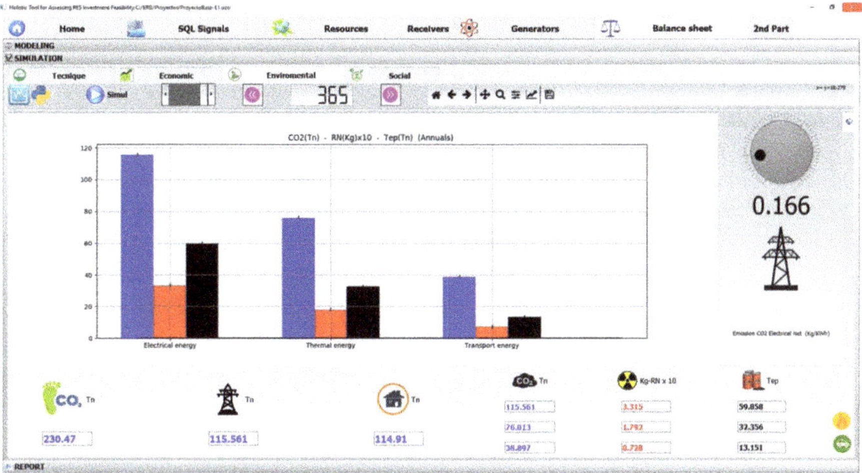

Figure 13. Window showing the emissions of CO_2 and tons of oil equivalent (toe) and nuclear energies.

3.3. Demand Profile Modeling

Three different types of energy consumers were considered in this application: electric consumption, thermal demand, and transportation issues using EVs, as shown in Figure 14.

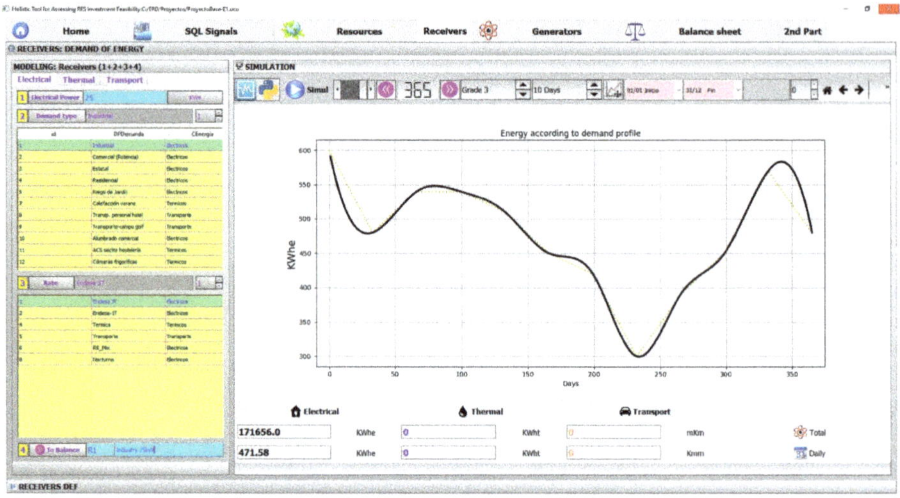

Figure 14. Window with the consumer demand profile.

Each of these consumptions has several characteristic patterns whose values were stored in various databases that can easily be uploaded and updated by means of the SQL features.

In addition to other sources of information, data from energy audits done in different types of companies can be easily adapted to the SQL format to obtain useful realistic profiles.

3.4. Energy Balance Calculation

On the one hand, the system contains the entire set of energy resources and generators database with their characteristics from primary energy transformation to the PCC. On the other hand, a set of demand profiles for different consumers is used. Therefore, combining all the information for the chosen items, a monthly energy balance report can be generated.

A setting window is used to configure the desired power from either a generating profile or a consuming one as it can be seen in Figure 15. This allows performing an energy balance monitoring profile, as depicted in Figure 16.

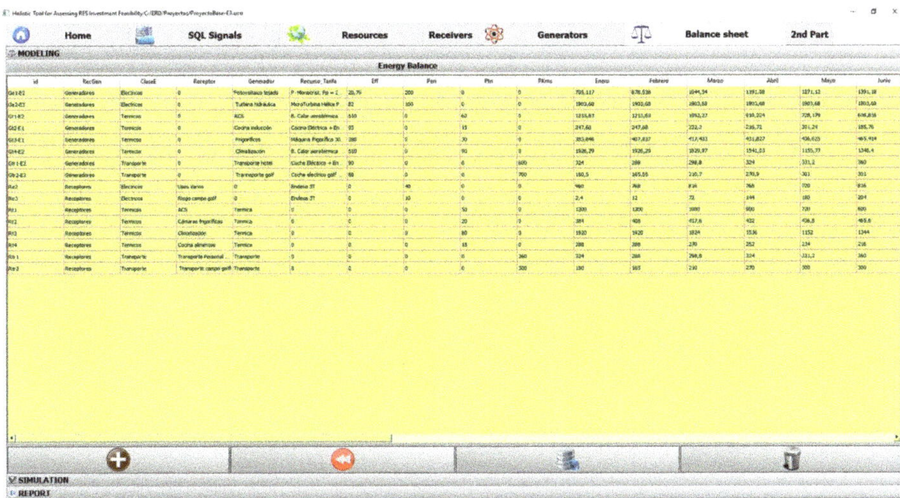

Figure 15. Window for configuring an energy balance profile.

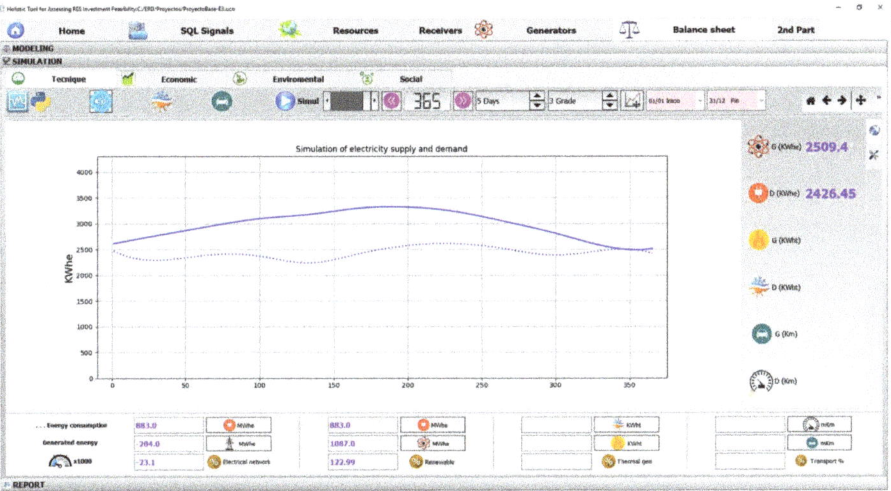

Figure 16. Window for monitoring the energy balance: electrical, thermal and transport.

As a result, an energy balance report is displayed in a window whose data can also be exported into spreadsheets, as shown in Figure 17. Moreover, this balance tab provides interactive dial controls for introducing variations in the parameters to establish different analysis conditions in simulation mode.

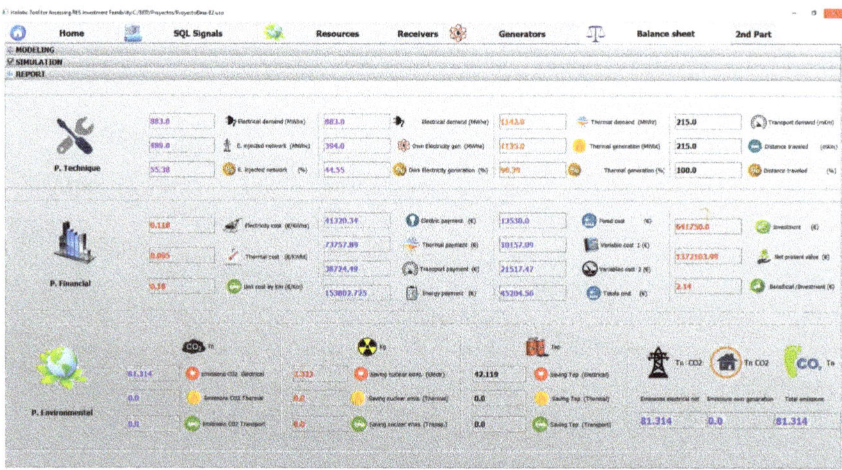

Figure 17. Final balance report spreadsheet.

4. An Illustrative Example of Application

This section is intended both to expose how to proceed with the proposed tool by an example and to provide a comparative analysis between two scenarios with different sets of devices installed in the same facility to generate a couple of contrasting reports.

4.1. Initial Settings

4.1.1. Overall Description

This is a sample study on the supply of electric power to a new hypothetical 115-room hotel and golf course which would be built in Sierra Morena, Cordova, Spain (37°58′16.7″N 4°48′36.8″W). The study considers that there is a nearby electrical network able to supply the whole power demanded by the hotel [25].

The power planned for the facility at project stage is as shown in Table 1.

Table 1. Power types & uses for the projected facility.

Type	Use	Power	Further Info
Electric	Lighting	10 kWe	
Electric	Multiple purposes	40 kWe	
Thermal	Water heating	50 kWt	
Thermal	Kitchen Cooking	15 kWt	3 × 5 kW stoves
Thermal	Refrigerators and freezers	20 kWt	
Thermal	HVAC [1]	80 kWt	
Electric	Golf course watering	10 kWe	1× pumping unit
Transport	Golf course transport	n.s. [2]	10 golf carts at 30 km per day each (300 km/day).
Transport	Staff transport	n.s. [2]	6 staff cars at 60 km per day (360 km/day).

[1] Heating, ventilating and air conditioning (HVAC). [2] Not specified (n.s.).

For the sake of simplicity, on the one hand, it will be considered that appropriate capacitor banks are installed to ensure that the power factor is the unit. On the other hand, absolute data on secondary losses in devices, cable lengths, etc. are not detailed, but have been considered in terms of sector statistics, otherwise, the present document would be oversized and possibly undermined its true objective.

4.1.2. Choosing Demand Profiles

Based on previous data obtained from an energy audit in an existing golf club and resort placed in Cordova's surrounding, demand profiles are chosen for each type of energy according to what it is displayed in Table 1 and exposed in Section 2. The graphs represent the demand per kW of power per day all year round. Each (shown in Figures 18–20) is built from a respective matrix of 12 monthly averaged values corresponding to a year. The displayed data graphs result from the spline interpolation with the Python *scipy* library.

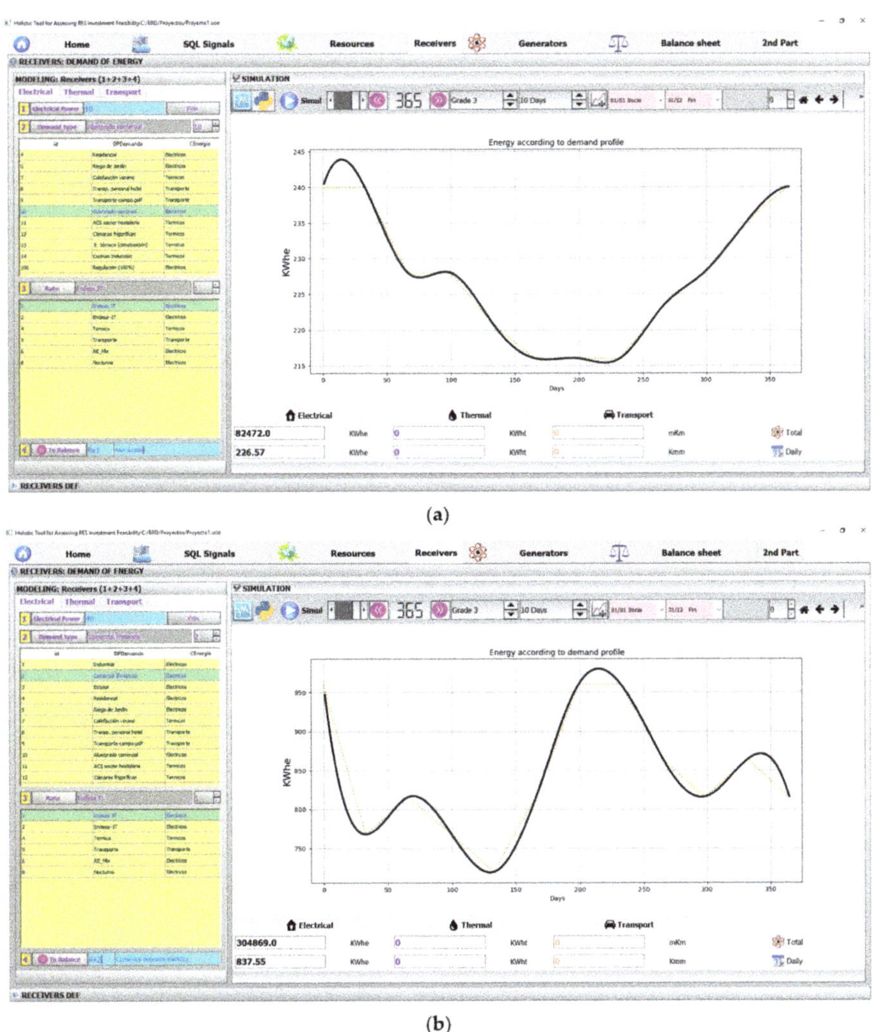

(a)

(b)

Figure 18. *Cont.*

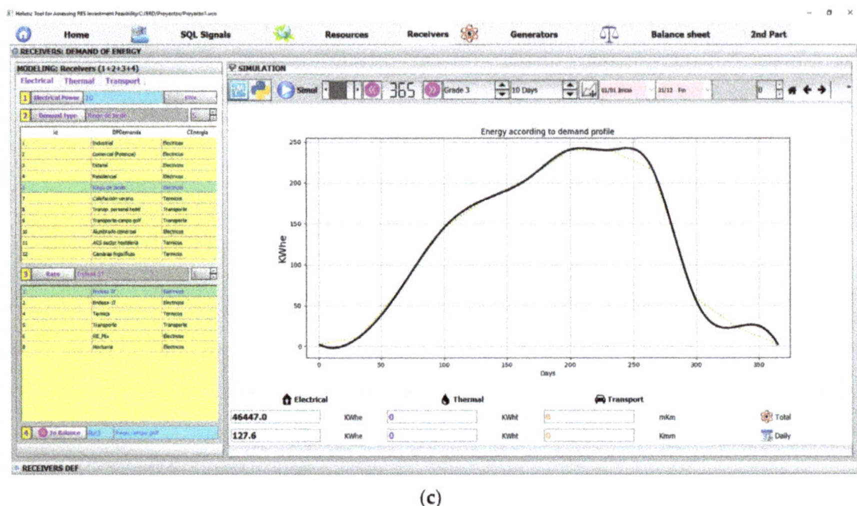

(**c**)

Figure 18. Electric receivers demand profiles: hotel lighting (**a**); plugs for multiple purposes (**b**); and pumps and valves for golf course watering (**c**).

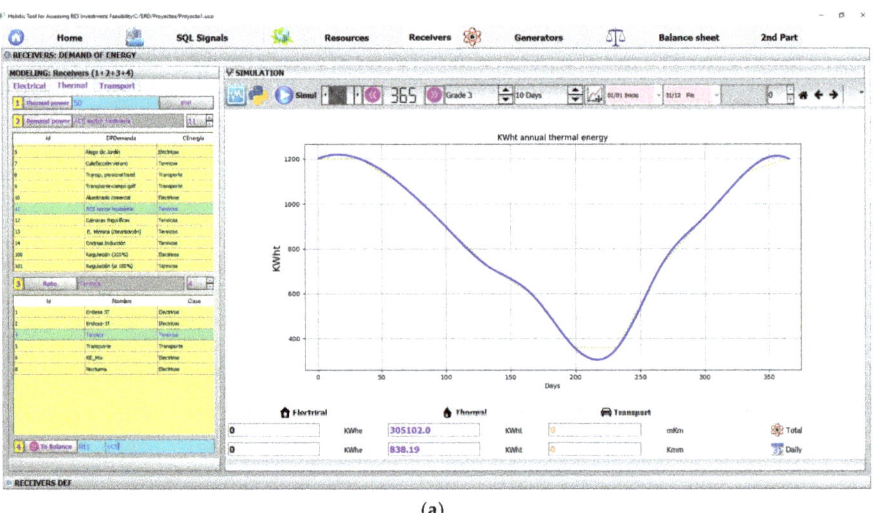

(**a**)

Figure 19. *Cont.*

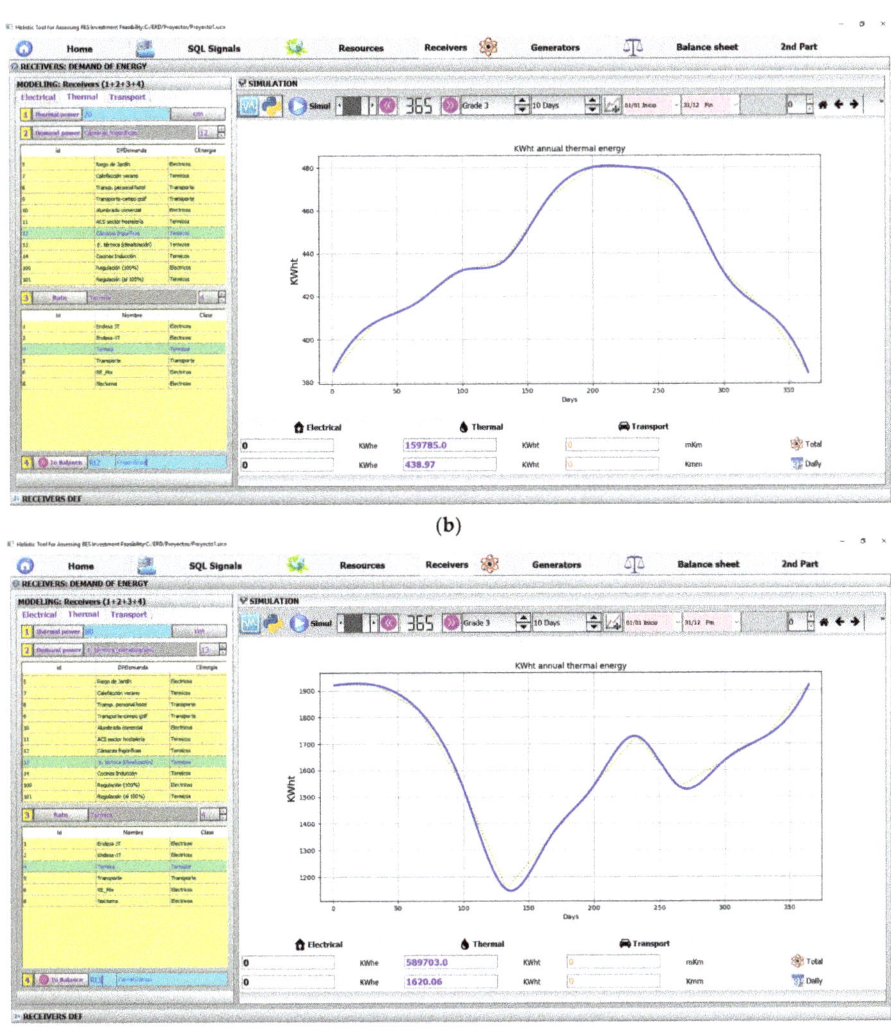

(b)

(c)

Figure 19. *Cont.*

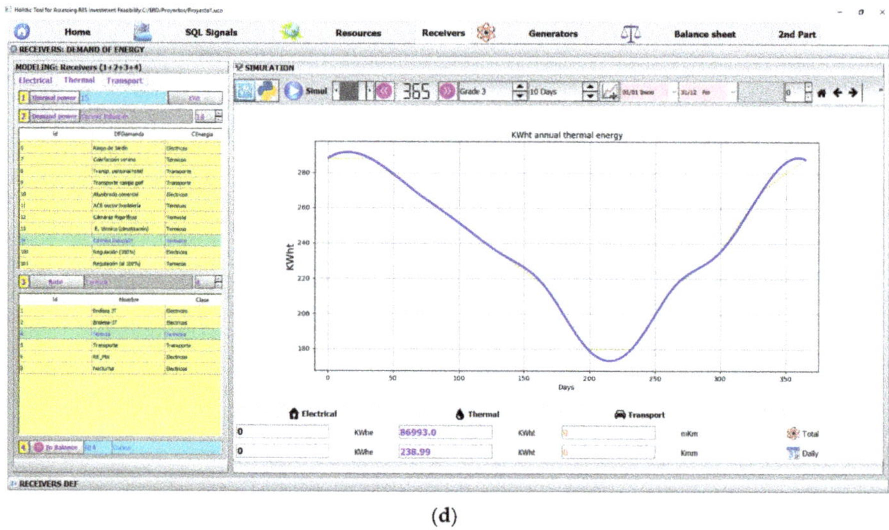

(**d**)

Figure 19. Thermal receivers demand profiles: water heating (**a**); refrigerators and freezers (**b**); HVAC (**c**); and the kitchen (**d**).

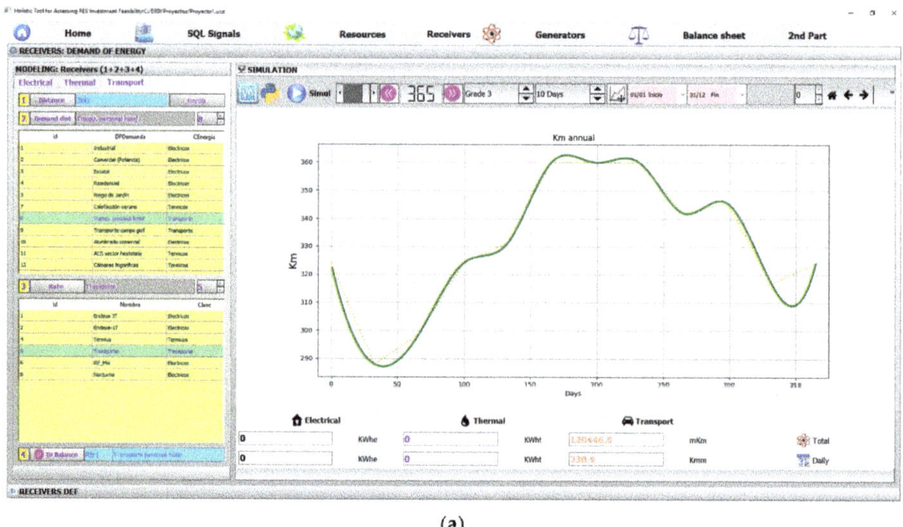

(**a**)

Figure 20. *Cont.*

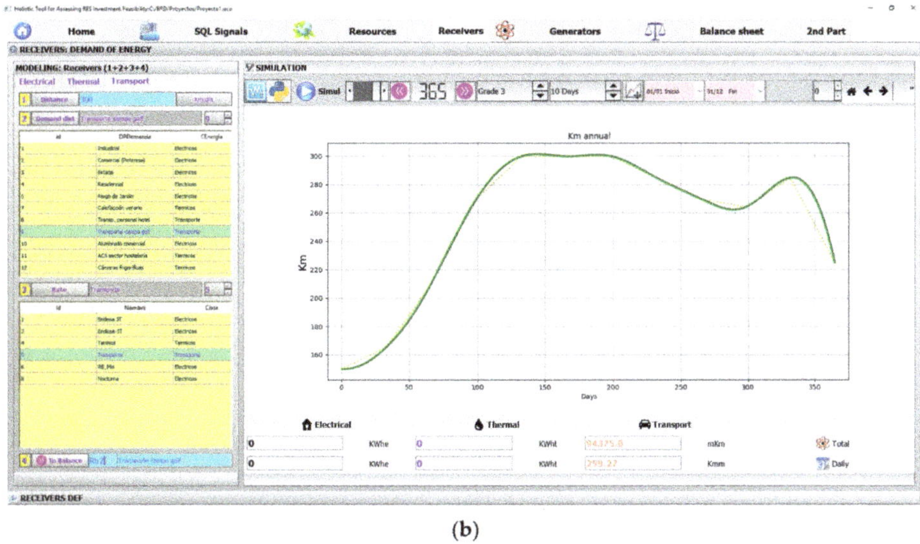

(b)

Figure 20. Transport receivers demand profiles: staff car fleet (**a**); and golf carts (**b**).

Thus, all receivers have already been included in the Balance Spreadsheet (Figure 21).

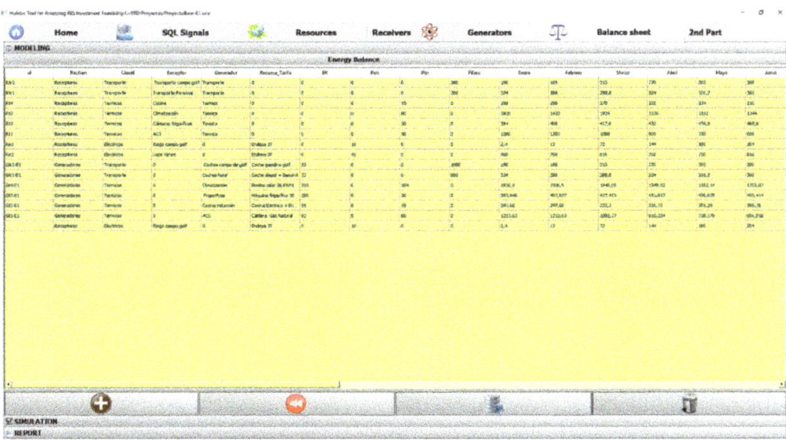

Figure 21. Receivers in Energy Balance Spreadsheet (detail of the tab).

4.2. Analyzing Different Scenarios

Once the demand profiles have been selected, the requested power is assigned to each group of receivers. To identify the receivers the prefix "R" is used and "G" for the generators; if the denomination is followed by "E1", it represents the first Scenario, and so on.

The experimental study will evolve through different supply Scenarios and the way in which the environment and the economy are affected will be shown.

The generators are selected from the "Energy balance" menu and the resource of each energy source is assigned as well. The prices of the resource, its maintenance, and the investment represent a part of the total cost amount to perform viability checking of the project, to what there must be added the losses for the machine and the transport efficiency, the thermal losses, etc. All these data are part of the SQL database of each generator and resource.

4.2.1. Scenario 1

Constraints: choosing generators and their respective profiles

Scenario 1 considers that all power is provided by the existing network of ENDESA (ENDESA is the local power utility in Southern Spain) with triple pricing [26]. Standard heat pumps will be used for HVAC, a natural gas boiler for water heating, the declared induction stoves for the kitchen and standard refrigerators and freezers will be installed. In addition, golf carts will be gas vehicles and the staff cars will be diesel.

Figure 22a shows the boiler graphs, in which the red line indicates the energy that the boiler can supply at full power, whereas the blue one represents the estimated demand profile. Proceeding in the same way, Figure 22b–d presents this information for the other generators in Scenario 1.

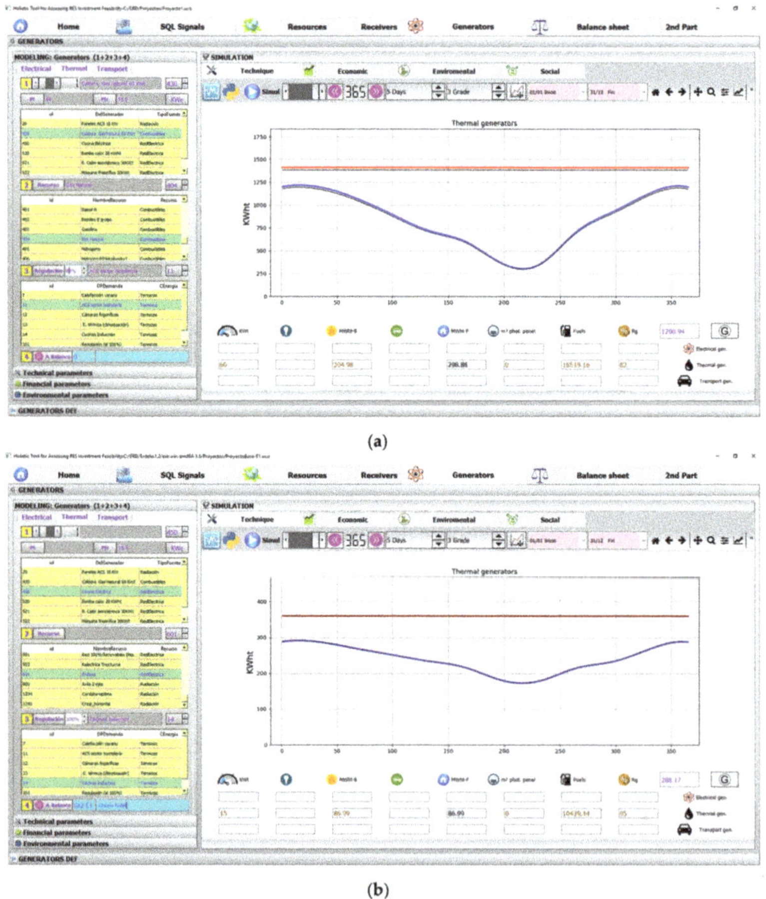

(a)

(b)

Figure 22. *Cont.*

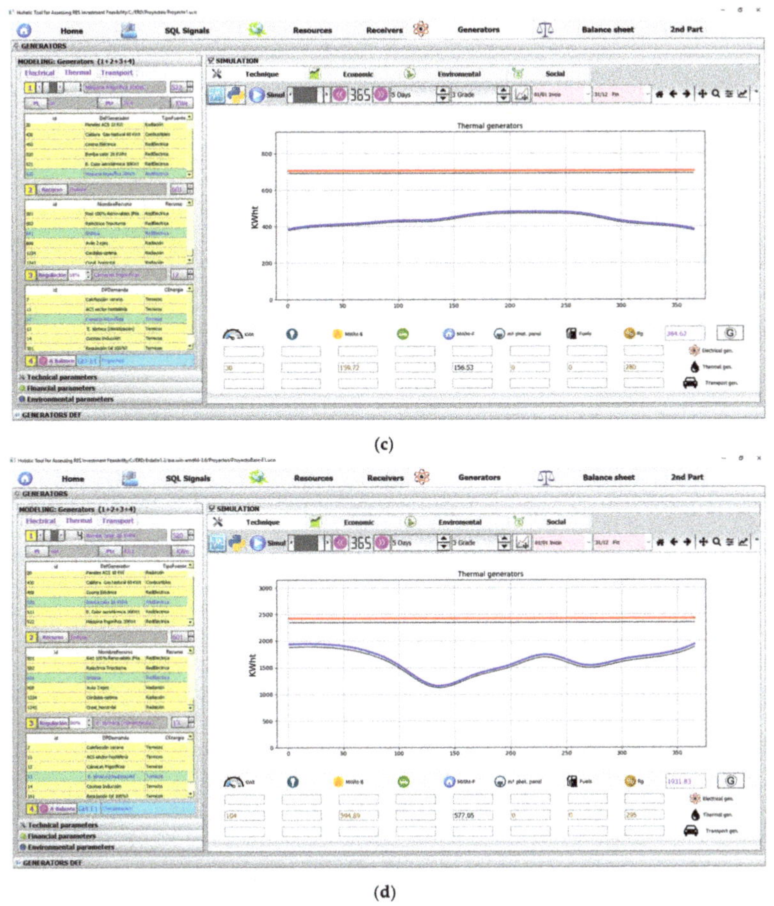

(c)

(d)

Figure 22. Thermal generating capacity (red trace) vs. demand profile (blue trace) for: boilers (**a**); kitchen induction stoves (**b**); refrigerators and freezers (**c**); and HVAC (**d**).

Now, the transport elements will be added. Thus, six cars for the hotel staff and 10 carts for the golf course will be considered for simultaneous use. Figure 23 depicts the capacity of generating energy for transport generators. Note that the gap between the capacity of generators and the requirements of the demand is larger than in previous analysis since the use of each vehicle is essentially discontinuous.

Once configured, the proposed consumption profile model can be simulated from a technical, economic and environmental point of view. Figure 24 displays all the elements involved in the simulation of the first Scenario of the proposed Balance Report Sheet.

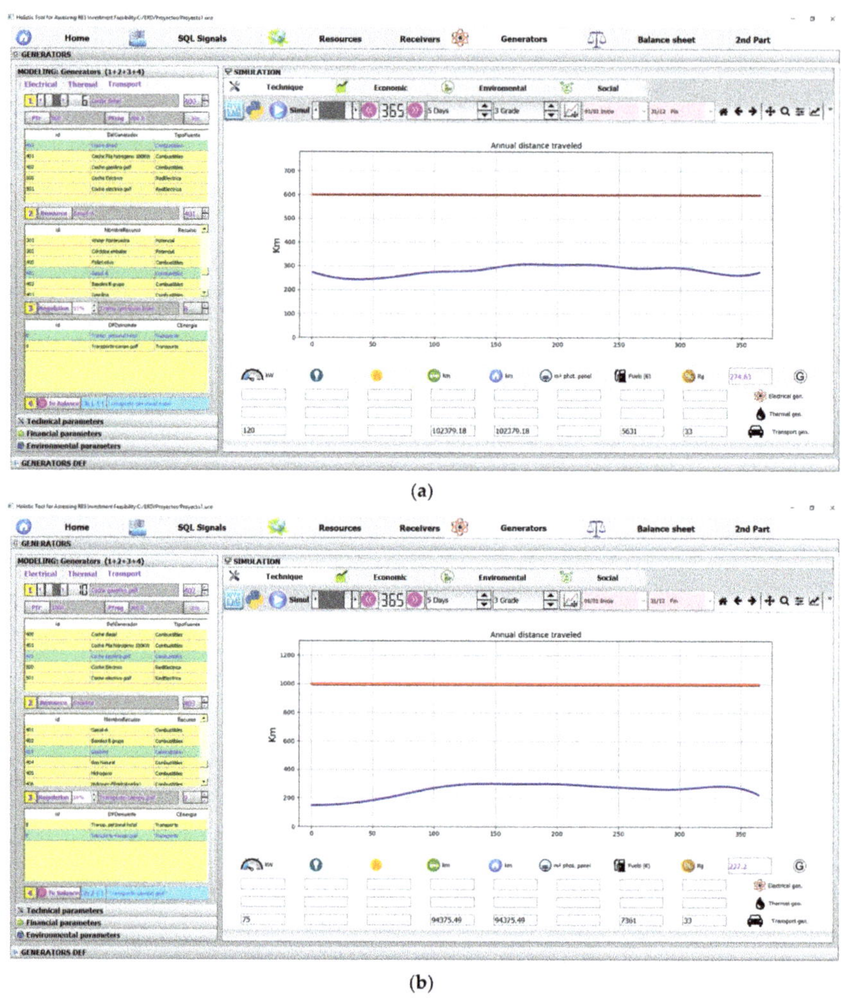

(a)

(b)

Figure 23. Transport generators generating capacity (red trace) vs. demand profile (blue trace) for: staff cars (**a**); and golf carts (**b**).

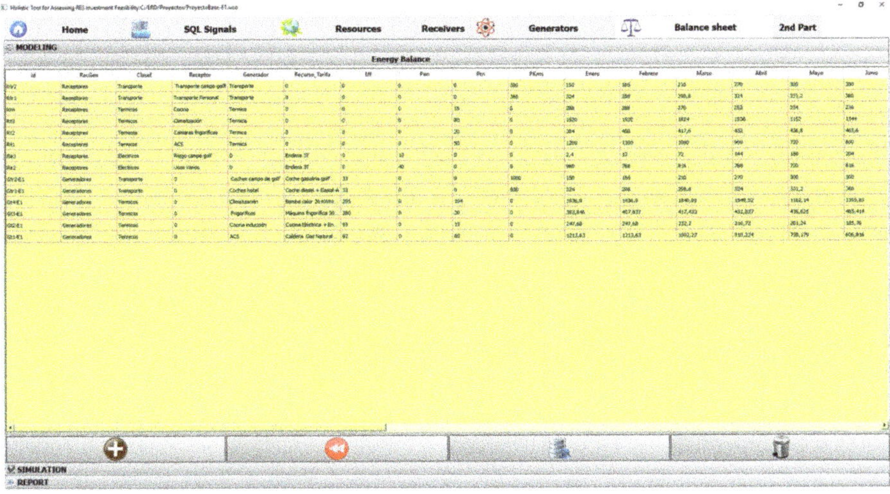

Figure 24. The balance of Scenario 1 (detail of the tab).

Technical Analysis

As stated in the previous section, all electric power is supplied by the grid (696 MWh, dotted line in Figure 25a) in this Scenario, while the generation itself remains null (solid line). On the right side of Figure 25, the instantaneous daily values are shown.

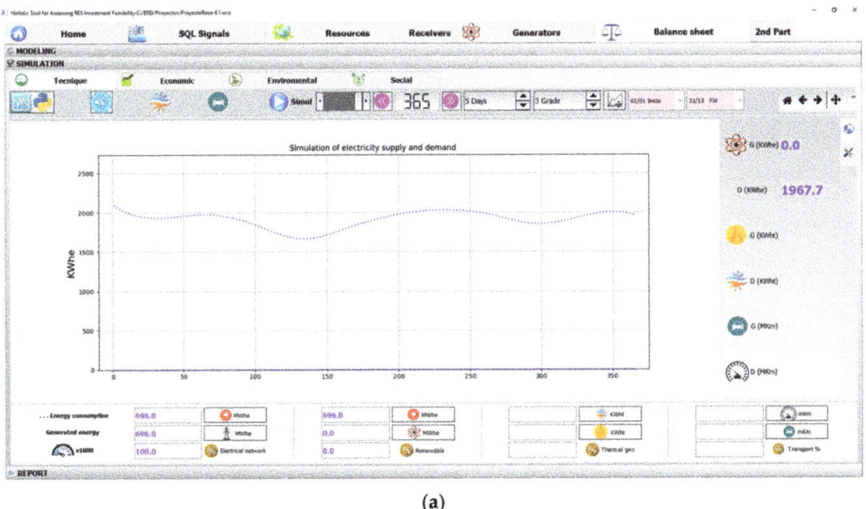

(a)

Figure 25. *Cont.*

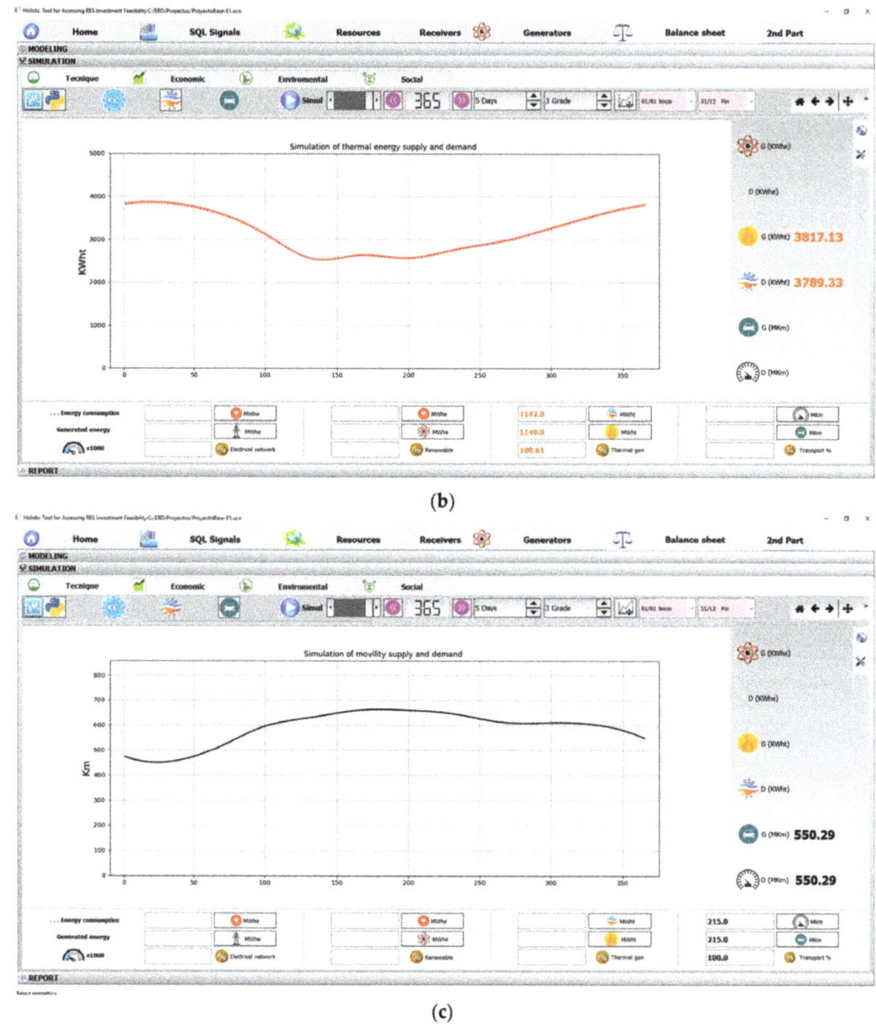

(c)

Figure 25. The perspective of the energy balance, Scenario E1 for: electrical energy (**a**); and thermal energy (**b**); and transport (**c**).

Furthermore, thermal energy (1142 MWht) is entirely supplied by the proposed generators (demand track has been 1% intentionally diverted in Figure 25b to show that they are overlapped).

Moreover, once the transport is analyzed from the technical point of view, it can be clearly seen in Figure 25c that it is 100% satisfied (215,000 km per year).

With these results, the installation is in working order, however, it is time to analyze its economic feasibility and CO_2 emissions.

Economic Analysis

As can be seen in the graph of accumulated cash flow (Figure 26), the investment from the financial point of view is profitable, given that the investment would be recovered after 14 years, with a benefit–investment ratio (BIR) of 0.21 €.

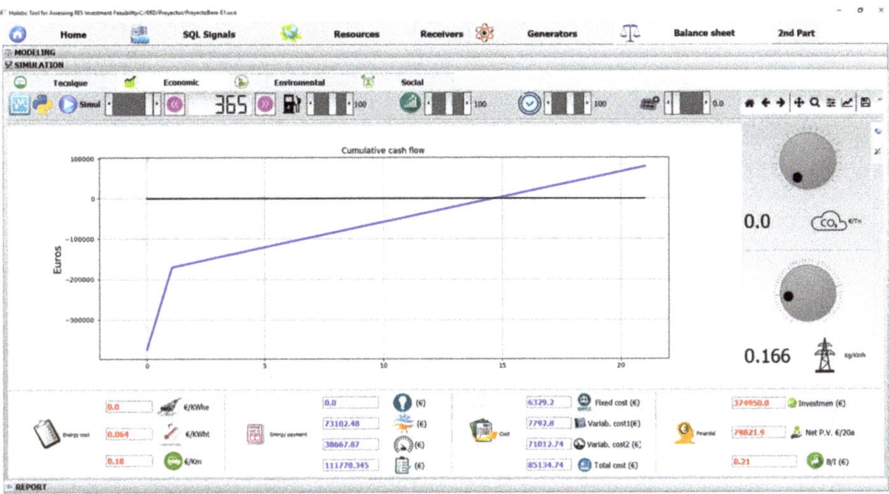

Figure 26. Economic analysis (cumulative cash flow).

Even though it might be feasible, the environmental aspect has not been considered, thus the project might become obsolete in a short period of time.

There are also several dial and scroll controls available to allow users to modify several economic parameters, especially those related to price and taxes, to perform basic sensitivity analyses. Figure 27a–c shows three illustrative cases in which investment sensitivity due to the increase in fuel price and emission issues [26,27], respectively, result in unprofitability within the considered 20 years horizon.

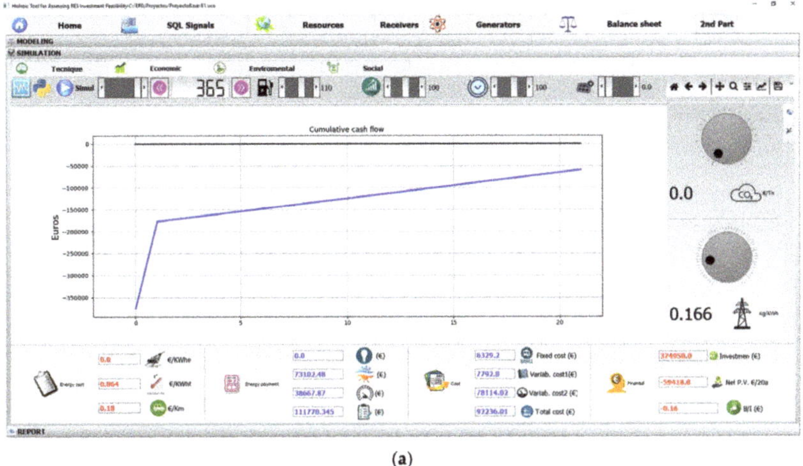

(a)

Figure 27. *Cont.*

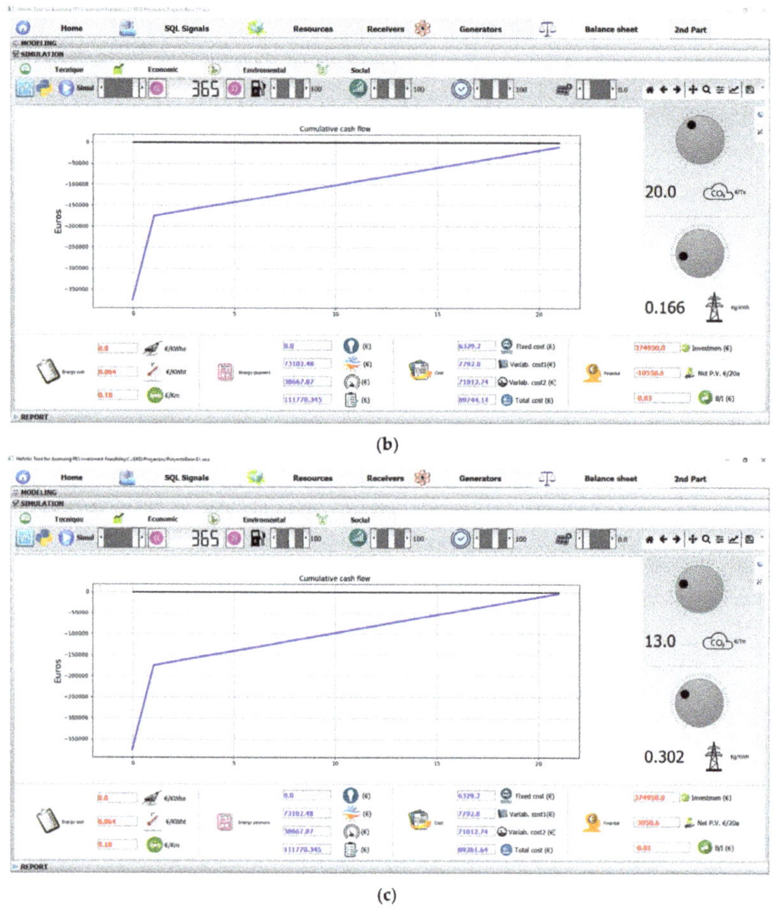

(b)

(c)

Figure 27. Sensitivity analysis with respect to: consumption (**a**); emission rights (**b**); and the CO_2 emission factor of the network and CO_2 emission rights (**c**).

The Benefit–Investment Ratio has lowered to negative and the recovery period is outside the predictable limits, which warns about the low flexibility of adaptation to likely events, and the need to make convenient decisions to prevent that. Otherwise, the project would come out of the profitability margin and would make it unviable.

Environmental Analysis

It is obvious that emissions in this Scenario are very high. In the lower left-hand side of Figure 25 are displayed the CO_2 emissions in tons due to the consumption of AC grid (129 t) and in total (244 t), as well as that of the own facility (114.91 t).

Below, on the right-hand side (in the graph, nuclear waste is multiplied by 10 for visibility purposes), energies are split up by type and their equivalent emissions in nuclear waste and toe are also described in Figure 28.

The dial in the simulation balance tab allows varying the CO_2 emission factor to check the project profitability through the evolution of the energy mix.

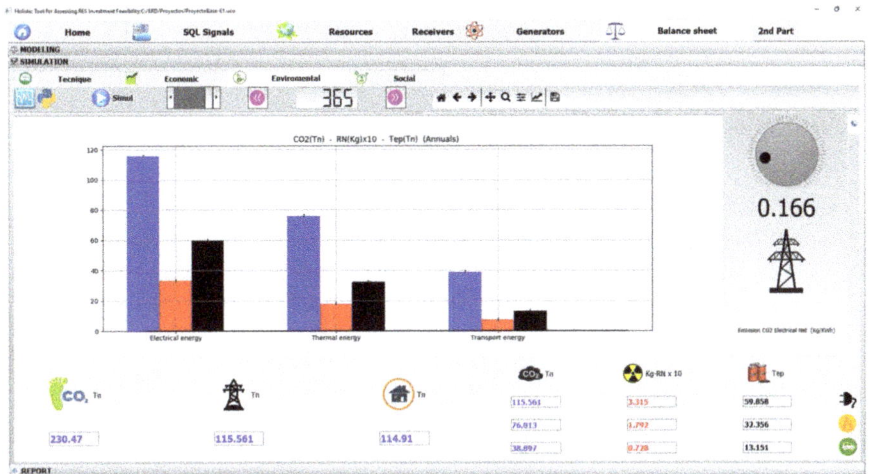

Figure 28. Sensitivity analysis with respect to emission rights.

4.2.2. Scenario 2

The investment improvement action aimed to assess its feasibility will be directed in two ways. Firstly, replacing combustion systems by those provided with alternative technologies. For instance, the gas boiler and standard air–air heat pumps with the most efficient aerothermal [28] heat pumps (air–water).

With regard to transport, internal combustion vehicles are going to be replaced with electric ones; given that the demand for travel is not affected by the autonomy of these vehicles, their efficiency is 90%, compared to 30% or 35% of combustion vehicles. Diesel and high-pressure gasoline vehicles are not only transport generators, they are also large emitters of CO_2 and NO_x and especially 2.5-micron particles are highly carcinogenic as they enter the cells more directly.

Despite its high price, the annual battery charge loss is close to 1%, which may consider a service-life of about 10–20 years depending on use. The average consumption is 20 kWh per 100 km and the battery is charged in 40 minutes at 80%, but for a full charge it takes 2 h, provided that the power of the charger allows for it.

Secondly, for this Scenario, a 200 kW photovoltaic system [29] will be installed on the roof, with a 35° slope and 10° East orientation. To do this, the user might select the section named "Solar Radiation Resources" and simulate the roof parameters, as shown in Figure 29a. Once it is done, a daily average of 5.66 KWh·per sq.m. and per day, with an increase of 8.49% over the horizontal radiation, is obtained. Furthermore, various inclination/orientation models are simulated to have real data regarding the convenience of using the roof of the building to place the panels, and the results are shown in Figure 29b.

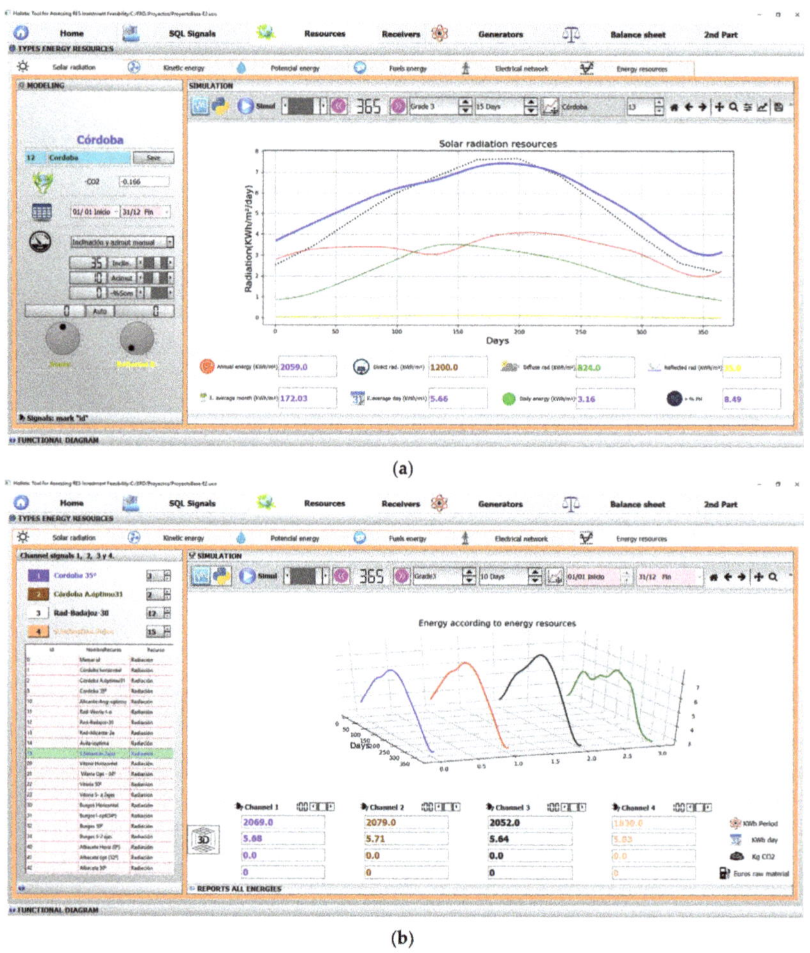

(a)

(b)

Figure 29. Solar radiation (**a**); and comparative analysis of solar radiation (**b**) in Cordova.

Note in Figure 29b that, even though the two-axis tracking (channel 4) gets much more radiation, the roof solution (traced in channel 1) should be kept because of its optimal slope (31°) and shade-free location. That is why a 200 kWe photovoltaic generator might be installed onto the roof, since there is enough roof available (962 sq.m. required), with monocrystalline panels, obtaining the results in Figure 30.

Upgrading to the new Scenario or model, the Balance to be considered is that shown in Figure 31.

Both the induction cooker and refrigerators and freezers should be kept because they are optimized to the current technology. The new results from the technical point of view are shown in Figure 32.

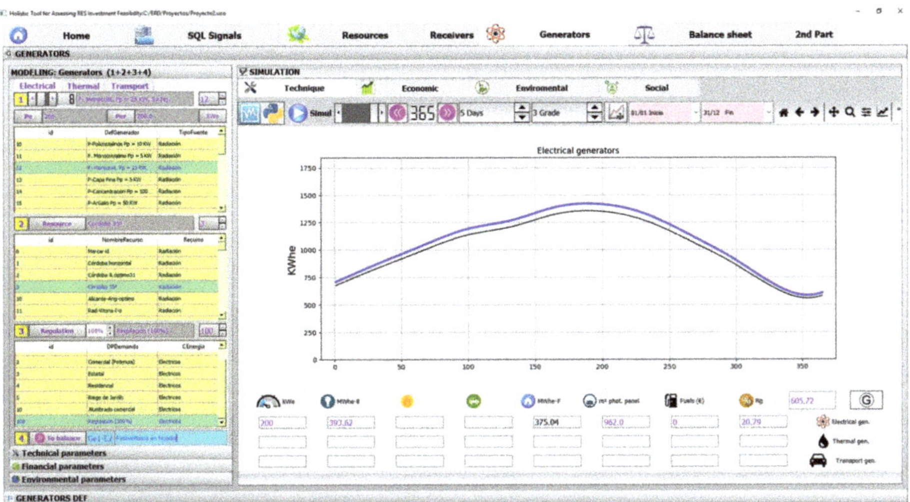

Figure 30. Photovoltaic generator of 200 kWe in Cordova at 35° slope and 10° East orientation.

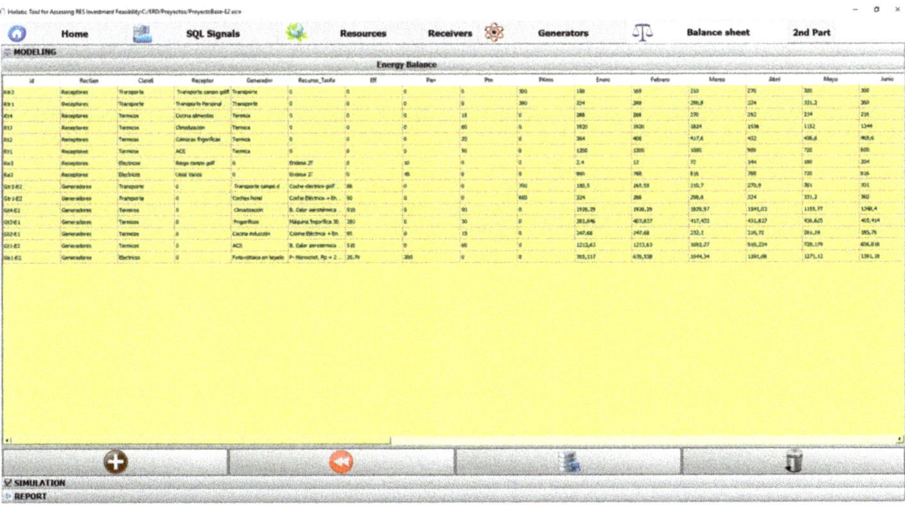

Figure 31. New nalance elements sheet in Scenario 2.

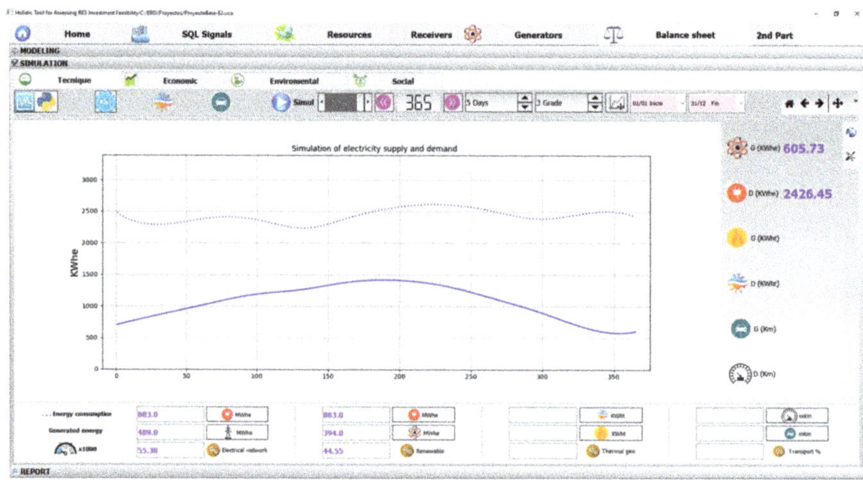

Figure 32. New balance from the electrical point of view.

It is not recommended to exceed 45% of the total demand since what is not consumed during sunny hours can be injected into the network (if you are a producer) at the minimum price since this case considers no accumulation.

Regarding the thermal and transport issues, they are technically the same as the previous curves, since the demands have not changed and the new receivers have been adjusted again according to the corresponding control curve.

From the economic point of view, the return on investment has been reduced to four years and BIR has been increased to 2.14 € (Figure 33a). It is mainly due to the energy provided by the photovoltaic system at zero fuel cost and the higher profitability of the electric car as opposed to the internal combustion, as well as that coming from the aerothermal pumps.

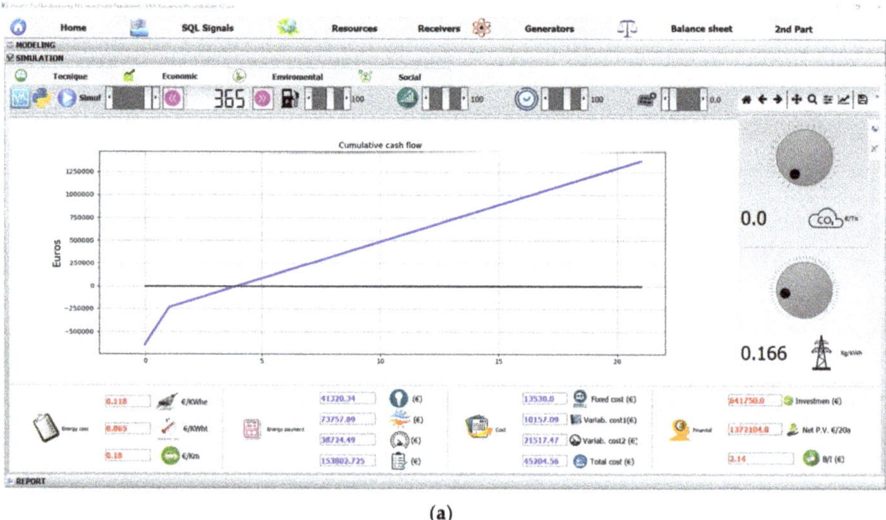

(a)

Figure 33. *Cont.*

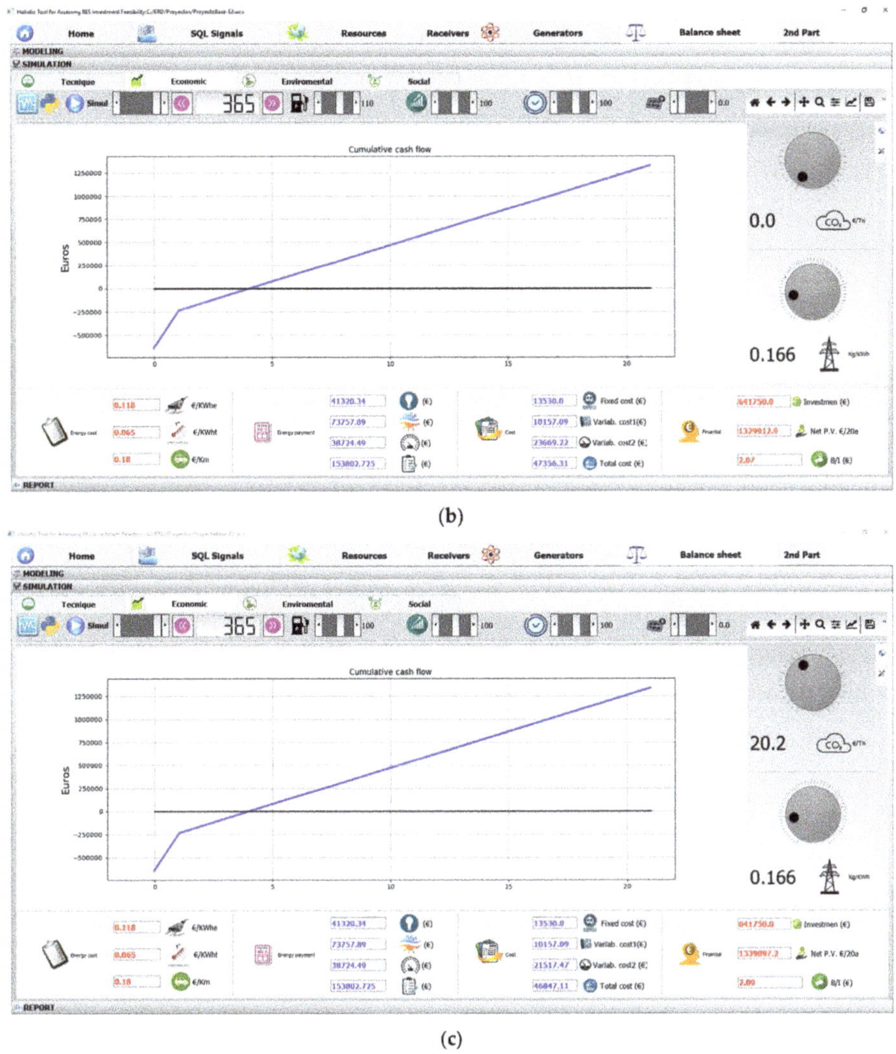

(b)

(c)

Figure 33. Accumulated movement of funds (**a**); sensitivity analysis with respect to consumption (**b**); and sensitivity analysis with respect to CO$_2$ emission rights (**c**) in Scenario 2.

Note that, when repeating the sensitivity analysis in which consumption was increased by 10%, the graph results as shown in Figure 33b, in which the values of profitability and return have barely moved, which means that the project is economically stronger now.

Simulating the model with the new CO$_2$ emissions and invoicing and keeping the emission rights at 20 € as above, the return and profitability remain almost the same (Figure 33c), mainly because the emission cost is negligible compared to the increase in profit. The project is now safer and more reliable with respect to potential threats in the coming years.

From an environmental perspective, the most noteworthy is the enormous reduction in CO$_2$ emissions, which has dropped by around 65%, from 230.47 t to 81.34 t (Figure 34). This is because the facility has stopped burning fuel and passed to the electricity grid the thermal energy generators, which has fewer emissions, has also been offset by a photovoltaic installation that produces 45% of consumption.

Having reached this point, to continue searching for a Zero Emissions [30,31] panorama, it is necessary to install an accumulation system and continue adding renewable energy, passing it on to manageable ones or adding fuel cells, which, as it is not mature technology, should be dismissed for the moment.

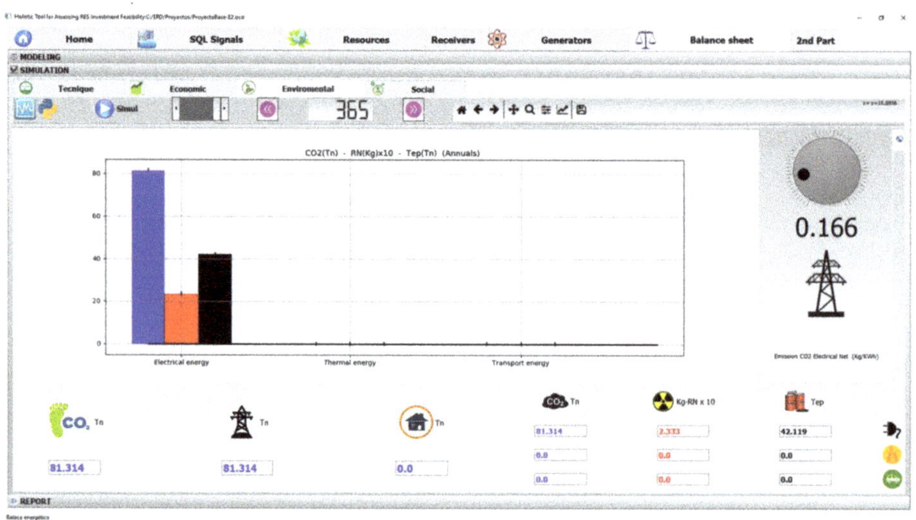

Figure 34. Environmental analysis in Scenario 2.

5. Conclusions

A first approach to an analysis tool has been developed that provides a holistic software environment, capable of generating technical, economic and environmental reports to support an investment decision.

To make the system useful, flexible and modular, an SQL database system has been programmed.

The current profiling methodology allows for the use of any modeling information related to RES, generators or consumption profiles by adapting the source data to a monthly average database structure, regardless of how the model was built or how the measurements were taken.

These databases can automatically be, and are presented as, one of the main lines of future development of the proposed tool.

Although no real investment examples or proven validation cases have been performed to assess the validity of the tool described in this paper, the process followed in the previous section proves that obtaining feasibility reports through the proposed tool is a user-friendly process. Moreover, the obtained reports allow for comparative analysis against different design alternatives, as can easily be seen from a fully completed theoretical case and the several simulations that have been carried out yielding consistent and promising results.

Author Contributions: J.M.F.A. and A.E. defined and programmed the software tool and performed the simulations. L.C. and A.M. reviewed and suggested improvements in the source code. A.E. developed behavioral models in SQL from public repositories; F.J.B.O. and E.J.P.G. developed the power receiver definition by adapting previous models to SQL databases; L.C. and A.M. did the same about the power generation model definition; and J.M.F.A., A.E. and E.J.P.G. developed models for generators and adapted them to SQL databases. All authors provided technical information on systems and equipment and the energy environment for the test scenarios and agreed to the selection of the case shown. J.M.F.A. and A.E. coordinated and wrote the paper. All authors supervised and approved the final version of the manuscript.

Conflicts of Interest: The authors declare no conflict of interest.

References

1. Salom, J.; Marszal, A.J.; Widén, J.; Candanedo, J.; Lindberg, K.B. Analysis of load match and grid interaction indicators in net zero energy buildings with simulated and monitored data. *Appl. Energy* **2014**, *136*, 119–131. [CrossRef]
2. Palacios-Garcia, E.J.; Moreno-Munoz, A.; Santiago, I.; Moreno-Garcia, I.M.; Milanés-Montero, M.I. Smart community load matching using stochastic demand modeling and historical production data. In Proceedings of the 2016 IEEE 16th International Conference on Environment and Electrical Engineering (EEEIC), Florence, Italy, 6–8 June 2016; pp. 1–6.
3. Kavgic, M.; Mavrogianni, A.; Mumovic, D.; Summerfield, A.; Stevanovic, Z.; Djurovic-Petrovic, M. A review of bottom-up building stock models for energy consumption in the residential sector. *Build. Environ.* **2010**, *45*, 1683–1697. [CrossRef]
4. Short, W.; Packey, D.J.; Holt, T.; U.S. Department of Energy. *A Manual for the Economic Evaluation of Energy Efficiency and Renewable Energy Technologies*; No. NREL/TP–462-5173; National Renewable Energy Lab: Golden, CO, USA, 1995.
5. Suganthi, L.; Samuel, A.A. Energy models for demand forecasting—A review. *Renew. Sustain. Energy Rev.* **2012**, *16*, 1223–1240. [CrossRef]
6. Solar-Estimate, Solar Calculator. Available online: http://www.solar-estimate.org (accessed on 30 November 2017).
7. EC. Institute for Energy and Transport, Photovoltaic Geographical Information System (PVGIS). 2010. Available online: http://re.jrc.ec.europa.eu/pvgis/apps4/pvest.php (accessed on 15 August 2017).
8. Energy Saving Trust. Tools and Calculators. Available online: http://www.energysavingtrust.org.uk/resources (accessed on 13 June 2017).
9. National Renewable Energy Laboratory (NREL). Solar and Wind Energy Resource Assessment (SWERA) Model. 2016. Available online: www.nrel.gov/analysis/models_tools.html (accessed on 30 May 2017).
10. Ciabattoni, L.; Grisostomi, M.; Ippoliti, G.; Longhi, S.; Mainardi, E. On line Solar Irradiation Forecasting by Minimal Resource Allocating Networks. In Proceedings of the 20th IEEE Mediterranean Conference on Control & Automation (MED 2012), Barcelona, Spain, 3–6 July 2012; pp. 1506–1511.
11. Ameli, S.M.; Agnew, B.; Potts, I. Integrated distributed energy evaluation software (IDEAS): Simulation of a micro-turbine based CHP system. *Appl. Therm. Eng.* **2007**, *27*, 2161–2165. [CrossRef]
12. Anastaselos, D.; Giama, E.; Papadopoulos, A.M. An assessment tool for the energy, economic and environmental evaluation of thermal insulation solutions. *Energy Build.* **2009**, *41*, 1165–1171. [CrossRef]
13. Axelsson, E.; Harvey, S.; Berntsson, T. A tool for creating energy market scenarios for evaluation of investments in energy intensive industry. *Energy* **2009**, *34*, 2069–2074. [CrossRef]
14. Grandjean, A.; Adnot, J.; Binet, G. A review and an analysis of the residential electric load curve models. *Sustain. Energy Rev.* **2012**, *16*, 6539–6565. [CrossRef]
15. Swan, L.; Ugursal, V. Modeling of end-use energy consumption in the residential sector: A review of modeling techniques. *Renew. Sustain. Energy Rev.* **2009**, *13*, 1819–1835. [CrossRef]
16. Richardson, I.; Thomson, M.; Infield, D.; Clifford, C. Domestic electricity use: A high-resolution energy demand model. *Energy Build.* **2010**, *42*, 1878–1887. [CrossRef]
17. Widén, J.; Wäckelgård, E. A high-resolution stochastic model of domestic activity patterns and electricity demand. *Appl. Energy* **2010**, *87*, 1880–1892. [CrossRef]
18. Palacios-Garcia, E.J.; Chen, A.; Santiago, I.; Bellido-Outeiriño, F.J.; Flores-Arias, J.M.; Moreno-Munoz, A. Stochastic model for lighting's electricity consumption in the residential sector. Impact of energy saving actions. *Energy Build.* **2015**, *89*, 245–259. [CrossRef]
19. Bellido-Outeirino, F.; Flores-Arias, J.; Linan-Reyes, M.; Palacios-garcia, E.; Luna-rodriguez, J. Wireless sensor network and stochastic models for household power management. *IEEE Trans. Consum. Electron.* **2013**, *59*, 483–491. [CrossRef]
20. Spanish Ministry of Industry Energy and Tourism. Resolution of the Directorate General for Energy Policy and Mines Published the Provisional List of Plants with Biodiesel Production Number Assigned to the Calculation of Compliance with Mandatory Targets. Published in the Spanish Official Bulletin Nr. 315.

Available online: http://www.minetad.gob.es/energia/en-US/novedades/Paginas/Propuesta-resolucion-provisional-biodiesel.aspx (accessed on 13 September 2017).

21. Ciabattoni, L.; Ferracuti, F.; Grisostomi, M.; Ippoliti, G.; Longhi, S. Fuzzy logic based economical analysis of photovoltaic energy management. *Neurocomputing* **2015**, *170*, 296–305. [CrossRef]
22. National Aeroanutics and Space Administration (NASA). Surface meteorology and Solar Energy. Available online: https://eosweb.larc.nasa.gov/cgi-bin/sse/sse.cgi?rets@nrcan.gc.ca (accessed on 24 June 2017).
23. Duffie, J.A.; Beckman, W.A. *Solar Engineering of Thermal Processes*, 4th ed.; Wiley-Interscience: New York, NY, USA, 2013; ISBN 978-0-470-87366-3.
24. Agüera Soriano, J. *Inconpressible Fluid Mechanics and Hydraulic Turbomachines*, 5th ed.; Ed. Ciencia 3 S.L.: Madrid, Spain, 2002; ISBN 84-95391-01-05.
25. Red Eléctrica de España. Available online: http://www.ree.es/es/ (accessed on 8 September 2017).
26. Tariff 3.0a, ENDESA. Available online: https://www.endesaclientes.com/empresas/catalogo/electricidad.html#subTabsProduct-2 (accessed on 12 September 2017).
27. International Energy Agency. Available online: http://www.iea.org (accessed on 19 March 2017).
28. Anonymous. What Is Aerothermia? (n.d.). Available online: http://www.toshiba-aire.es/que-es-aerotermia (accessed on 12 October 2017).
29. Anonymous. Solar Panels (n.d.). Available online: https://www.tesla.com/solarpanels?energy_redirect=true (accessed on 8 October 2017).
30. National Energy Commission. Report on the Proposal for a Royal Decree Establishing the Regulation of the Administrative, Technical and Economic Conditions of the Electricity Supply Modality with Net Balance. 2012. Available online: http://www.efimarket.com/wp/wp-content/uploads/2012/04/cne09_12.pdf (accessed on 24 October 2017).
31. Spanish Ministry of Industry Energy and Tourism. Royal Decree 900/2015 of 9 October 2015 Regulating the Administrative, Technical and Economic Conditions Governing the Supply of Electricity for Own Consumption and the Production of Electricity for Own Consumption. Published in the Spanish Official Bulletin Nr. 243 10/10/2015. Available online: https://www.boe.es/boe/dias/2015/10/10/pdfs/BOE-A-2015-10927.pdf (accessed on 24 October 2017).

 *sustainability*

Article

Transforming Data Centers in Active Thermal Energy Players in Nearby Neighborhoods

Marcel Antal, Tudor Cioara *, Ionut Anghel, Claudia Pop and Ioan Salomie

Computer Science Department, Technical University of Cluj-Napoca, Memorandumului 28, 400114 Cluj-Napoca, Romania; marcel.antal@cs.utcluj.ro (M.A.); ionut.anghel@cs.utcluj.ro (I.A.); claudia.pop@cs.utcluj.ro (C.P.); ioan.salomie@cs.utcluj.ro (I.S.)
* Correspondence: tudor.cioara@cs.utcluj.ro; Tel.: +40-264-202-352

Received: 22 January 2018; Accepted: 18 March 2018; Published: 23 March 2018

Abstract: In this paper, we see the Data Centers (DCs) as producers of waste heat integrated with smart energy infrastructures, heat which can be re-used for nearby neighborhoods. We provide a model of the thermo-electric processes within DCs equipped with heat reuse technology, allowing them to adapt their thermal response profile to meet various levels of hot water demand. On top of the model, we have implemented computational fluid dynamics-based simulations to determine the cooling system operational parameters settings, which allow the heat to build up without endangering the servers' safety operation as well as the distribution of the workload on the servers to avoid hot spots. This will allow for setting higher temperature set points for short periods of time and using pre-cooling and post-cooling as flexibility mechanisms for DC thermal profile adaptation. To reduce the computational time complexity, we have used neural networks, which are trained using the simulation results. Experiments have been conducted considering a small operational DC featuring a server room of 24 square meters and 60 servers organized in four racks. The results show the DCs' potential to meet different levels of thermal energy demand by re-using their waste heat in nearby neighborhoods.

Keywords: data centers; waste heat re-use; thermal energy flexibility; thermal profile adaptation; neural networks; computation fluid dynamics

---

1. Introduction

Energy efficiency in Data Centers (DCs) is of utmost importance as they are among the largest consumers of energy and their energy demand is rapidly increasing due to the increasing digitization of human activities. The cooling system is the second largest consumer in a DC after the Information Technology (IT) equipment and deals with maintaining the proper temperature for servers' safe operation. Even in the best-designed DCs, the cooling system can consume up to 37% of the total electricity [1]. The main cause of this high energy consumption is that most nowadays, DCs use low temperature set points to cool down the server room, minimizing the risk of overheating the servers. Nevertheless, it has been shown in [2] that a temperature between 15 °C and 32 °C does not affect the proper operation of servers. Thus, methods for decreasing the energy consumption of the cooling system had emerged considering actions such as increasing the temperature of the air pumped in the server room by air conditioning units or minimizing the air volume pumped in the room. However, approaches bring the risk of creating hot spots in the server room, resulting in the emergency stop or even total damage of the servers. To avoid such unpleasant situations, workload allocation methods that consider the heat distribution resulting from the execution of tasks by the servers have been proposed in [3,4].

Lately, with the advent of smart city concept to urban development, the DCs have become fundamental in providing the technological foundation to process the huge amount of data the smart cities are generating [5]. Thus, many DCs are located in urban environments providing theological support for the implementation of various smart city services. One way to achieve cost-effectiveness and/or energy efficient solutions is the integration of DCs with the smart cities' energy infrastructure and utilities. The H2020 CATALYST project [6] vision is that DCs have the potential of becoming active and important stakeholders in a system-level thermal energy value chain. Being electrical energy consumers as well as heat producers, the DC can be successfully integrated into both electrical and thermal energy grids and the waste heat generated by the IT components can be effectively re-used either internally for space heating and/or domestic or district heating network operators. As a result, the DCs will gain financial benefits from cooperating under various modalities with smart energy grids by becoming intelligent hubs at the crossroad of three energy networks: electrical energy network, heat network and data network. Thus, in our vision waste heat reuse is expected to become a considerable financial revenue stream for DCs, trading off additional costs for waste heat regeneration (e.g., heat pumps) with incremental revenues from waste heat valorization.

This paper addresses the issue of transforming the DCs into active players in the local heat grid by proposing an optimization methodology that will allow them to adapt their heat generation profile by optimal setting operational parameters of the cooling system (airflow and air temperature) and intelligent workload allocation to avoid hot spots. Our approach to DCs wasted heat recovery problem will allow for exploiting and adapting DCs' internal yet latent thermal flexibility and feed heat (as hot water for example) to the nearby neighborhood on demand. The methodology is based on a thermo-electrical mathematical model of a virtualized DC equipped heat reuse technology and relates the servers' utilization to the server room temperature set points as well as the cooling system energy consumption to the amount heat amount and quality generated by the heat pump. Because the thermal processes within the server room are highly complex, models based on Computational Fluid Dynamics (CFD) techniques and neural networks are used to estimate the available thermal energy flexibility of the DC. In summary, the paper brings the following contributions:

- A thermo-electrical model that correlates the workload distribution on servers with the heat generated by the IT equipment for avoiding hot spot appearance.
- Techniques for assessing the DCs' thermal energy flexibility using CFD simulations and neural networks.
- An optimization technique for DCs equipped with heat reuse technology that enables the adaptation of thermal energy profile according to the heat demand of the nearby buildings.
- Experimental evaluation of the defined models and techniques on a simulated environment.

The rest of the paper is structured as follows: Section 2 presents the related work in DC heat reuse, Section 3 presents a model for estimating the heat generated by server in correlation with the workload distribution, Section 4 describes the CFD and neural networks techniques used to assess the DC thermal flexibility, Section 5 presents results obtained for a small-scale test bed DC, while Section 6 concludes the paper and shows the future work.

2. Related Work

Few approaches can be found in state of the art literature addressing the thermal energy flexibility of DCs and reuse of waste heat in nearby neighborhoods and buildings [7,8]. The hardware development in the last years has led to the increase of the power density of chips and of servers' density in DCs. As the servers' design improves for allowing them to operate at increasingly higher temperatures and their density in server rooms will continue to rise, the DCs will be transformed in producers of heat [9]. This generates a cascading energy loss effect in which the heat generated by servers must be dissipated and gets wasted while the cooling system needs to work at even higher capacity leading to even increased levels of energy consumption. From the perspective of utilities and district heating, there is a clear identified need for intelligent heat distribution systems that can leverage on re-using the heat generated on demand by third parties such DCs [10]. The goal is to provide heat-based ancillary services leveraging on forecasting techniques to determine water consumption patterns [11,12]. Modeling and simulation tools for the transport of thermal energy within district heating networks are defined in [13,14] to assess benefits and potential limitations of re-using the wasted heat. They can help communities and companies to make preliminary feasibility studies for heat recovery and make decisions and market analysis before actual project implementation. Besides increasing the efficiency of district heating, the heat reuse will also contribute to the reduction of emissions during the peak load production of thermal energy which is being typically produced using fossil fuels.

There are two big issues with re-using the heat generated by DCs in the local thermal grid: the relatively low temperatures in comparison with the ones needed to heat up a building and the difficulty of transporting heat over long distances [15]. In Wahlroos et al. [16], the case study of DCs located in Nordic countries is analyzed because in these regions the heat generated by DCs is highly demanded by houses and offices, while due to specific climate conditions free cooling can be used for most of the year. The study concluded that the DCs can be a reliable source of heat in a local district heating network and may contribute to the costs reduction of the heat network with 0.6% up to 7% depending on the DC size. The low quality of heat (i.e., low temperature or unstable source) is identified as the main barrier in DCs' waste heat reuse in [17], thus an eight-step process to change this was proposed. It includes actions like selecting appropriate energy efficiency metrics, economic and CO_2 analysis, implementing systematic changes and investment in heat pumps. A comprehensive study of reusing the waste heat of DCs within the London district heat system is presented in [18]. In the UK, DCs consume about 1.5% of the total energy consumption, most of this energy being transformed in residual heat, which, with the help of heat pumps that can boost its temperature, can be used in heating networks leading to potential carbon, cost, and energy savings. An in-depth evaluation of the quantity and quality of heat that can be captured from different liquid-based cooling systems for CPUs (Central Processing Units) on blade servers is conducted [19]. The study determines a relation between CPU utilization, power consumption and corresponding CPU temperatures showing that CPUs running at 89 °C would generate high quality heat to be reused. In [20] the authors propose a method to increase the heat quality in air-cooled DCs by grouping the servers in tandem to create cold aisles, hot aisles and super-hot aisles. Simulations showing a decrease in DC energy consumption with 27% compared to a conventional architecture and great potential for obtaining high quality heat to be reused.

Nowadays the DCs have started to use heat pumps to increase the temperature of waste heat, making the thermal energy more valuable, and marketable [21]. With the help of heat pumps, the heat generated by servers can be transferred to heat the water to around 80 degrees Celsius, suited for longer distance transportation in the nearby houses. In [22] the authors evaluate the main heat reuse technologies, such as plant or district heating, power plant co-location, absorption cooling, Rankine cycle, piezoelectric, thermoelectric and biomass. In [23] a study of ammonia heat-pumps used for simultaneous heating and cooling of buildings is conducted. They are analyzed from a thermodynamic point of view, computing the proper scaling with the building to be fit in, showing great potential. The authors of [7] present an infrastructure created to reuse the heat generated by a set of servers running chemistry simulations to provide heat to a greenhouse located within the Notre Dame University. It consists of an air recirculation circuit that takes the hot air from the hot aisle containment and pumps it in the space needed to be heated. The main issue with this approach is that the heat is provided at the exhaust temperature of the servers is relatively low making it unsuited for long distance transportation. In [24,25] thermal batteries are attached to the heat pumps to increase the quality of reused heat. The thermal battery consists of two water tanks, one for storing cold water near the evaporator of the heat pump, and one for storing hot water near the heat exchanger of the heat pump. The two tanks allow the heat pump to store either cool or heat. The authors conclude that such approach proves to be more economically efficient than traditional electrical energy storage devices. In [26] an analytical model of a heat pump equipped with thermal tanks is defined and used to study the thermodynamics of the system within a building. The model was used to simulate the behavior of a building during a 24-h operational day, analyzing the sizes of the storage tanks to satisfy the heating and cooling demands. A set of measurable performance indices such as Supply/Return Heat Index, Return Temperature Index and Return Cooling Index are proposed in [27–29] to measure the efficiency of the server rack thermal management systems. They conclude that the rack's thermal performance depends on its physical location within the server room and the location of air conditioning units, and that cold aisle containment increases the energy efficiency of the DC.

Thermal processes simulations within the server room are essentials for increasing the DC's thermal flexibility without endangering the operation of servers. It is well known that by increasing the temperature in the server room of a DC, hot spots might appear near highly utilized equipment, leading to emergency shutdown or total damage of this equipment [30,31]. Thus, management strategies need to cope with these situations by simulating in advance workload deployment scenarios and avoiding dangerous situations. CFD simulation techniques are traditionally used for evaluating the thermal performance DCs [32–34]. Using CFD software tools (e.g., OpenFoam [35], BlueTool [36], etc.), a detailed infrastructure of the DC with air flows and temperature distributions can be simulated [37]. However, the main disadvantage of these kind of tools is the time overhead, most of them being unsuitable for real-time management decisions that can arise in a dynamic DC environment. Depending on the complexity of the simulated environment, a CFD simulation can take from few minutes up to a few hours and can consume a lot of hardware resources. Techniques to reduce the execution time are developed and are classified into white-box or black-box approaches. The white-box simulations are based on a set of equations that describe the underlying thermal physical processes which are parametrized with real or simulated data describing the modeled DC [3,38–41]. The black-box simulations use machine learning algorithms which are trained to predict the thermal behavior of the DC of without knowing anything about the underlying physical processes [25,29,42–44].

Our paper builds upon existing state of the art proposing a set of optimization techniques for DCs equipped with heat reuse technology addressing the problem of waste heat reuse by exploiting the thermal energy generation of servers within the server room without endangering their operation. The proposed solution combines existing cooling system models with workload allocation strategies and heat reuse hardware to explore the potential of DC thermal energy flexibility for providing heat to nearby neighborhoods. Our paper goes beyond the state of the art by proposing a black-box time-aware model of the DC thermodynamic behavior that can compute the evolution of the temperatures within the server room for future time intervals. A simulation environment based on a CFD tool is proposed to generate training data for the neural network within the black box model. Once defined and trained, the neural network is integrated into a proactive DC management strategy that can control both workload allocation and cooling system settings, to adapt the heat generation to meet various heating demands and to avoid the hot spots forming.

3. DC Workload and Heat Generation Model

In this section, we introduce a mathematical model to relate the DC workload and associated electrical energy consumption with the generated thermal energy that must be dissipated by the cooling system (waste heat), as shown in Figure 1. The model's objective is to provide the support necessary for the DC infrastructure management software in taking workload scheduling decisions as well as cooling system operation optimization decisions such that the DC thermal profile is adapted to meet the heat demand coming from nearby neighborhoods. The model is based on and enhances the previous one proposed by us in [45], this time focusing on the thermo-electrical model of the server room and energy dynamics by providing mathematical formalisms that relate the deployment of the VMs on the servers with the electrical energy consumption and thermal energy generation of the IT equipment. Furthermore, in this paper the relations between the IT equipment heat generation and cooling system thermal energy dissipation capabilities are presented, also considering the impact of the volume and temperature of the input airflow in the case of an air-cooled server room.

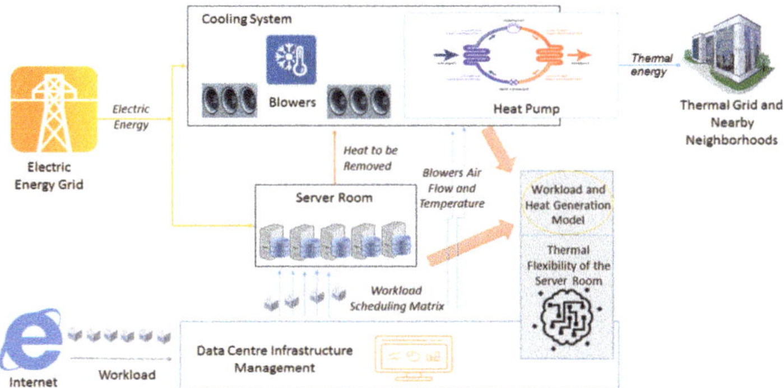

Figure 1. Data Center (DC) workload, electric and thermal energy exchange.

To execute the clients' workload, the IT equipment from the server room consumes electrical energy which is transformed into heat that must be dissipated by the cooling system. If the DC is equipped with heat reuse technology, the heat pump can perform both cooling of the server room and waste heat reuse. We consider a cloud-based DC hosting applications for its clients in VMs. When the number of user requests for an application increases, the cloud platform load balancer automatically scales up or down the resources allocated to VMs. Thus, the workload is modeled as a set of VMs each having virtual resources allocated:

$$\begin{cases} Workload = \{VM_1, VM_2, \dots, VM_M\} \\ VM_j = (CPU_j, RAM_j, HDD_j), \quad 0 \le j \le M \end{cases} \tag{1}$$

The server room of the DC to be composed of set of N homogeneous servers defined according to relation (2) the main hardware resources of the server being the number of CPU cores, the RAM and the HDD capacity.

$$\begin{cases} Server_{Room} = \{Server_1, Server_2, \dots, Server_N\} \\ Server_i = (CPU_i, RAM_i, HDD_i), \quad 0 \le i \le N \end{cases} \tag{2}$$

We define an allocation matrix of the VMs on servers, $\epsilon R^{N * M}$, where the element $S[Server_i][VM_j]$ is defined according to relation (3) and specifies if VM_j is allocated to server $Server_i$:

$$S[Server_i][VM_j] = \begin{cases} 1, \text{ if } VM_j \text{ is allocated to } Server_i \text{ during time window } T \\ 0, \text{ otherwise} \end{cases} \tag{3}$$

Using the S matrix, we can compute the utilization rate of server $Server_i$ over the time window of length T by summing the hardware resources used by the VMs allocated on this server. For example, in case of the CPU it is denoted as μ_{server_i}, and calculated using the following relation:

$$\mu_{server_i} = \frac{\sum_{j=1}^{M} S[i][j] \times CPU_{VM_j}}{CPU_{Server_i}} \tag{4}$$

Knowing the utilization of the server, one can approximate the power consumed by the server using Equation (5) proposed in [3], that defines a linear relationship between the server utilization, the idle power consumption P_{idle} and the maximum power consumption of the server P_{max}:

$$P_{server}(\mu_{server}) = P_{idle} + \mu_{server} \times (P_{max} - P_{idle}) \tag{5}$$

The *IT* equipment energy demand over the time interval T can be computed as:

$$E_{IT}(T) = \int_{t}^{t+T} P_{IT}(t)dt \ \text{ where } P_{IT} = \sum_{i=1}^{n} P_{server}(\mu_{server_i}) \tag{6}$$

Most of the electrical energy consumed by the servers is converted to thermal energy that must be dissipated by the electrical cooling system for keeping the temperature in the server room below a defined set point:

$$Heat_{Removed}(t) = E_{IT}(t) \tag{7}$$

The electrical cooling system is the largest energy consumer of the DC after the *IT* servers. Following our assumption of being equipped with heat reuse pumps, the cooling system must extract the excess heat from the server room and dissipate it into the atmosphere. By operating in this way there are two sources of energy losses: first the electrical energy needed to extract the heat from the server room and second the thermal energy dissipated. The main idea to increase efficiency is to use the heat pump to capture the excess heat and reuse it for nearby neighborhoods.

The heat pump operation is modeled leveraging two main parameters [8,24–26]: the COP for cooling, denoted as $COP_{cooling}$ and the COP for heating, denoted as $COP_{heating}$. The first parameter is calculated as the ratio between the heat absorbed by the heat pump, denoted $Heat_{Removed}(t)$, and the work done by the compressor of the heat pump to transfer the thermal energy, in this case, represented by the electrical energy consumed by the heat pump compressor denoted as $E_{cooling}(t)$:

$$COP_{cooling} = \frac{Heat_{Removed}(t)}{E_{cooling}(t)} \tag{8}$$

The second parameter, the COP for heating, the relation between the heat produced by the heat pump and supplied in the nearby neighborhoods, $Heat_{Suplied}(t)$, and the electrical energy consumed by the heat pump to extract the heat, $E_{cooling}(t)$.

$$COP_{heating} = \frac{Heat_{Supplied}(t)}{E_{cooling}(t)} + 1 \tag{9}$$

Using the thermodynamics law of heat capacity, the heat removed from the server room in a specific moment of time t can be computed as the product between the airflow through the room, $air_{flow}(t)$, the specific heat of air, c_{air}, and the difference between the temperature of the air entering the room, $T_{IN}(t-dt)$, and exiting the room, $T_{OUT}(t)$, where dt is the time needed by the air to pass through the room:

$$Heat_{Removed}(t) = air_{flow}(t) \times c_{air} \times [T_{OUT}(t) - T_{IN}(t-dt)] \tag{10}$$

The air traveling through the server room takes the heat from the servers, increasing its temperature from $T_{IN}(t-dt)$ to $T_{OUT}(t)$ during a period dt. If during this interval, the heat removed by the air passing through the room is equal to the electrical energy consumed by the servers during this interval, then, according to relation (7) the temperature inside the room remains constant. Otherwise, the temperature inside the room can fluctuate, increasing or decreasing.

It is difficult to determine a relation between the input and output airflow temperature on one side and the server room temperature on the other side in different interest points because the thermodynamics of the room must be considered. Consequently, we have defined a function T_{SIM} that takes as inputs the set of servers in the server room ($Server_{Room}$), the heat produced by each server, H_{Server_i} (considered equal to the energy consumed), the input airflow and temperature T_{IN}, air_{flow} and computes the output temperature of the air leaving the room, T_{OUT}, and the temperatures in critical points of the server room, $T_{Server_i}^{Surround}$, after a time period t:

$$T_{SIM}(Server_{Room},\ H_{Server_i},\ T_{IN},\ air_{flow}, t) \rightarrow (T_{OUT},\ T_{Server_i}^{Surround}) \tag{11}$$

For a given configuration of the DC and setting of the cooling system, the T_{SIM} function computes the temperatures in the room after a time t. To function properly, the temperatures of the critical DC equipment must not exceed the manufacturer threshold:

$$T_{Server_i}^{Surround} < Threshold_{Server} \forall\, i \in \{1 \dots N\} \tag{12}$$

Because the aim of this paper is to define a proactive optimization methodology for DC waste heat reuse on-time window of length T, we choose to construct a discrete model of the thermodynamic processes described above and to split the time window $[0 \dots T]$ in T equidistant time slots. Consequently, the values of the control variables of the cooling system equipped with heat reuse pumps as well as the workload allocation matrix will be computed. The optimization problem is defined in Figure 2.

Input:

$\qquad Server_{Room} = \{Server_1, Server_2, \dots Server_N\}$

$\qquad Workload = \{VM_1, VM_2, \dots, VM_M\}$

$\qquad Heat_{Demand}[T], T$ the heat demand time interval

Output (control variables):

$\qquad S[Server_i][VM_j],\ air_{flow},\ T_{IN}$ for each $t \in T$

Objective function:

$\qquad \min \text{Distance}(Heat_{Demand}, Heat_{Supplied})$

Given the constraints:

C1: $E_{DC}(t) = E_{IT}(t) + E_{Cooling}(t)$

C2: $E_{IT}(t) = \int_t^{t+1} P_{IT}(x)\, dx$

C3: $E_{Cooling}(t) = \dfrac{Heat_{Removed}(t)}{COP_{cooling}} = \dfrac{E_{IT}(t)}{COP_{cooling}}$

C4: $Heat_{Removed}(t) = air_{flow}(t) \times c \times [T_{OUT}(t) - T_{IN}(t)]$

C5: $Heat_{Supplied}(t) = (COP_{Heating} - 1) \times \dfrac{Heat_{Removed}(t)}{COP_{Cooling}}$

C6: $T_{SIM}(Server_{Room}, H_{Server_i}, T_{IN}, air_{flow}, t) \rightarrow (T_{OUT}, T_{Server_i}^{Surround})$

C7: $T_{Server_i}^{Surround} < Threshold_{Server} \forall\, i \in \{1..N\}$

C8: $P_{IDLE} \leq P_{Server}(\mu_{Server_i}) \leq P_{MAX}$

C9: $T_{ROOM}^{MIN} < T_{OUT}(t) < T_{ROOM}^{MAX} \forall\, t \in \{1..T\}$

C10: $T_{COOLING}^{MIN} < T_{IN}(t) < T_{COOLING}^{MAX} \forall\, t \in \{1..T\}$

C11: $air_{flow}^{MIN} < air_{flow}(t) < air_{flow}^{MAX} \forall\, t \in \{1..T\}$

Figure 2. DC heat generation optimization problem for reusing in nearby neighborhoods.

The inputs are represented by the DC server room configuration, the workload that has to be deployed during and executed during the time window $[0 \dots T]$ as well as the heat demand profile $Heat_{Demand}[T]$ (i.e., requested by the neighborhood) that has to be supplied $Heat_{Supplied}$ by the DC computed using the model presented above. The outputs of the optimization problem are the scheduling matrix VMs to the servers, the airflow and air temperature of the cooling system for each time slot of the optimization window $[0 \dots T]$.

The defined optimization problem is a Mixed-Integer-Non-Linear Program (MINLP) [46] due to the nonlinearities of the objective function and the integer variables of the scheduling matrix. It can be shown that the resulting decision problem is NP-Hard [47], thus we have used a multi-gene genetic algorithm heuristic to solve the optimization problem, similar with the one presented by us [48]. The solution of the algorithm is represented by three chromosomes corresponding to the three unknowns of the optimization problem described in Figure 2. The chromosome corresponding to the S workload scheduling matrix is an integer vector of size M, equal to the number of VMs. The chromosomes corresponding to the air_{flow} and T_{IN} variables are two real number arrays of length T, equal to the length of the optimization time window. As a result, we deal with a total number of M integer and $2 \times T$ real unknowns. The fitness function is defined as the Euclidean distance between the heat profile of the DC corresponding to a solution and the heat demand. When evaluating each solution, the thermal profile of the DC is computed during the time window $[0 \ldots T]$ by calling the T_{SIM} function to compute the server temperatures. Thus, the run-time complexity of the genetic algorithm heuristic is:

$$O(No_{iterations} \times Population_{size} \times (T \times O(T_{SIM}) + T + M) \tag{13}$$

where $No_{iterations}$ represents the number of iterations of the genetic a algorithm and $Population_{size}$ the size of chromosomal population. The T_{SIM} function implementation and time complexity is detailed in Section 4.

4. Thermal Flexibility of Server Room

An important component in adapting the DC thermal energy profile to meet a certain demand is the thermal energy flexibility of server room. Our objective is to estimate the available thermal energy flexibility considering different schedules of the workload to be executed and different settings of the cooling system (i.e., T_{SIM} function behavior). This will allow us to obtain a high server utilization ratio and increase for limited time periods the temperature set points in the server room (i.e., allowing heat to accumulate) while being careful not to endanger the proper operation of IT equipment (i.e., servers' temperature thresholds $T_{Server_i}^{Surround}$). Moreover, this will optimize the operation of the heat pump which will increase with more efficiency the temperature of waste heat leaving the server room, T_{OUT}, (i.e., due to higher temperature baseline) making the thermal energy more valuable and marketable.

To estimate the thermal flexibility of the server room we have used CFD techniques to simulate the thermodynamic processes in the server room in relation to workload allocation on servers. We have considered a generic configuration of the server room as the one presented in Figure 3.

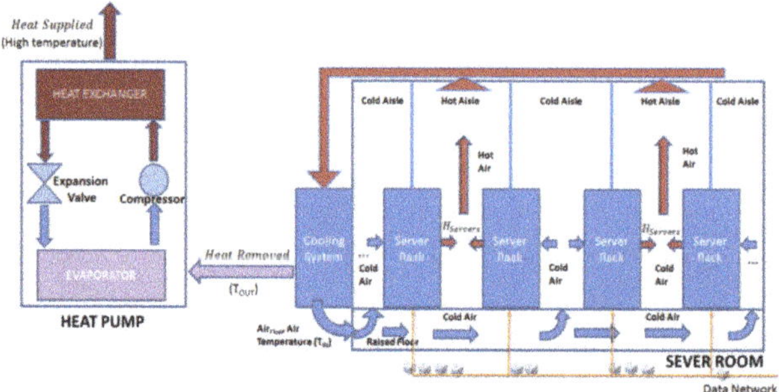

Figure 3. Server room configuration used in Computational Fluid Dynamics (CFD)-based thermal flexibility estimation.

The server room is modeled as a parallelepiped with physical dimensions defined in tri-dimensional space expressed in meters (length, width, and height), number of homogenous blocks in which the server room volume will be divided, used for computing the simulation parameters, initial temperature at the beginning of the simulation, air substance characteristics and input airflow velocity vector and temperature. The solver will calculate the physical parameters of the simulation in each defined homogeneous block, the simulation accuracy depending on the size and number of these blocks. If they are smaller, the simulation is more accurate but takes longer to finish the calculations. Inside the server room, the server racks are defined as parallelepipeds with coordinates in tri-dimensional space having the reference frame the server room (*X*-coordinate, *Y*-coordinate, *Z*-coordinate), physical dimensions of each server rack expressed in meters (length, width and height), and average temperature of each rack due to server loads (H_{Server}).

Cold air is pumped in the room through raised floor perforated tiles and hot air is extracted through the ceiling. The heat reuse and cooling of the server room is performed using a heat pump similar to the solution presented in [9] featuring coefficients of performance for cooling and heating. The heat pump transfers the heat absorbed from the DC server room to the thermal grid by using a refrigerant-based cycle to increase its temperature and consuming electrical energy. The cooling system parameters are the input air velocity vector, defined as a vector in tri-dimensional space showing the direction and the speed of the air, the origin of the air velocity vector defined as a position in the server room space, and finally the temperature of the air supplied in the server room.

The control variables of the simulation are the two cooling system parameters (input airflow air_{flow} and temperature T_{IN}) and the matrix of VMs scheduling on servers (*S*) from which the rack temperature is derived. From a thermodynamic point of view, the system operates as follows: cold air enters the cold aisle of the server room through the floor perforated tiles, passes through the racks, removing the heat generated by the electrical components. The hot air is eliminated behind the racks in the warm aisle. Due to its physical properties, the hot air rises, being absorbed by the ceiling fans and redirected to the main cooling unit containing cold water radiators that take over the heat from the air. The excess thermal energy from the air is passed to the water from the radiators that increases its temperature, while the air lowers its temperature and is pumped back to the room. The warm water is sent to the evaporator part of the heat pump that extracts the excess heat, lowering its temperature. The cold water is then pumped again back to the radiators, while the thermal energy extracted is transferred to the heat exchanger of the heat pump that heats water at the other end to temperatures as high as 80 degrees Celsius. Figure 4 presents the server room thermodynamic processes simulation view using OpenFoam [35].

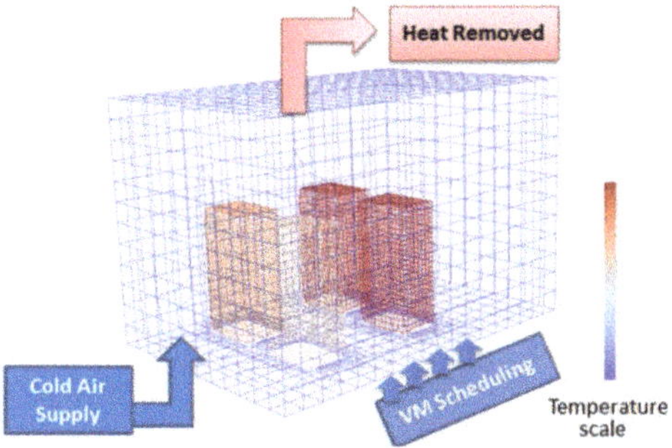

Figure 4. Server room thermal flexibility simulation.

The advantage of using such a CFD-based simulation for determining the temperatures in the server room is the accuracy of the results, with the main drawback being the long execution time that makes it unsuitable for real-time iterative computations needed when solving the optimization problem described in Section 3. Thus, we have used deep neural networks-based technique to learn the T_{SIM} function from a large number of samples, while using the OpenFoam CFD-based simulation to generate the training dataset (see Figure 5).

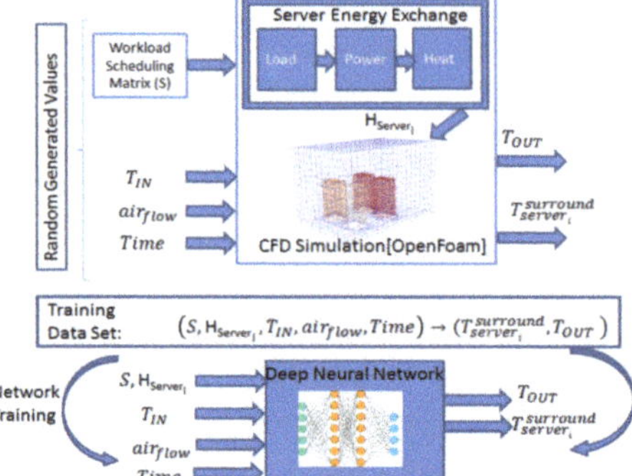

Figure 5. Estimating server room thermal flexibility-CFD simulation and deep neural networks.

We have chosen the deep neural networks because they provide low time overhead during runtime (i.e., after training) and because the relations between the inputs and the outputs are non-linear and highly complex. The inputs of the neural network is represented by: (i) the matrix $S \in R^{N*M}$ representing the VMs allocation on servers together with the array H_{Server_i} representing the estimated heat generated by each of the N servers in the server room as result of executing the allocated workload; (ii) the airflow (air_{flow}) and temperature of the air fed (T_{IN}) into the room by the cooling system and (iii) the time interval for thermal flexibility estimation. From a structural point of view, the network used is a fully-connected one having $N + 3$ inputs and $N + 1$ outputs and two hidden layers. The neurons use Rectified Linear Unit (ReLU) activation functions and for training the network the ADAM optimizer [49] is used with the Mean-Square-Error (MSE) loss function.

The time complexity of the T_{SIM} function is different in the two implementation cases. In the first case when CFD is used for implementation, the time complexity depends on the number of iterations of the simulation process and the complexity of an iteration. At its turn, the complexity of an iteration depends on several parameters, the most important being the accuracy of the surfaces modeled, while the number of iterations varies according to the convergence time of the model. In our implementation, a step of the simulation took about 1.5 s, and simulations converged in 100 up to 500 steps, so the simulation time could take up to 12 min. In the second case when neural networks have been used the complexity depends on the number of neurons of the network. However, the neural network call from Java environment was measured to be between 1 and 2 ms, with an average of 1.73 ms which represents a significant improvement in comparison to the CFD implementation.

5. Validation and Results

We have conducted numerical simulation-based experiments to estimate the potential of our techniques for exploiting the DCs thermal energy flexibility in reusing waste heat in nearby neighborhoods. The simulation environment is leveraging on a nonlinear programming solver to deal with the DC heat generation optimization as presented in Figure 2. For estimating the thermal flexibility of the server room the CFD-based techniques for simulating the thermodynamic processes involved were implemented using OpenFoam Open Source CFD Toolbox. It provides a set of C++ tools for developing numerical-based software simulations and utilities for solving continuous mechanics problems, including CFD related ones. Among the many solvers available in OpenFoam libraries we choose the BouyantBoussnesqSimpleFoam solver, which is suitable for heat transfer simulations and uncompressible stationary flow. The OpenFoam simulation process involves the following steps: (i) setting the parameters of the physical process underlying the simulation; (ii) setting the computational and space parameters of DC server room used in simulation and (iii) unning the simulation and add the results to the neural network training data set. The neural network was implemented in Keras [50] with a Tensorflow [51] backend and was loaded in a Java application containing the solver for the optimization problem. For loading the network in Java, the Deeplearining4J framework [52] was used, that converts a JSON model of the network exported from Keras to the Java implementation.

In our experiments, we have modeled a DC server room of 6 m long, 4 m width and 4 m height, having a surface of 24 square meters and a volume of 96 cubic meters. The server room contains 4 standard 42U racks of servers (see Table 1) with the following dimensions: width −600 mm and depth-1000 mm, with both front and rear doors. The racks contain 10U Blade Systems servers with full-height and half-height blade servers featuring multiple power sources: 2U servers with 2 redundant power supplies or 1U servers with a single power supply.

Table 1. Server Room Equipment Description.

Rack Type	Total Rack Idle Power [W]	Total Rack Max Power [W]	# Servers	Server Type	Server Idle Power [W]	Server Max Power [W]	# CPU Cores
HP Bladesystem C7000	2000	3000	10	IBM Blade	200	300	4
IBM Bladecenter S	4000	6000	20	IBM Blade	200	300	4
HP Bladesystem C7000	2550	4350	15	HP DL 360 G7	170	290	4
IBM Rack	5200	7200	20	IBM X3650 M4	260	360	4

The considered electrical cooling system (see Table 2) is based on indirect free cooling technology and consists on a set of redundant external chillers. Inside the server room, cooling is done by pumping cold air through the raised floor perforated to the server racks. The server racks are grouped in cool aisles, as shown in Figure 3.

Table 2. Electrical cooling system characteristics.

Energy Consumption	Heat Generation	$COP_{cooling}$	$COP_{heating}$	Min Air_{flow}	Max Air_{flow}	Min T_{IN}	Max T_{IN}
5 kWh	7.5 kWh	3.8	2.3	~0.6 kg/s	~6 kg/s	14 °C	30 °C

The server racks are rear-to-back, being isolated from each other. Cold air enters a lane through the slopes of the floor, passes through the racks, taking over the heat of the electrical components. The hot air is discharged behind the racks in the warm corridor. Due to its physical properties, the hot air rises, being absorbed by the ceiling fans and redirected to the main cooling unit containing cold water radiators that take over the heat from the air as well as electric compressors used to cool the water. The heat reuse and cooling of the server room is performed using a heat pump having the characteristics presented in Table 3. There are two heat exchanges on the server room side of the cooling process: heat exchange between hot air extracted from the server room by the cooling system and cold water inside the radiators and heat exchange between cold air that is pumped into the room and IT servers to be cooled.

Table 3. Heat exchange circuits characteristics.

Cooling System	Cooling Capacity (Operational Level)	24 kWh (Normal Operation)	4 kWh (Minimum)	48 kWh (Maximum)
	Water Flow	4120 L/h	687 L/h	8240 L/h
Water Circuit	Input Air Temperature	10 °C	10 °C	10 °C
	Output Air Temperature	15 °C	15 °C	15 °C
	Airflow	9340 m³/h	1600 m³/h	18,680 m³/h
Air Circuit	Input Air Temperature	17 °C	17 °C	17 °C
	Output Air Temperature	24 °C	24 °C	24 °C
	Air Mass	~3.18 kg/s	~0.6 kg/s	~6.5 kg/s

To train the neural network, a data set was generated by running 15,000 OpenFoam simulations considering different values of the input parameters such as workload scheduling matrix S, input airflow and temperature (T_{IN} and air_{flow}). The simulation results were generated as tuples of the form $< Timestamp, < T_{OUT}, T_{rack-1}, T_{rack-2}, T_{rack-3}, T_{rack-4} > >$, where *Timestamp* is the simulation time instance when results are generated, T_{OUT} represents output temperature of the air leaving the room, and $T_{rack-1}, T_{rack-2}, T_{rack-3}, T_{rack-4}$ represent the surrounding temperatures of the 4 racks of server. Thus, a time log showing the progress of the server room temperatures is created containing more than 50,000 different data samples. The generated data set, together with the values of OpenFoam simulation input parameters is converted into a format accepted by the neural network, being split in 80% training data samples and 20% as test data samples for the neural network. When evaluating the precision of the temperature predictions using the trained neural network against a number of reference OpenFoam simulations we obtain the mean square error (MSE) and mean absolute percentage error (MAPE) values shown in Table 4. They are computed using relations (14) and (15) where: n is the number of samples considered, $Temp_{Reference}$ is the reference temperature of one of the parameters generated from OpenFoam simulations (i.e., T_{OUT}, $T_{rack-1}, T_{rack-2}, T_{rack-3}, T_{rack-4}$, etc.) and $Temp_{Predicted}$ is the temperature predicted by our neural network model.

$$MSE = \frac{1}{n} \times \sum_{i=1}^{n} \left(Temp_{Reference} - Temp_{Predicted} \right)^2 \tag{14}$$

$$MAPE = \frac{100}{n} \times \sum_{i=1}^{n} \frac{|Temp_{Reference} - Temp_{Predicted}|}{Temp_{Reference}} \tag{15}$$

Table 4. Temperature Prediction based on Neural Network Model Evaluation.

Metric	T_{out}	T_{rack-1}	T_{rack-2}	T_{rack-3}	T_{rack-4}	T_{out}
MSE	0.19	2.21	0.08	0.53	0.11	0.11
MAPE	1.48%	3.26%	0.93%	2.26%	1.02%	1.05%

The temperatures are measured on the Celsius scale and vary from 18 to 60 degrees, thus an error of 2% means a temperature miss prediction of about 1 degree Celsius. As it can be noticed from Table 4, the mean absolute error percentage of the predictions is less than 2% for the rack temperature prediction, meaning that the temperatures are predicted with an accuracy of 1 degree Celsius for the racks. The prediction error is the cost of cutting down the OpenFoam simulation time to less than 2 milliseconds needed to predict the temperatures using neural networks, being a reasonable tradeoff in case of DC optimization where multiple evaluations have to be carried out fast to determine an optimal parameter setting for DC operation.

5.1. Workload Scheduling and Temperature Set Points

This section evaluates the relation between the DC workload scheduling on servers (S allocation matrix control variable defined in Section 3), heat generation in the server room and temperature set points. We aim to determine workload deployment strategies on servers which will allow us to maximize the average temperature in the server room without impacting the IT server operation through overheating (i.e., hot spots avoidance). To estimate the heat generated by the servers as result of workload execution we have used the relation between CPU (Central Processing Unit) usage and temperature presented in Table 5. The values are reported in the Processors Thermal Specifications sheets [53–55].

Table 5. CPU Usage vs. Temperature.

Server Type	Processor Family	P_{MAX} [W]	Temperature vs. Power Relation	Typical Thermal Profile
IBM Blade	Intel Xeon E5-2600	135 W	$CPU_{Temp} = 0.205$ $*CPU_{Power} + 48.9$	
HP DL 360 G7	Intel Xeon 5600	130 W	$CPU_{Temp} = 0.190$ $*CPU_{Power} + 55.7$	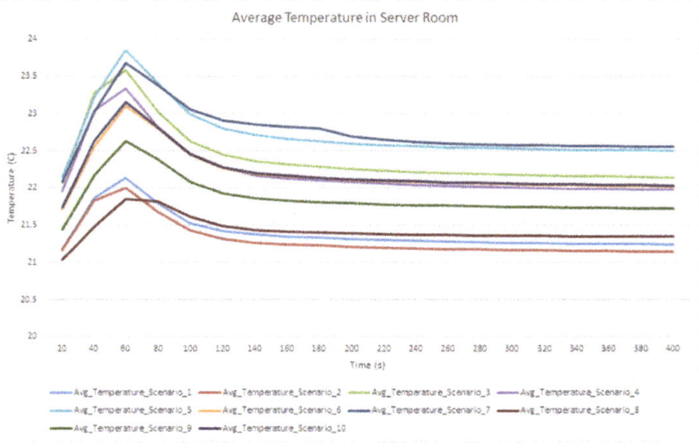
IBM X3650	Intel Xeon X5400	120 W	$CPU_{Temp} = 0.221$ $*CPU_{Power} + 43.5$	

We have defined and used a set of 10 simulation scenarios. The cooling system is set to pump air in the room at a speed of 0.1 m/s and a volume of 3 kg/s at a temperature set point of 20 °C. The initial room temperature is also 20 °C. A set of 50 up to 65 synthetic VMs having 1 CPU core, 1 GB of RAM and 30 GB for HDD must be deployed in the virtualized environment running on the server racks. Each VM will run a synthetic workload that will keep its CPU between 80% and 100%.

Table 6 matrices of the VMs allocation on servers have been considered using two main workload scheduling strategies: either group the VMs on servers leading to an increased CPU usage (>50%) and increased CPU temperatures (Scenarios 1–3) or spread the VMs among more servers within each rack and have a lower CPU usage (<40%) and a lower CPU temperature (Scenarios 4–10). After simulating each workload scheduling configuration for 400 s, enough time to reach a thermal equilibrium for most scenarios, the average and maximum temperatures are plotted in Figure 6, respectively Figure 7.

Table 6. Workload Deployment in the DC and corresponding CPU usages.

Scenario	Rack 1		Rack 2		Rack 3		Rack 4		Total VMs Workload
	VMs	CPU Usage	VMs	CPU Usage	VMs	CPU Usage	VMs	CPU Usage	
1	0	0	64	80	0	0	0	0	64
2	0	0	56	70	0	0	0	0	56
3	20	50	0	0	30	50	0	0	50
4	0	0	32	40	24	40	0	0	56
5	12	30	24	30	18	30	0	0	54
6	10	25	20	25	0	0	20	25	50
7	10	25	20	25	15	25	20	25	65
8	8	20	16	20	12	20	16	20	52
9	8	20	32	40	12	20	0	0	52
10	0	0	24	30	12	20	24	30	60

Figure 6. Average temperature variation in server room.

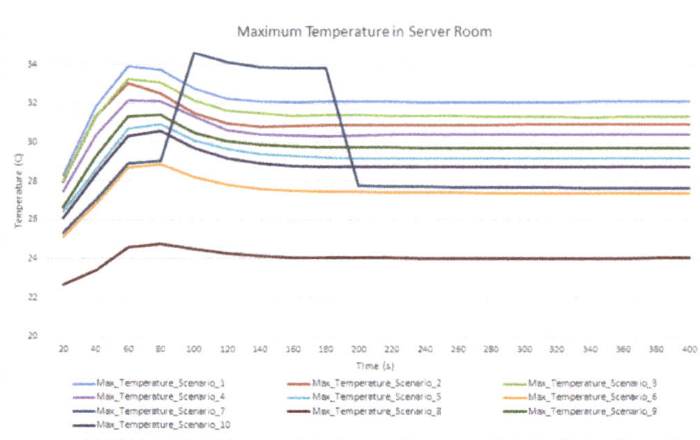

Figure 7. Maximum temperature variation in the server room.

As it can be seen from Figure 6, the highest average temperature in the server room is reached in the Scenarios 5–8, which disperse the workload among more servers with a lower utilization. The allocation technique that consolidates the workload on servers with high CPU usage, such as scenarios 1 and 2 show a smaller average temperature in the server room. Furthermore, by analyzing Figure 7 that displays the peak temperature in the server room, the situation is reversed. The deployment strategies in which the workload is consolidated on servers with high CPU usage exhibit the largest peak temperatures while the scenarios where the workload is spread on more servers with lower CPU usage show a lower maximum temperature in the server room. Even if at first sight these results are counter-intuitive, they can be explained. By distributing the workload on more servers with lower CPU usage, more thermal energy is released, because according to Equation (5) from Section 3, a server consumes electrical energy even in idle state, energy that will be converted in thermal energy which in turn will lead to a temperature increase in the server room. Thus, when the workload is consolidated on the minimum number of servers that run at high CPU usage, less electrical energy is consumed, and thus less thermal energy is eliminated in the room, a situation illustrated by a lower temperature in these scenarios. However, in these cases, because the energy is concentrated in a small number of points in the server room, hot spots might occur, as illustrated in Figure 7 by higher temperatures achieved by deployment strategies that overload the servers to use fewer resources.

In conclusion to transform the DCs in active thermal energy players which may provide heat in nearby neighborhoods the deployment strategies that maximize the average room temperature and minimize the maximum room temperatures are desired, for two reasons. Firstly, by having a higher average temperature in the server room, the exhaust air from the room is hotter and more heat can be recovered for heat reuse. Secondly, by minimizing the peak temperatures from the room, hotspots can be easily avoided in case of cooling system adjustments, such as pre- and post-cooling presented in the following section.

5.2. DC Thermal Profile Adaptation

This section aims at evaluating the DC capability of adapting its thermal energy generation to match a given heat demand by using only the cooling system control variables: input air temperature (T_{IN}) and input airflow (air_{flow}). By adjusting these two variables, the temperature in the server room can be increased for limited time periods allowing the DC to reuse the heat generating more efficiently through the heat pump and provide it to the nearby neighborhood. We have created four scenarios with a different combination of control variables adjustment (see Table 7). For each of them we have evaluated the DC thermal flexibility over a period of two hours, with a five-minute time step, considering the same workload allocation matrix on servers (about 25% of the servers used).

Table 7. Scenarios defined and control variables settings.

Scenarios	Heat Demand Matching (Before Optimization)	Control Variable		Heat Demand Matching (After Optimization)
		T_{IN}	air_{flow}	
Post-cooling Scenario 1	63%	Constant at 20 °C	First decrease and then increase	98%
Post-cooling Scenario 2	79%	First increase and then decrease	Constant at 2.89 kg/s	97%
Pre-cooling Scenario 3	78%	Constant at 20 °C	First increase and then decrease	98%
Pre-cooling Scenario 4	86%	First decrease and then increase	Constant at 2.89 kg/s	99%

Scenarios 1 and 2 investigate the server room "post-cooling" mechanism as a source of thermal energy flexibility. In Scenario 1 the electrical cooling system usage is reduced for a short period of time to accumulate residual heat that can be later used to meet the heat demand. The cooling system input airflow (air_{flow}) is steadily decreased in the first 35 min, as shown by the yellow line in Figure 8-left, leading to an increase in the maximum temperature in the server room, as shown in Figure 8-right. Then, the airflow starts to steadily increase and to eliminate the excess residual heat from the server room, leading to a temperature decrease below the normal operating temperature shown by the blue line in Figure 8-right. Finally, in the last 15 min, the airflow is brought back to normal values leading to a rebalance of the server room operating temperature. By using this intelligent airflow management technique, the DC heat generation matches more than 98% the given heat demand profile.

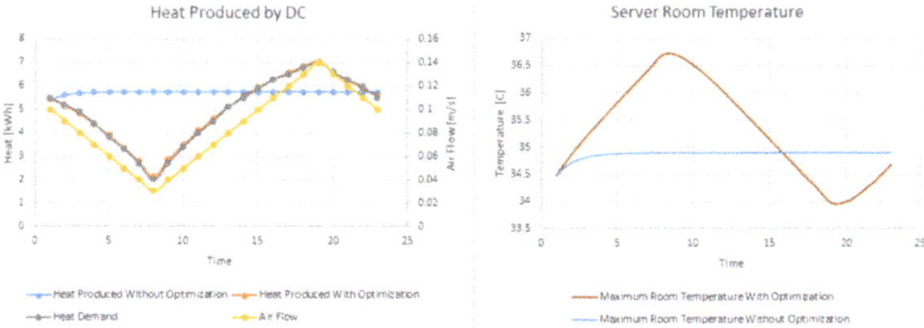

Figure 8. DC thermal profile adaptation in Scenario 1 (**left**—heat produced by the DC; **right**—server room maximum temperature).

In Scenario 2 we evaluate the DC thermal energy profile adaption by modifying the cooling system input air temperature (T_{IN}). Initially, the input airflow temperature is increased, as shown by the yellow line in Figure 9-left, a process that leads to an increase of the overall maximum server room temperature, is shown by the red line in Figure 9-right. The heat accumulates in the first hour in the server room. In the second hour, the input air temperature is steadily decreased below the normal operating set point temperature of 22 °C, shown by the blue line in Figure 9-left, increasing the amount of thermal energy removed from the server room. This thermal energy is transported to the heat pump that converts it to heat that can be reused to meet the nearby neighborhood heat demand profile with a matching degree of 97%.

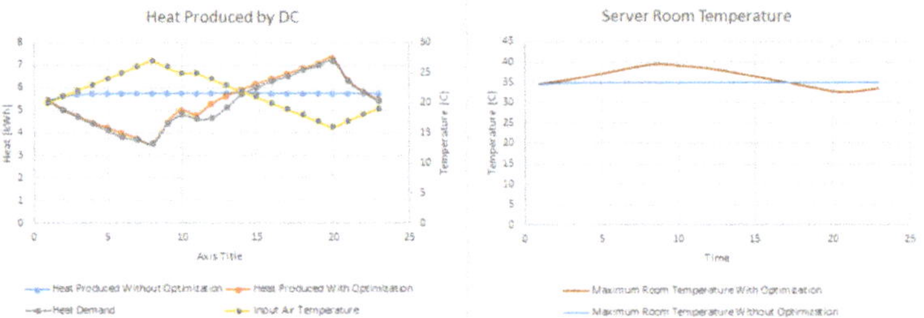

Figure 9. DC thermal profile adaptation in Scenario 2 (**left**—heat produced by the DC; **right**—server room maximum temperature).

Scenarios 3 and 4 investigate the server room pre-cooling mechanism as a source of thermal energy flexibility. In scenario 3, as shown in Figure 10-left and Figure 10-right this is achieved toggling the airflow (air_{flow}) in the server room. To meet the increased heat demand of 7 kWh in the first 30 min of the simulation, the airflow is increased accordingly leading to an excess heat removal from the server room, as shown in Figure 10-right. Next, the heat demand is decreasing to 3 kWh, less than the normal 5.5 kWh heat generation of the servers. To meet this low heat demand, the cooling system airflow is drastically decreased, thus less heat is removed from the server room and the temperature increases. Finally, the airflow is brought back to normal values bringing back the temperature from the server room to the initial values. In this case, the heat demand is matched with an accuracy of 98%.

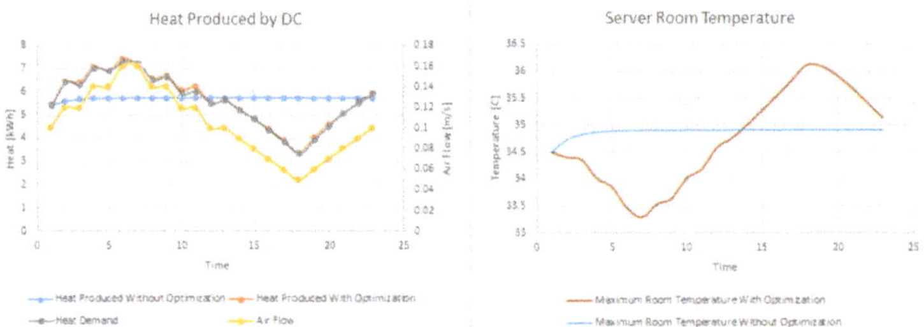

Figure 10. DC thermal profile adaptation in Scenario 3 (**left**—heat produced by the DC; **right**—server room maximum temperature).

Scenario 4 presented in Figure 11 shows the DC response to an increased heat demand by adjusting the cooling system input air temperature (T_{IN}). In the first 30 min of the response period, the heat demand increases above 7 kWh, much higher than the 5.5 kWh normal thermal generation of the servers. To collect the necessary thermal energy, the cooling system decreases the input air temperature in the server room, removing more heat than generated, leading to a decrease in the server room temperature, as shown in Figure 11-right. In the second part of the response period, the heat demand decreases, thus the input air temperature is adjusted accordingly to remove less heat from the server room. As an impact, the average server room temperature increases and exceeds the normal operating temperature by 3°. In the last 15 min of the response, the DC adjust again the input air temperature to bring back the server room to normal operating parameters. In this case, the heat demand is matched with an accuracy of 99%.

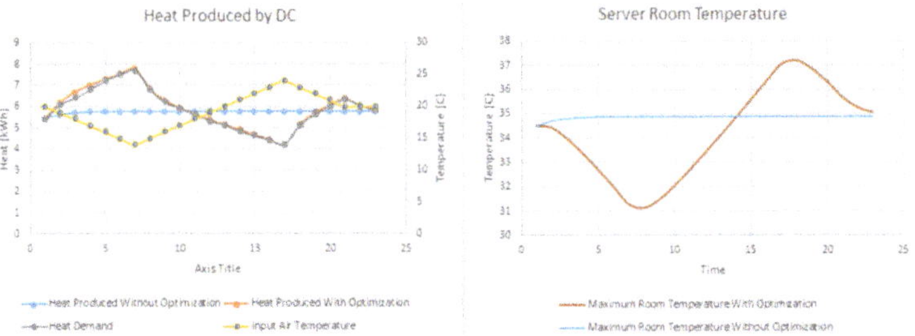

Figure 11. DC thermal profile adaptation in Scenario 4 (**left**—heat produced by the DC; **right**—server room maximum temperature).

6. Conclusions

In this paper, we have proposed techniques that will allow the DCs equipped with heat reuse technology to adapt their thermal energy profile and provide heat to nearby neighborhoods on demand. We have defined a mathematical model for describing the thermodynamics processes inside the server room and used it to decide on the optimal shifting DCs thermal energy profile to meet expected heat demands. CFD and neural networks have been employed to estimate the thermal flexibility of the DC considering the cooling system air temperature and flow level as control variables as well as workload allocation on servers to avoid hot spots. The validation conducted using the characteristics of a small DC was promising the results showing the DCs potential in acting as thermal energy players in their nearby neighborhoods. In our tests, the DC could meet the heat demand with an accuracy of over 90% in all the defined scenarios. As future work, we will investigate new business scenarios for DC operation which will allow them to cooperate and jointly exploit their thermal flexibility in providing a more stable heat supply and as result to gain new revenue streams. At the same time, we will address the case of fully decentralized DCs and supercomputers case which may provide free and efficient heating for residential buildings.

Acknowledgments: This work has been conducted within the CATALYST project Grant number 768739, co-funded by the European Commission as part of the H2020 Framework Programme (H2020-EE-2016-2017) and it was partially supported by a grant of the Romanian National Authority for Scientific Research and Innovation, CNCS/CCCDI—UEFISCDI, project number PN-III-P2-2.1-BG-2016-0076, within PNCDI III.

Author Contributions: Marcel Antal and Tudor Cioara designed the DC Workload and Heat Generation Model; Ionut Anghel and Ioan Salomie defined the thermal flexibility techniques; Claudia Pop, Marcel Antal and Tudor Cioara implemented the simulation based prototype; Claudia Pop, Tudor Cioara and Ioan Salomie designed and performed the experiments; Tudor Cioara, Ionut Anghel and Marcel Antal wrote the paper.

Conflicts of Interest: The authors declare no conflict of interest.

References

1. Five Strategies for Cutting Data Center Energy Costs through Enhanced Cooling Efficiency. Available online: http://www.emersonnetworkpower.com/documentation/en-us/brands/liebert/documents/white%20papers/data-center-energy-efficiency_151-47.pdf (accessed on 2 December 2017).
2. Data Center Environmental Control. Available online: https://en.wikipedia.org/wiki/Data_center_environmental_control (accessed on 2 December 2017).
3. Tang, Q.; Mukherjee, T.; Gupta, S.K.S.; Cayton, P. Sensor-based Fast Thermal Evaluation Model for Energy Efficient High-Performance Datacenters. In Proceedings of the Fourth International Conference on Intelligent Sensing and Information Processing, Bangalore, India, 15 October–18 December 2006; pp. 203–208.
4. Das, R.; Yarlanki, S.; Hamann, H.; Kephart, J.O.; Lopez, V. A Unified Approach to Coordinated Energy-Management in Data Centers. In Proceedings of the 7th International Conference on Network and Service Management (CNSM), Paris, France, 24–28 October 2011; pp. 1–5.

5. MacKinnon, C. Foundations of the Smart City. Available online: http://www.datacenterdynamics.com/content-tracks/core-edge/foundations-of-the-smart-city/99167.fullarticle (accessed on 10 January 2018).
6. H2020 Catalyst Project. Available online: http://project-catalyst.eu/ (accessed on 15 January 2018).
7. Brenner, P.; Go, D.B.; Buccellato, A.P.C. Data Center Heat Recovery Models and Validation: Insights from Environmentally Opportunistic Computing. In Proceedings of the ASHRAE Winter Conference Technical Program, Dallas, TX, USA, 26–30 January 2013.
8. Sarkar, J.; Bhattacharyya, S.; Ramgopal, M. Performance of a transcritical CO_2 heat pump for simultaneous water cooling and heating. *Int. J. Appl. Sci. Eng. Technol.* **2010**, *6*, 57–64.
9. Mitchell, R.L. Data Center Density Hits the Wall. Computerworld Magazine 2010. Available online: http://www.computerworld.com/article/2522601/it-management/data-center-density-hits-the-wall.html (accessed on 15 December 2017).
10. Gelažanskas, L.; Gamage, K.A.A. Forecasting hot water consumption in residential houses. *Energies* **2015**, *8*, 12702–12717. [CrossRef]
11. Department for Environment, Food & Rural Affairs (DEFRA). *Measurement of Domestic Hot Water Consumption in Dwellings*; Report; DEFRA: London, UK, 2008.
12. Parker, D.S.; Fairey, P.W. Estimating daily domestic hot-water use in North American homes. *ASHRAE Trans.* **2015**, *121*, 258.
13. Arce, I.H.; López, S.H.; Perez, S.L.; Rämä, M.; Klobut, K.; Febres, J.A. Models for fast modelling of district heating and cooling networks. *Renew. Sustain. Energy Rev.* **2018**, *82*, 1863–1873. [CrossRef]
14. Goumba, A.; Chiche, S.; Guo, X.; Colombert, M.; Bonneau, P. Recov'Heat: An estimation tool of urban waste heat recovery potential in sustainable cities. *AIP Conf. Proc.* **2017**, *1814*, 020038.
15. Waste Heat Recovery: Technology and Opportunities in U.S. BCS Incorporated Report. Available online: https://www1.eere.energy.gov/manufacturing/intensiveprocesses/pdfs/waste_heat_recovery.pdf (accessed on 15 December 2017).
16. Wahlroos, M.; Pärssinen, M.; Manner, J.; Syri, S. Utilizing data center waste heat in district heating—Impacts on energy efficiency and prospects for low-temperature district heating networks. *Energy* **2017**, *140*, 1228–1238. [CrossRef]
17. Wahlroos, M.; Pärssinen, M.; Rinne, S.; Syri, S.; Manner, J. Future views on waste heat utilization—Case of data centers in Northern Europe. *Renew. Sustain. Energy Rev.* **2018**, *82*, 1749–1764. [CrossRef]
18. Davies, G.F.; Maidment, G.G.; Tozer, R.M. Using data centres for combined heating and cooling: An investigation for London. *Appl. Therm. Eng.* **2016**, *94*, 296–304. [CrossRef]
19. Haywood, A. Modeling, Experimentation, and Analysis of Data Center Waste Heat Recovery and Utilization. Ph.D. Thesis, Arizona State University, Tempe, AZ, USA, May 2014.
20. Taniguchi, Y.; Suganuma, K.; Deguchi, T.; Hasegawa, G.; Nakamura, Y.; Ukita, N.; Aizawa, N.; Shibata, K.; Matsuda, K.; Matsuoka, M. Tandem equipment arranged architecture with exhaust heat reuse system for software-defined data center infrastructure. *IEEE Trans. Cloud Comput.* **2017**, *5*, 182–192. [CrossRef]
21. Monroe, M. How to Reuse Waste Heat from Data Centers Intelligently. Data Center Knowledge 2016. Available online: http://www.datacenterknowledge.com/archives/2016/05/10/how-to-reuse-waste-heat-from-data-centers-intelligently (accessed on 8 January 2018).
22. Ebrahimi, K.; Jones, G.F.; Fleischer, A.S. A review of data center cooling technology, operating conditions and the corresponding low-grade waste heat recovery opportunities. *Renew. Sustain. Energy Rev.* **2014**, *31*, 622–638. [CrossRef]
23. Stene, J. Design and Application of Ammonia Heat Pump Systems for Heating and Cooling of Non-Residential Buildings. In Proceedings of the 9th International IEA Heat Pump Conference, Zurich, Switzerland, 20–22 May 2008.
24. Dharkar, S.; Kurtulus, O.; Groll, E.A. Analysis of a Data Center Liquid-Liquid CO_2 Heat Pump for Simultaneous Cooling and Heating. Paper 1419. In Proceedings of the 2014 15th International Refrigeration and Air Conditioning Conference, West Lafayette, IN, USA, 14–17 July 2014.
25. Blarke, M.B.; Yazawa, K.; Shakouri, A.; Carmo, C. Thermal battery with CO_2 compression heat pump: Techno-economic optimization of a high-efficiency Smart Grid option for buildings. *Energy Build.* **2012**, *50*, 128–138. [CrossRef]
26. Houbak-Jensen, L.; Holten, A.; Blarke, M.B.; Groll, E.A.; Shakouri, A.; Yazawa, K. Dynamic Analysis of a Dual-Mode CO_2 Heat Pump With Both Hot and Cold Thermal Storage. In Proceedings of the ASME 2013 International Mechanical Engineering Congress and Exposition, San Diego, CA, USA, 15–21 November 2013; Volume 8B.

27. Nada, S.A.; Said, M.A.; Rady, M.A. CFD investigations of data centers' thermal performance for different configurations of CRACs units and aisles separation. *Alex. Eng. J.* **2016**, *55*, 959–971. [CrossRef]

28. Nada, S.A.; Elfeky, K.E.; Attia, A.; Alshaer, W. Experimental parametric study of servers cooling management in data centers buildings. *Heat Mass Transf.* **2017**, *53*, 2083–2097. [CrossRef]

29. Kumar, V.A. Real Time Temperature Prediction in a Data Center Environment using an Adaptive Algorithm. Master's Thesis, University of Texas at Arlington, Arlington, TX, USA, December 2013.

30. Paradis, P.L.; Ramdenee, D.; Ilinca, A.; Ibrahim, H. CFD modeling of thermal distribution in industrial server centers for configuration optimisation and control. *Int. J. Simul. Process Model.* **2014**, *9*, 151–160.

31. Wibron, E. CFD Modeling of an Air-Cooled Data Center. Master's Thesis, Department of Applied Mechanics Chalmers, University of Technology Gothenburg, Gothenburg, Sweden, 2015.

32. TileFlow Data Center CFD Modeling Software. Available online: http://tileflow.com/ (accessed on 2 December 2017).

33. FP7 CoolEmAll Project. Available online: http://cordis.europa.eu/project/rcn/100127_en.html (accessed on 2 December 2017).

34. CoolSim—Data Center CFD Modeling and Design Software for Airflow Modeling and Management. Available online: http://www.coolsimsoftware.com/ (accessed on 2 December 2017).

35. OpenFOAM. Available online: https://www.openfoam.com/ (accessed on 2 December 2017).

36. Gupta, S.; Varsamopoulos, G.; Haywood, A.; Phelan, P.; Mukherjee, T. *Bluetool: Using a Computing Systems Research Infrastructure Tool to Design and Test Green and Sustainable Data Centers. Handbook of Energy-Aware and Green Computing*, 1st ed.; Ahmad, I., Ranka, S., Eds.; Chapman & Hall/CRC: Boca Raton, FL, USA, 2012.

37. Patankar, S.V. Airflow and cooling in a data center. *J. Heat Transf.* **2010**, *132*, 073001. [CrossRef]

38. Jonas, M.; Gilbert, R.R.; Ferguson, J.; Varsamopoulos, G.; Gupta, S.K.S. A Transient Model for Data Center Thermal Prediction. In Proceedings of the 2012 International Green Computing Conference (IGCC), San Jose, CA, USA, 4–8 June 2012; pp. 1–10.

39. Zavřel, V.; Barták, M.; Hensen, J.L.M. Simulation of a Data Center Cooling System in an Emergency Situation. In Proceedings of the 8th IBPSA-CZ Conference Simulace Budov a Techniky Prostřed, Prague, Czech Republic, 6–7 November 2014.

40. Fouladi, K.; Wemhoff, A.P.; Llanca, L.S.; Abbasi, K.; Ortega, A. Optimization of data center cooling efficiency using reduced order flow modeling within a flow network modeling approach. *Appl. Therm. Eng.* **2017**, *124*, 929–939. [CrossRef]

41. De Boer, G.N.; Johns, A.; Delbosc, N.; Burdett, D.; Tatchell-Evans, M.; Summers, J.; Baudot, R. Three computational methods for analysing thermal airflow distributions in the cooling of data centers. *Int. J. Numer. Methods Heat Fluid Flow* **2017**, *28*, 271–288. [CrossRef]

42. Kansara, N.; Katti, R.; Nemati, K.; Bowling, A.P.; Sammakia, B. Neural Network Modeling in Model-Based Control of a Data Center. In Proceedings of the International Electronic Packaging Technical Conference and Exhibition, San Francisco, CA, USA, 6–9 July 2015.

43. Li, L.; Liang, C.J.M.; Liu, J.; Nath, S.; Terzis, A.; Faloutsos, C. ThermoCast: A Cyber-Physical Forecasting Model for Datacenters. In Proceedings of the 17th ACM SIGKDD International Conference on Knowledge Discovery and Data Mining, San Diego, CA, USA, 21–24 August 2011; pp. 1370–1378.

44. Wang, L.; Laszewsk, G. Task scheduling with ANN-based temperature prediction in a data center: A simulation-based study. *Eng. Comput.* **2011**, *27*, 381–391. [CrossRef]

45. Antal, M.; Cioara, T.; Anghel, I.; Pop, C.; Salomie, I.; Bertoncini, M.; Arnone, D. DC Thermal Energy Flexibility Model for Waste Heat Reuse in Nearby Neighborhoods. In Proceedings of the Eighth International Conference on Future Energy Systems, Hong Kong, China, 16–19 May 2017; pp. 278–283.

46. Zhang, S.; Xia, Y. Solving nonlinear optimization problems of real functions in complex variables by complex-valued iterative methods. *IEEE Trans. Cybern.* **2018**, *48*, 277–287. [CrossRef] [PubMed]

47. Belotti, P.; Kirches, C.; Leyffer, S.; Linderoth, J.; Luedtke, J.; Mahajan, A. Mixed-integer nonlinear optimization. *Acta Numer.* **2013**, *22*, 1–131. [CrossRef]

48. Antal, M.; Cioara, T.; Anghel, I.; Pop, C.; Tamas, I.; Salomie, I. Proactive Day-Ahead Data Center Operation Scheduling for Energy Efficiency. In Proceedings of the IEEE 13th International Conference on Intelligent Computer Communication and Processing, Cluj-Napoca, Romania, 7–9 September 2017.

49. Duchi, J.; Hazan, E.; Singer, Y. Adaptive subgradient methods for online learning and stochastic optimization. *J. Mach. Learn. Res.* **2011**, *12*, 2121–2159.

50. Keras: The Python Deep Learning Library. Available online: https://keras.io/ (accessed on 2 December 2017).
51. Tensorflow Software Library for Machine Intelligence. Available online: https://www.tensorflow.org/ (accessed on 2 December 2017).
52. Deeplearning4j: Deep Learning for Java. Available online: https://deeplearning4j.org/ (accessed on 2 December 2017).
53. Intel Xeon X5400. Available online: https://www.intel.com/assets/PDF/datasheet/318589.pdf (accessed on 15 January 2018).
54. Intel Xeon E5-2600. Available online: https://www.intel.co.jp/content/dam/www/public/us/en/documents/guides/xeon-e5-v3-thermal-guide.pdf (accessed on 15 January 2018).
55. Intel Xeon 5600. Available online: http://downloadt.advantech.com/ProductFile/PIS/96mpxe-2.4-12m13t1/Product%20-%20Datasheet/96MPXE-2.4-12M13T1_Datasheet20170315091546.pdf (accessed on 15 January 2018).

Article

Socio-Cultural Impact of Energy Saving: Studying the Behaviour of Elementary School Students in Greece

Sideri Lefkeli [1], Evangelos Manolas [1], Konstantinos Ioannou [2] and Georgios Tsantopoulos [1,*]

[1] Department of Forestry and Management of the Environment and Natural Resources, Democritus University of Thrace, 193 Pantazidou Street, 68200 Orestiada, Greece; roylaleu@hotmail.gr (S.L.); emanolas@fmenr.duth.gr (E.M.)

[2] Hellenic Agricultural Organization "DEMETER", Forest Research Institute, Vasilika, 57006 Thessaloniki, Greece; ioanko@fri.gr

* Correspondence: tsantopo@fmenr.duth.gr; Tel.: +30-25520-41118

Received: 24 January 2018; Accepted: 3 March 2018; Published: 7 March 2018

Abstract: Education makes it possible for students to become familiar with the rational management of energy as well as learn to implement energy saving practices in their everyday life. The study of certain student characteristics helps in the direction of applying strategies of behavioural change. The aim of this research is to record the knowledge and attitudes of elementary school students in the Prefecture of Evros with regard to energy saving. The collection of research data was done through the use of a structured and anonymous questionnaire with closed questions. The method used for the collection of the research data was cluster sampling. This involved 17 elementary schools of the continental part of the prefecture. 612 questionnaires were completed by students of the 5th and 6th grade of these schools. The evaluation of the research data showed that 69.6% of the students think that the most appropriate house temperature is 20°C with 79.1% of the students keeping the thermostat switched off while the house is aired. With regard to the use of TV, stereo, play station and PC the research showed that 93.8% of the students switch off the above devices when these are not in use. In parallel, 86.6% of the respondents usually or always switch off the lights when coming out of a room and 46.2% of the students use energy saving bulbs. Also, 93% of the students recycle because they believe that doing so contributes to the protection of the environment while 41% always chooses to walk to school. With regard to the significance of reasons concerning energy saving 85.9% thinks that energy saving is important to very important for reducing environmental pollution.

Keywords: irrational energy management; environmental education; energy education programs; active participation of students

---

1. Introduction

The competitive behaviour of modern man led to his isolation from the natural environment and the degradation of natural resources. The use of renewable energy sources is one of the most important elements of sustainable development, which also constitutes the most effective solution for combating environmental problems [1]. The biggest part of the energy used comes from fossil fuels such as petroleum, coal and natural gas. As the energy requirements of countries continue to increase, the consumption of fossil fuels also continues to increase. The uncontrolled consumption of fossil fuels leads to the release of pollutant gases which impact negatively land and marine ecosystems. However, the finite nature of fossil fuels and the pollution of the natural environment constitute key determinants for the development and exploitation of new alternative forms of energy. Renewable sources of energy are infinite and environmentally friendly because they contribute to agricultural development, to the promotion of variety in energy sources and to the minimization of the danger resulting from the availability of nuclear weapons [2]. Their use brings some advantages which contribute to the reduction of land and marine pollution. Renewable energy sources constitute the most attractive choice for satisfying energy demands. They can also be used without the fear of depletion [3].

Interest on renewable energy sources started in the 1970s mainly as a result of the petroleum crises of the era but also as a result of the degradation of the environment and of the quality of life from the use of traditional energy sources. Renewable energy sources were particularly costly in the beginning because at that time renewable energy applications were at an experimental stage. However, today renewable energy sources are taken into account in the official energy plans of developed countries and although they still constitute a very small percentage of energy production, nevertheless, steps are taken for further use of such sources. It needs to be noted that the cost of applying renewable energy solutions is continuously reduced in the last twenty years. In particular, wind energy and biomass can now compete with traditional sources such as coal and nuclear energy [4].

Energy saving is an efficient way of confronting our energy and ecological problem in order to achieve sustainability. Energy saving means reduction in the amount of energy consumed in a process or system, or by an organization or society, through economy, elimination of waste, and rational use. Sustainability is the ability of a process or human activity to meet present needs but maintain natural resources and leave the environment in good order for future generations. Change of attitude and behaviour with regard to the sustainable use of energy is necessary. It is important that citizens and particularly students understand the significance of rational energy management so that they can create positive attitudes towards the management of natural resources. Education on energy issues plays a key role in the formation of student behaviour. The application of initiatives regarding energy saving within school units can only bring benefits and lead towards reduction of energy cost [5]. Educational institutions are the most appropriate places in which students are taught energy conservation and involved in activities regarding rational energy management. Students are given opportunities to appreciate activities regarding energy saving and disseminate what they learnt in their wider social environment [6]. The environmental education strategies applied constitute a significant educational process which strengthens student awareness of environmental issues [7].

Factors which influence the attitudes of students towards the environment are the features of their family environment, their energy education at school and social interaction [8,9]. What needs to be emphasized is the importance of the social context of the adolescent, and the necessity to take this into account as a channel which amplifies the impact of specific environmental education strategies [10].

Through the activities of an appropriate educational curriculum the students can revise his personal values, understand the good called energy and learn to use it rationally [3,11]. The aim of this paper is to investigate the behaviour of elementary school students with regard to energy management. This paper is expected to show the influence of the social context of students on environmental protection. In particular, the aims of the research are to investigate the everyday habits of students with regard to energy management, to recycling processes, and to how they commute from house to school. Also, an important aim of the research is to study the reasons for which students think energy saving is important. At the same time, it is important to highlight the importance of the stimuli students have received through the material they have been taught from the physical sciences curriculum, their participation in environmental education programs as well as their wider social environment.

2. Energy Saving in the Educational System

The concept of energy is taught in the Greek educational system since the end of the 1970s [12]. This interest is due to the response of the educational system to the energy crisis which appeared and marked the beginning of the 1970s. In recent years, in the curriculum of elementary education, energy became independent as a concept. However, its importance for everyday issues and problems was also highlighted (energy saving, renewable and non-renewable energy sources). In particular, the Cross Curricular/Thematic Framework (C.C.T.F) aims that students acquire a holistic view of the concept of energy and become familiar with the scientific way of thinking and methodology (observation, collection of information, formulation of hypotheses, analysis and interpretation of data, extraction of conclusions) [12].

From early grades students learn to construct simple energy systems which they continue to do until senior grades where they calculate the energy footprint of their homes. In the last grades of elementary school students are capable to use the term "energy", the terms "energy storage' and 'energy transfer" and, in addition, they are in a position, to suggest solutions regarding the consumption of less energy in meeting their needs and desires. Physical science school books provide knowledge with regard to restricted energy reserves and attempt to orient students towards the adoption of simple everyday habits so that they can contribute to the restriction of energy waste. Conscious behaviour is a product of the learning process and for this reason "energy education" is important. Energy education can help students form an objective view regarding the use of energy in their everyday life [13].

Today students receive several stimuli with regard to energy management and have formulated conceptual representations on the issue of energy through their experiences from the school program and everyday interaction. The Cross Curricular/Thematic Framework (C.C.T.F) and the program of studies emphasize the use of renewable sources of energy and promote the significance of environmentally friendly behaviour. It is important that students are prepared for the exploitation of renewable energy sources [14] and for this reason the content of educational programs should contribute to the creation of a positive attitude towards the management of the natural environment [15].

Through environmental education programs students acquire knowledge and develop the skills they need to defend the environment in everyday life. They formulate a new code of behaviour which is based on energy saving practices. The result is that students become familiar with electrical energy management, participate in the recycling process and avoid the use of a car for going to school. In Jordan, students are aware of the use of renewable energy sources and show a positive attitude and willingness towards using them [2]. Similar behaviour is shown by school students in Turkey who are supporters of renewable energy sources because they believe that the energy which such sources can provide is adequate [16]. Students in Finland participate in sustainability activities and learn through them about natural resources management and energy saving [17]. The study of student behaviour in schools in Taiwan shows that energy saving is linked with their everyday life [18]. Also, research conducted on 2400 students in Taiwan showed that students are encouraged to develop innovative ideas on energy through programs of energy education implemented in their schools [19]. Also, the study of the views of students in N. California and E. Massachusetts, regarding management of energy, shows that students after participation in an energy education program, have improved their home energy saving behaviour [20].

The study of research data regarding female students (also members of the Scouts association), 8–14 years old, showed that after their involvement in a program of energy saving, have improved their energy saving behaviour. These students usually deactivate electric devices, adopt desirable behaviour with regard to heating practices in winter, and with regard to energy management, they exchange information with other people [21]. Similar behaviour is shown by students in an elementary school in North-eastern USA, who, after, their participation in an energy management program, reduced their energy consumption of energy for more than 15% [22]. Research carried out with high school students in Grevena in Greece showed that these students have adopted a positive attitude towards energy management. 66.67% of the sample turn off the lights when coming out of the room, 44.98% always switches off television and personal computer using the central button of operation and 57% never leaves the windows open when the heating or air-conditioning is activated. 65.46% of the students asked usually walk to school from home [23].

With regard to the management of electricity from students in the Santa Elena high school in the Phillipines, during the academic year 2013–2014, it is found that an over-all weighted mean of 2.15 was computed, which signifies that the respondents usually switch off the lights when leaving a room. The responses about water and energy conservation consist of four items, with emphasis on the following: avoid the waste of water by using glass, when brushing teeth and using basin when washing dishes; switching-off the lights and the fan in the classroom if not in use; using solar flashlight and calculator; and providing support for local programs on conservation through conserving electricity at home and at school [24]. Svolis says that the students of the 4th high school of Lamia in Greece, after involvement with the issue of rational energy management, aim to use alternative forms of energy and suggest ways to reduce energy consumption both at home and at school [25]. Vasili's research on 50 students of preschool age in Patra and Kyparissia in Greece shows that 98% of these students deactivates the lights when leaving the room while 72% of the sample uses both sides of paper [26].

Students must be aware of the importance of recycling so that they can develop a positive attitude and successfully participate in recycling activities [27]. Research carried out in schools in Ireland reveals that students of "green schools" participate in high percentage in the process of recycling. These students recognize the need to separate garbage, contributing to the reduction of the quantity of garbage, have the opportunity to think of the consequences of their actions on the environment and, with regard to recycling as well as the adoption of rational environmental practices, they function as transmitters in their family environment [28].

Elementary school students in Chalkidiki in Greece collect batteries and take them to recycling points, a fact which confirms that students are aware and educated on the significance of recycling [29]. Research concerning students 6-10 years old, confirm the views of the students asked on the significance of recycling with regard to the reduction of garbage, the conservation of natural resources and the economy [30]. Research on the views of 41 students, 6–12 years old, on recycling, shows that students recognize its effectiveness while a percentage of the sample usually recycles at home [31]. Similar behaviour is adopted by 14 elementary school students who hold ecocentric views on recycling [32].

However, research carried out on preschool and school students of four public schools in Volos with differing social and economic background, provides information on the definition of recycling, its process as well as the views of students regarding benefits from recycling. The elaboration of research data shows that the two age groups are used to recycling only paper or batteries at home, without, however, realizing to a significant degree the benefits of recycling. In particular, preschool students find it difficult to separate garbage according to the material they are made of [33]. The investigation of the attitudes of students regarding energy saving in schools in Spain and Mexico shows that students in Spain are famous regarding recycling behaviour while students in Mexico use mass transportation for their movement from and to school [34].

Research by Panter et al. showed that half of the students usually walk to school or alternatively use a bicycle [35]. Napier et al [36] found that students walk to school if this is near home since their parents believe that their children are not in danger, while 75% of the students in San Fransisco, for reasons of convenience, and for saving time, go to school by car [37]. 50% of students in New Zealand, due to danger in their neighbourhoods, go to school by car [38]. According to Beck and Greenspan the most common means of transportation to school is the car (46.3%), next is the school bus (39.6%) and last in the hierarchy is walking (14.2%) [39].

With regard to energy management students are receivers of many stimuli and can realize the significance of environmentally friendly behaviour [40]. According to Kandpal and Broman, students are aware of the basic principles of using energy and prepare for strategies which exploit renewable forms of energy [14]. Research on energy management on students, shows that they can understand the good called energy and through "energy education" learn to use it rationally [11].

Saving energy must become an everyday practice for every student. It is important that we use strategies for making students aware for the seriousness of the issue and also involve students in appropriate activities [41]. According to Coker et al. the provision of knowledge on energy management and experiential learning can form active and conscientious students [42]. It is also important that the students/future citizens are equipped with environmental knowledge which will lead to the development of sustainable solutions.

In order to secure environmental quality it is important to adopt a way of life which will be based on: the reduction of household energy consumption, the reduction of CO_2 emissions, the development of more green areas, the reduction of pollution, the development of environmental programs at school and knowledge retrieved from newspapers, books and journals Recycling of goods, less use of electricity, less use of cars and the creation of devices with energy efficiency could seriously contribute to the protection of the environment [43].

3. Methodology

The research area included the geographical borders of Elementary Education of the Prefecture of Evros which belongs to the administrative region of Eastern Macedonia and Thrace. The Directorate of Primary Education of the Prefecture of Evros supervises and cooperates with 162 schools (kindergartens and elementary schools). It is comprised of five municipalities and, in particular, the Municipality of Alexandroupolis, the Municipality of Orestiada, the Municipality of Didymoteicho, the Municipality of Soufli and the Municipality of Samothraki. One of the responsibilities of Elementary Education is the organizing of school activities. Each year the teachers of these schools, in cooperation with students, design and realize various programs. The programs are categorized in three categories according to their theme: cultural, health and environmental. These programs are supervised by the teacher of each class and the district supervisor for school programs and activities. The questionnaire was divided into three sections. The first section was about questions with regard to the contribution of renewable energy sources in confronting environmental problems, the emotions of students concerning climate change and the significance of developing in the future technologies designed to produce electrical energy from energy sources which are not harmful to the natural environment. The second section was about student daily habits with regard to electrical energy management, the recycling process, the use of means of transportation as well as the study of reasons with regard to the significance of energy saving. The third section was about the social characteristics of students.

For the collection of research data regarding the views and attitudes of 5th and 6th grade students of the Prefecture of Evros a structured questionnaire was used because it was regarded as the most appropriate tool in order to achieve the aim of the research. Questionnaires are used to collect data by asking people to answer to the same group of questions. They are usually used in the framework of research strategies which aim to collect data on the views, behaviors, characteristics, attitudes etc. Although there is a variety of definitions, we use the questionnaire as a general term which refers to techniques of data collection where every respondent answers the same group of questions in a pre-determined sequence. The advantages of collecting information through the use of questionnaires are that these are less costly, can be sent to a large number of people, they are easy to create and use, the respondents are free to express their opinion (lack of direct communication), the ways of analyzing the material are standardized, the researcher cannot influence the answers, the questionnaires are the less time-consuming method [44].

The sample was comprised of 17 elementary schools out of 53 which operate in the area. 612 questionnaires were completed by students of 5th and 6th grade. The reliability of the particular research questions is mainly based on the experience of the senior researcher in the particular research area. The senior researcher who conducted this research had sufficient educational experience both with regard to teaching at this educational level and with regard to issues concerning the area under investigation. In addition, the senior researcher had devoted sufficient time with regard to the issue under investigation both with regard to previous research projects in the particular area as well as with regard to data collection. In addition, this questionnaire was based on the available literature and was given for checking/corrections to ten educators of schools in which the research was to be conducted. Finally, after the completion of the questionnaire a pilot research was conducted to 25 students. The final questionnaire was formulated after completion of this pilot research. The sample was comprised of 17 elementary schools out of 53 which operate in the area and 612 questionnaires were completed by students of 5th and 6th grade. For purposes of approval for carrying out the research the procedure followed was the procedure set by the Pedagogical Institute of Greece which supervises all research in Greek schools. The collection of data was done the period May–June 2016 and for the evaluation of data descriptive statistics were used, Friedman's non-parametric test and the test of independence χ^2.

4. Results

The results came from the content, the concepts and the logical sequence of questions in the questionnaire. In particular, from the evaluation of the research data the following themes were created: "management of household energy", "recycling", "movement of students and means of transport", and "saving energy". In particular, the focus was on the regulation of house heating, the management of house electrical energy, the operation of the heating or cooling system as well as on the condition of electrical devices when these are not used. The focus was also on habits of saving paper and recycling, on the most important reasons of energy saving and particular attention was paid to the way students commute from their house to school. With regard to the social and economic characteristics of the students asked, 38.9% of fathers of students are civil servants while 26.3% of fathers are self-employed. 14.7% are private employees and 13.4% farmers and stock farmers while 3.3% of fathers are unemployed and 3% pensioners. With regard to mothers of students 25.9% are civil servants and 21% private employees. 27.5% of mothers only do house work and 13.4% are free lance professionals. A very small percentage of mothers (2.7%) are farmers and stock breeders. 47.0% of men and 44.5% of women are graduates of upper secondary schools (lyceums). The percentage of fathers who are university graduates is 20.6% while the percentage of mothers who are university graduates is 26.6%. 10.9% of men and 9.3% of women are graduates of technical and vocational lyceums. The percentage of fathers who are graduates of lower secondary schools is 7.9% while the percentage of mothers with the same level of education is 5.5%. The percentage of fathers who are elementary school graduates is 4.4% and the percentage of mothers who are elementary school graduates is 3.5%. The percentage of parents who are graduates of Technological Education Institutes is 7% for fathers and 7.9% for mothers. The percentage of parents who are holders of a graduate degree is 2.2% for fathers and 2.7% for mothers. With regard to the sex and grade of students (Tables 1 and 2) 44.3% of students are girls and 55.7% are boys. In addition, 54.7% of students are at the 5th grade and 45.3% of students are at the 6th grade. With regard to regulating home temperature almost seven out of ten students (69.6%) think that the most appropriate temperature is 20 degrees.

Table 1. Sex (Percentage %).

Sex	Percentage %
Girl	44.3
Boy	55.7

Table 2. Grade (Percentage %).

Grade	Percentage %
5th	54.7
6th	45.3

Table 3 shows that 68.4% of the students think that it is preferable for a person to wear more clothes when he/she is cold rather than increase the heating. 17.4% of students have regulated home temperature at the level above 20 degrees and 13% at a level below 19 degrees (Table 4).

Table 3. The first thought of pupils when they are very cold while the hearing is on (Percentage %).

Increase the Heating	Wear More Clothes
31.6	68.4

Table 4. Regulation of home temperature (Percentage %).

Below 19 Degrees	20 Degrees	More than 20 Degrees
13.0	69.6	17.4

Table 5 shows that almost eight out of ten students (79.1%) have the thermostat switched off when they open the windows in order to renew the air of the house. Next the test of independence X^2, in relation to the position of the thermostat, was carried out while the house is aired and this if the position of the thermostat is different in relation to grades and sex.

Table 5. Position of the thermostat when the house is aired (Percentage %).

Heating/Thermostat Switched on	Heating/Thermostat Switched off
20.9	79.1

According to Tables 6 and 7 the statistical test of independence χ^2 showed that there is dependence between grade and sex and the answers students gave for the variable "Position of the thermostat when the house is aired". In particular, when the house is aired, girls are more careful compared to boys.

Table 6. Relationships between school grade and views of pupils with regard the position of the thermostat when the house is aired.

| | Grade | | | |
| | 5th | | 6th | |
When they open the window in order to renew the air of the house	Frequency	Percentage %	Frequency	Percentage %
Heating/thermostat switched on	84	24.3	48	16.8
Heating/thermostat switched off	262	75.7	238	83.2
Total	346	100.00	286	100.00

$$\chi^2 = 10.322, p = 0.001 \ (p < 0.005).$$

Table 7. Relationships between sex of pupils and views of pupils with regard to position of thermostat when the house is aired.

| | Sex | | | |
| | Boy | | Girl | |
When they open the window in order to renew the air of the house	Frequency	Percentage %	Frequency	Percentage %
Heating/thermostat switched on	90	25.6	42	15.0
Heating/thermostat switched off	262	74.4	238	85.0
Total	352	100.00	280	100.00

$$\chi^2 = 10.541, p = 0.001 \ (p < 0.005).$$

75.5% of students (Table 8) say that they have the windows closed when the heating system is in operation in winter or alternatively when the cooling system is in operation in summer. Almost four out of ten students usually use the air-conditioner (40.5%) or the fan (40.8%) (Table 9). 18.7 usually lower the tents or close the blinds of the windows.

Table 8. View of the students regarding the position of the windows when the heating system is in operation in winter or alternatively when the cooling system is in operation in summer (Percentage %).

Closed	Open
75.5	24.5

Table 9. Habits in dealing with heat (Percentage %).

Use the Air-Conditioner	Use the Fan	Lower the Tents and Close the Blinds of the Windows
40.5	40.8	18.7

With regard to the use of television, stereo, play station-computer game console, almost nine out of ten students (93.8%) switch off the above devices from the central button (Table 10). 87.7% of students switch off the television from the central button while 11.2% of students leave on standby position. 92.9% of students show the same behaviour with regard to the stereo, and the same happens for 93.8% of students with regard to the play station. Almost eight out of ten students (89.7%) usually switch off the personal computer from the central button while 6.8% of students leave the personal computer on standby position.

Table 10. State of devices when not in use (Percentage %).

	Switch off from Central Button	Leave on Standby	Leave Switched on
Television	87.7	11.2	1.1
Stereo	92.9	6.3	0.8
Play station—computer game console	93.8	4.6	1.6
Personal computer	89.7	6.8	3.5

Table 11 shows the habits of students with regard to the use of paper, the recycling of batteries and the management of electricity. Almost eight out of ten students (86.6%) usually or always switch off the lights when coming out of a room. Also, 47.3% of pupils always or usually recycle batteries either at home or at school while 46.2% of students use energy saving light bulbs. Almost five out of ten students always make rational use of paper. In particular, 57.8% of students usually or always use used paper for notes etc and 54.5% always use both sides of the paper when printing or photocopying documents.

Table 11. Saving habits with regard to energy, paper and recycling (Percentage %).

	Always	Usually	Sometimes	Rarely	Never	I Do Not Know
You use in your house energy saving light bulbs	26.6	19.6	14.4	9.7	8.7	21.0
You switch off lights when coming out of a room	66.5	20.1	7.8	3.5	1.6	0.6
Use used paper for notes etc.	32.0	25.8	19.8	13.1	8.1	1.3
You use both sides of paper when you print or photocopy documents	31.6	22.9	14.6	11.4	9.8	9.7
You recycle batteries at home or at school	28.2	19.1	13.9	17.2	16.9	4.6

Friedman's statistical test was used to examine the potential existence of a statistical difference between the views of students concerning saving habits with regard to paper, energy and recycling (Table 12). According to the results of the above mentioned test, the first energy saving habit, with mean rank 3.97, is that students switch off the lights when coming out of a room. Their second habit, with mean rank 3.03, is to use used paper for notes etc. Their third habit, with mean rank 2.83, is to use both sides of the paper when they print or photocopy documents. The students'fourth habit, with mean rank 2.64, is to recycle batteries at home or at school. In the last position, with mean rank 2.53, is the habit of the students to use energy saving light bulbs at home.

Table 12. Application of Friedman's test regarding saving habits with regard to energy, paper and recycling.

	Mean Rank
You use in your house energy saving light bulbs	2.53
You switch off the lights when coming out of a room	3.97
You usually use used paper for notes etc.	3.03
You use both sides of the paper when printing or photocopying documents	2.83
You recycle batteries at home or at school	2.64

$N = 632$, Chi-Square = 411.416, $df = 4$, Asymp. Sig. 0.000.

With regard to the views of students on recycling (Table 13) nine out of ten students recycle. In particular, 93% of students recycle because they believe that they contribute to the protection of the environment. The percentage of students who do not recycle is 7% and this because they believe that they do not contribute much to the protection of the environment.

Table 13. Recycling and protection of the environment (Percentage %).

I Recycle Because I Contribute to the Protection of the Environment	I Do Not Recycle Because I Think That Recycling Does Not Contribute Much to the Protection of the Environment
93.0	7.0

Table 14 presents the way students move from home to school. In particular, five out of ten students (51%), regarding the route from home to school, never use a car or rarely move by car. In parallel, eight out of ten students rarely or never use public transport in order to go to school. 41% of the students asked always go to school on foot and 23.6% of students report that they usually or always use their bicycle as a means of transport from home to school.

Table 14. Mode of transport to school (Percentage %).

	Always	Usually	Sometimes	Rarely	Never
By car	23.3	12.2	13.6	29.6	21.4
By public transport—school bus	12.3	4.4	2.8	8.4	72.0
On foot	27.4	13.6	11.6	17.2	30.2
By bicycle	10.8	12.8	10.9	18.0	47.5

In order to investigate the possibility of existence of statistical difference between the views of students and how they move to school, Friedman's statistical test was applied. According to the results of the above mentioned test, students never move by car (mean rank 2.85). In particular, students, with mean rank 2.83, choose to go to school on foot, and, also, with mean rank 2.40, prefer to use their bicycle. Regarding means of transport the last position in the hierarchy, with mean rank 1.92, is the school bus, which rarely or never is used by students (Table 15).

Table 15. Application of Friedman's statistical test regarding the mode of movement of pupils to school.

	Mean Rank
By car	2.85
By public transport—school bus	1.92
On foot	2.83
By bicycle	2.40

$N = 632$, Chi-Square $= 270.717$, $df = 3$, Asymp. Sig. 0.000.

With regard to the importance of reasons concerning energy saving (Table 16) almost nine out of ten students think that it is important to very important to engage in energy saving in order to reduce environmental pollution. 74.2% of students think saving money for financial reasons is important while 68.7% thinks that saving energy is important in order not to exhaust natural resources.

Table 16. Importance of reasons for energy saving (Percentage %).

	Unimportant	Of Little Importance	Moderate	Important	Very Important
In order to save money	4.9	6.6	14.2	34.3	39.9
In order not exhaust natural resources	3.8	8.9	18.7	35.3	33.4
In order to reduce environmental pollution	3.2	4.4	6.5	24.2	61.7

Friedman's statistical test was used to examine the potential existence of a statistical difference between the views of the students regarding importance of reasons for saving energy, Friedman's test was applied (Table 17).

Table 17. Application of Friedman's statistical test regarding reasons for energy saving.

	Mean Rank
In order to save money	1.91
In order not to exhaust natural resources	1.78
In order to reduce environmental pollution	2.31

$N = 632$, Chi-Square $= 137.495$, $df = 2$, Asymp. Sig. 0.000.

According to the results of the above mentioned test, the reduction of environmental pollution, with mean rank 2.31, is a reason of primary importance for energy saving. Next in importance, with mean rank 1.91, is saving money and in the last position, with mean rank 1.78, is not to exhaust natural resources.

5. Discussion

Understanding the behaviour of students regarding energy saving is important for purposes of promoting energy saving. Together with an appreciation of the benefits for energy saving, understanding the attitude of the students is something which will help in making the appropriate political and educational decisions.

With regard to the first theme "management of household energy" the students asked think that the most appropriate house temperature is 20 degrees and declare that they prefer to wear extra clothes when they are cold. The students usually air their house when the thermostat is switched off. In addition, they close the windows when the heating system is in operation in winter and when the cooling system is in operation in summer. In parallel, students in N. California save more household energy [20]. In addition, students in North-eastern USA have also managed to reduce household energy [22].

During summer time, the majority of students use the air-conditioner or the fan and very few of them use tents. The majority of the sample always switches off electrical devices from the central button. They do the same with regard to lights when coming out of a room. Students in Grevena in Greece behave the same way and always deactivate electrical devices and lights when coming out of a room [23]. The same behaviour, with regard to the management of electricity applies to students in the Phillipines [24]. At this point, it is important to note that girls are more careful than boys with regard to household energy management, something which is verified in the research by Puttick et al. [21].

With regard to the second theme "recycling", the majority of students in the sample mainly recycle batteries just as students in Chalkidiki do [28,29,31] while, they simultaneously, support the view that recycling contributes to the protection of the environment. The same view is held by students in other research projects who acknowledge the significance of recycling for the reduction of garbage and the saving of natural resources [30,32]. Except recycling batteries, the same pupils have learnt to use used paper for notes etc or use both sides of paper, when they need to print or photocopy documents, something which proves their environmentally-friendly behaviour.

The study of the theme "movement of students and means of transport" shows that students rarely use a car or public transport in order to go to school. On the other hand, the majority of these students goes to school either on foot or by bicycle, just as it is proved from other research data which show that students go to school either on foot or by bicycle [35–37]. On the contrary, students in New Zealand, due to danger in their neigbourhoods, usually go to school by car [38].

The study of the fourth theme "saving energy" shows that the students of this sample think of energy saving as very important because they believe that it can first contribute to the reduction of pollution of the natural environment and second to saving money. This view has been adopted by students of schools in Finland, who have learnt through activities how to manage natural resources or as well as save energy [17]. In addition, taking initiatives in schools can only bring educational benefits [5], just as the content of educational programs may change behaviour towards the management of the natural environment [40].

6. Conclusions

This paper investigated the behaviour of elementary school students in the distant area of the Prefecture of Evros on a series on issues related to the management of household energy, recycling, movement of students to school and the importance of saving energy so that those who formulate and implement policy can use the findings of this research effectively. The habits of students regarding energy management show that students are careful and sensitive to such issues, which means that they have acquired these skills through their "energy education". Studentscan develop environmentally-friendly behaviours through environmental education programs [45] something which is shown by the habit of students to recycle and properly manage used paper. Environmental education plays a key role in the everyday life of students [46] just as it does in the program of studies. It is worth noting that schools gradually change and aim to promote investigative and experiential learning processes with the ultimate aim to create environmentally responsible citizens.

Finally, this research could be repeated and address lower grade students and thus make obvious their habits with regard to energy management. In addition, this research could be expanded, modified and focus on linking knowledge about the environment with different courses in the program of studies.

Author Contributions: Sideri Lefkeli and Georgios Tsantopoulos designed the research; Konstantinos Ioannou and Evangelos Manolas analyzed the data; Georgios Tsantopoulos and Evangelos Manolas wrote the paper.

Conflicts of Interest: The authors declare no conflict of interest.

References and Note

1. Dincer, I. Renewable energy and sustainable development: A crucial review. *Renew. Sustain. Energy Rev.* **2000**, *4*, 157–175. [CrossRef]
2. Zyadin, A.; Puhakka, A.; Ahponen, P.; Cronberg, T.; Pelkonen, P. School students' knowledge, perceptions, and attitudes toward renewable energy in Jordan. *Renew. Energy* **2012**, *45*, 78–85. [CrossRef]
3. Karatayev, M.; Clarke, M.L. A review of current energy systems and green energy potential in Kazakhstan. *Renew. Sustain. Energy Rev.* **2016**, *55*, 491–504. [CrossRef]
4. Sideridou, E.D.; Achilias, D.S.; Bikiaris, D. *Fuel-Lubricants*; Zitis Publications: Thessaloniki, Greece, 2011.
5. Castleberry, B.; Gliedt, T.; Greene, J.S. Assessing drivers and barriers of energy-saving measures in Oklahoma's public schools. *Energy Policy* **2016**, *88*, 216–228. [CrossRef]
6. Kuzume, K.; Tabusa, T.; Sawa, H. Electric Power Saving Awareness System at School Using ICT. In Proceedings of the MATEC Web of Conferences 55, Asia Conference on Power and Electrical Engineering Bangkok, Bangkok, Thailand, 20–22 March 2016.
7. Simsekli, Y. An Implementation to Raise Environmental Awareness of Elementary Education Students. *Procedia Soc. Behav. Sci.* **2015**, *191*, 222–226. [CrossRef]
8. Goldstein, N.J.; Cialdini, R.B.; Griskevicius, V. A Room with a Viewpoint: Using Social Norms to Motivate Environmental Conservation in Hotels. *J. Consum. Res.* **2008**, *35*, 472–482. [CrossRef]
9. Grønhøj, A.; Thøgersen, J. Action speaks louder than words: The effect of personal attitudes and family norms on adolescents' pro-environmental behaviour. *J. Econ. Psychol.* **2012**, *33*, 292–302. [CrossRef]
10. Duarte, R.; Escario, J.J.; Sanagustín, M.V. The influence of the family, the school, and the group on the environmental attitudes of European students. *Environ. Educ. Res.* **2017**, *23*, 23–42. [CrossRef]
11. Dias, R.A.; Mattos, C.R.; Balestieri, J.A.P. The limits of human development and the use of energy and natural resources. *Energy Policy* **2006**, *34*, 1026–1031. [CrossRef]
12. Andritsos, N. *Energy and Environment*; University of Thessaly: Volos, Greece, 2008.
13. DeWaters, J.; Qaqish, B.; Graham, M.; Powers, S. Designing an energy literacy questionnaire for middle and high school youth. *J. Environ. Educ.* **2013**, *44*, 56–78. [CrossRef]
14. Kandpal, T.C.; Broman, L. Renewable energy education: A global status review. *Renew. Sustain. Energy Rev.* **2014**, *34*, 300–324. [CrossRef]
15. Çelikler, D.; Aksan, Z. The development of an attitude scale to assess the attitudes of high school students towards renewable energy sources. *Renew. Sustain. Energy Rev.* **2016**, *54*, 1092–1098. [CrossRef]
16. Kilinç, A.; Stanisstreet, M.; Boyes, E. Incentives and disincentives for using renewable energy: Turkish students' ideas. *Renew. Sustain. Energy Rev.* **2009**, *13*, 1089–1095. [CrossRef]
17. Uitto, A.; Saloranta, S. The relationship between secondary school student's environmental and human values, attitudes, interests and motivations. *Procedia Soc. Behav. Sci.* **2010**, *9*, 1866–1872. [CrossRef]
18. Lee, L.S.; Lin, K.Y.; Guu, Y.H.; Chang, L.T.; Lai, C.C. The effect of hands-on "energy-saving house" learning activities on elementary school students' knowledge, attitudes, and behavior regarding energy saving and carbon-emissions reduction. *Environ. Educ. Res.* **2013**, *19*, 620–638. [CrossRef]
19. Lee, L.S.; Lee, Y.F.; Altschuld, J.W.; Pan, Y.J. Energy literacy: Evaluating knowledge, affect, and behavior of students in Taiwan. *Energy Policy* **2015**, *76*, 98–106. [CrossRef]
20. Boudet, H.; Ardoin, N.M.; Flora, J.; Armel, K.C.; Desai, M.; Robinson, T.N. Effects of a behaviour change intervention for Girl Scouts on child and parent energy-saving behaviours. *Nat. Energy* **2016**, *1*. [CrossRef]
21. Puttick, G.; Kies, K.; Garibay, C.; Bernstein, D. Learning and behavior change in a Girl Scout program focused on energy conservation: Saving energy to "save the planet". *J. Sustain. Educ.* **2015**, *8*. Available online: http://www. jsedimensions.org/wordpress/wp-content/uploads/2015/01/Puttick-et-al-JSE-Vol-8-Jan-2015.pdf (accessed on 7 May 2017).
22. Craig, C. Electricity Generation, Electricity Consumption, and Energy Efficiency in the United States: A Dual Climatic-Behavioral Approach. Ph.D. Thesis, University of Arkansas, Fayetteville, AR, USA, 2016.
23. Ntona, E.; Arabatzis, G.; Kyriakopoulos, G.L. Energy saving: Views and attitudes of students in secondary education. *Renew. Sustain. Energy Rev.* **2015**, *46*, 1–15. [CrossRef]
24. Cruz, J.P. Students' Environmental Awareness and Practises: Basis for Development of Advocacy Program. *Int. J. Educ. Stud.* **2017**, *9*, 29–40.

25. Svolis, I. Consumerism, Energy, Garbage and Environment. In Proceedings of the 2nd Conference of School Environmental Education Programs, Athens, Greece, 15–17 December 2006; Available online: http://kpe-kastor.kas.sch.gr/kpe/yliko/sppe2/sppe/PDFs/1574-1580_sppe.pdf (accessed on 4 May 2017).

26. Vasili, C. An Investigation of Environmental Attitudes of Children in Pre-School Age. Master's Thesis, University of Patras, Patra, Greece, 2016.

27. Md Zain, S.; Basri, N.E.A.; Mahmood, N.A.; Basri, H.; Zakaria, N.; Elfithri, R.; Ahmad, M.; Ghee, T.K.; Shahudin, Z. Recycling practice to promote sustainable behavior at University Campus. *Asian Soc. Sci.* **2012**, *8*, 163–173.

28. Byrne, S.; O'Regan, B. Attitudes and actions towards recycling behaviours in the Limerick, Ireland region. *Resour. Conserv. Recycl.* **2014**, *87*, 89–96. [CrossRef]

29. Gkliaos, K. *Recycling Batteries*; Office of Environmental Education, Directorate of Elementary Education of the Prefecture of Chalkidiki: Polygyros, Greece, 2003.

30. Palmer, J.; Grodzinska Jurczak, M.; Suggate, J. Thinking about Waste: Development of English and Polish Children's Understanding of Concepts Related to Waste Management. *Eur. Early Child. Educ. Res. J.* **2003**, *11*, 117–139. [CrossRef]

31. Honig, A.S.; Mennnerich, M. What does "go green" mean to children? *Early Child Dev. Care* **2013**, *183*, 171–184. [CrossRef]

32. Kara, G.E.; Hande Aydos, E.; Aydın, Ö. Changing Preschool Children's Attitudes into Behavior towards Selected Environmental Issues: An Action Research Study. *Int. J. Educ. Math. Sci. Technol.* **2015**, *3*, 46–63. [CrossRef]

33. Iliopoulou, I. Views of pre-school and early school age on recycling: Research on pupils in Volos. *Res. Educ.* **2016**, *5*, 148–164.

34. Vicente-Molina, M.A.; Fernández-Sáinz, A.; Izagirre-Olaizola, J. Environmental knowledge and other variables affecting pro-environmental behaviour: Comparison of university students from emerging and advanced countries. *J. Clean. Prod.* **2013**, *61*, 130–138. [CrossRef]

35. Panter, J.R.; Jones, A.P.; Van Sluijs, E.M.F.; Griffin, S.J. Neighborhood, Route, and School Environments and Children's Active Commuting. *Am. J. Prev. Med.* **2010**, *38*, 268–278. [CrossRef] [PubMed]

36. Napier, M.A.; Brown, B.B.; Werner, C.M.; Gallimore, J. Walking to school: Community design and child and parent barriers. *J. Environ. Psychol.* **2011**, *31*, 45–51. [CrossRef]

37. McDonald, N.C.; Aalborg, A.E. Why Parents Drive Children to School: Implications for Safe Routes to School Programs. *J. Am. Plan. Assoc.* **2009**, *75*, 331–342. [CrossRef]

38. Mitchell, H.; Kearns, R.A.; Collins, D.C.A. Nuances of neighbourhood: Children's perceptions of the space between home and school in Auckland, New Zealand. *Geoforum* **2007**, *38*, 614–627. [CrossRef]

39. Beck, L.F.; Greenspan, A.I. Why don't more children walk to school? *J. Saf. Res.* **2008**, *39*, 449–452. [CrossRef] [PubMed]

40. Çelikler, D.; Aksan, Z. The opinions of secondary school students in Turkey regarding renewable energy. *Renew. Energy* **2015**, *75*, 649–653. [CrossRef]

41. Zerva, A.; Tsantopoulos, G. Views and attitudes of public opinion on climate change on an international and national scale. In *Themes in Forestry and Management of the Environment and Natural Resources, International Environmental Politics: Encounters with the Future*; Manolas, E.I., Protopapadakis, E.D., Tsantopoulos, G.E., Eds.; Democritus University of Thrace: Orestiada, Greece, 2013; Volume 5, pp. 75–89.

42. Çoker, B.; Çatlıoğlu, H.; Birgin, O. Conceptions of students about renewable energy sources: A need to teach based on contextual approaches. *Procedia Soc. Behav. Sci.* **2010**, *2*, 1488–1492. [CrossRef]

43. Manolas, E. *Climate Change: Challenges for the 21st Century, Themes in Forestry and Management of the Environment and Natural Resurces, Vol. 7: Climate Change: Interdisciplinary Approaches*; Manolas, E.I., Protopapadakis, E.D., Eds.; Department of Forestry and Management of the Environment and Natural Resources, Democritus University of Thrace: Orestiada, Greece, 2015; Volume 7, pp. 161–168.

44. Lagoumitzis, G.; Vlachopoulos, G.; Koutsogiannis, K. Methods of Research in Health Sciences. 2015. Available online: https://repository.kallipos.gr/bitstream/11419/5356/1/00_master_document%20corrected%20links-KOY.pdf (accessed on 16 May 2017).

45. Sipsas, A.; Lekka, A.T.M.; Pagge, G. Educational games with computer and communication technologies for the development of children's environmental consciousness. *Sci. Ann. Dep. Pre-Sch. Age* **2013**, *6*, 267–279.

46. Zsóka, Á.; Szerényi, Z.M.; Széchy, A.; Kocsis, T. Author's personal copy Greening due to environmental education? Environmental knowledge, attitudes, consumer behavior and everyday pro-environmental activities of Hungarian high school and university students. *J. Clean. Prod.* **2013**, *48*, 128–138. [CrossRef]

 *sustainability*

Article

Public Perceptions and Willingness to Pay for Renewable Energy: A Case Study from Greece

Stamatios Ntanos [1,*], **Grigorios L. Kyriakopoulos** [2], **Miltiadis Chalikias** [3], **Garyfallos Arabatzis** [1] and **Michalis Skordoulis** [1]

1   Department of Forestry and Management of the Environment and Natural Resources, School of Agricultural and Forestry Sciences, Democritus University of Thrace, 68200 Orestiada, Greece; garamp@fmenr.duth.gr (G.A.); mskordoulis@gmail.com (M.S.)

2   School of Electrical and Computer Engineering, National Technical University of Athens, 15780 Zografou, Greece; gregkyr@chemeng.ntua.gr

3   Department of Business Administration, School of Business and Economics, Piraeus University of Applied Sciences, 12244 Aigaleo, Greece; mchalikias@hotmail.com

*   Correspondence: sdanos@ath.forthnet.gr or sdanos@puas.gr; Tel.: +30-210-5381275

Received: 31 January 2018; Accepted: 1 March 2018; Published: 3 March 2018

Abstract: The purpose of this study is to discover the factors shaping public opinion about renewable energy sources and investigate willingness to pay for expansion of renewable energy sources in the electricity mix. Data was collected through a questionnaire applied in Nikaia, an urban municipality of Greece. The respondents have a positive attitude towards renewable energy systems. Most of them have good knowledge of solar and wind energy systems and are using solar water heating, while several respondents own a solar PV system. Environmental protection is seen as the most important reason for investing in a renewable energy system. Willingness to pay for a wider penetration of RES into the electricity mix was estimated to be 26.5 euros per quarterly electricity bill. The statistical analysis revealed the existence of a relationship between RES perceived advantages and willingness to pay for renewable energy. Furthermore, by using a binary logit model, willingness to pay was found to be positively associated with education, energy subsidies, and state support.

Keywords: renewable energy sources; social acceptance; WTP; CVM; logit regression

---

1. Introduction

Life is directly linked to the quality of the natural environment and the availability of natural resources. Environment and life are interdependent concepts. Maintaining a balance in the world ecosystem is a basic prerequisite for preserving life. The atmosphere of our planet is a valuable and sensitive resource to be protected. On the contrary, undesirable inflows into the ecosystem, caused by anthropogenic activity, can shake this harmony and degrade living conditions [1]. Human influence on the environment is increasing due to mass production of technological goods [2], intensification of agriculture [3], the rapid rate of urbanization, and growing demand of fossil fuels for energy and transport [4]. According to data from the International Energy Agency, between 1971–2014, global primary energy consumption has increased by 2.5 times, as from 5.5 GTOE in 1971 to 13.7 GTOE for 2014 [5]. Over the same period, carbon dioxide emissions (hereinafter CO_2) have doubled. Climate change poses a significant environmental, social, and economic threat [6]. The increase in anthropogenic carbon emissions is linked to global warming. Scientists point out that CO_2 concentration in the atmosphere has significantly increased over the last century, compared to relatively stable levels of the pre-industrial era [7]. Since 1751, about 400 billion tons of coal have been released into the atmosphere due to fossil fuel combustion and cement production. Half of those CO_2 emissions were added since the late 1980s [8]. There are scientific publications from the early 1970s, calling for actions to control CO_2 emissions. Dyson [9], estimated that fossil fuel combustion adds 5×109 tons of CO_2 annually, of which, about half remain in the atmosphere. Bach [10] reported that the upward trend in global population, energy consumption, and economic activity contributes to climate change. He estimated that the average temperature would rise by about 1.5 °C to 3 °C by 2050, due to the increase of anthropogenic CO_2 emissions. Garret [11], calculated that the global temperature will increase between 2 °C (optimistic scenario) and 4 °C (pessimistic scenario), by the year 2100, compared to the average temperature of the Industrial Revolution era. Recent estimates of the Intergovernmental Conference on Climate Change point out that we are close to exceeding the 2 °C global warming threshold [12].

The phenomenon of the ever-increasing environmental burden due to the rising trend in energy demand has turned environmental research interest on energy management and renewable energy sources. There has already been a remarkable shift of developed countries towards green growth due to their commitment to the Kyoto Protocol and the Paris Agreement [13], which is favored by broad public access to environmental information [14]. There is a need for wider penetration of renewable energy sources (hereinafter RES), to achieve the Paris Agreement target of limiting temperature increase to only 1.5 °C above the pre-industrial levels. Therefore, in the context of the implementation of 'sustainable development', green investments in the energy sector have significantly evolved, especially during recent years. Greece, for the period of 2006–2016, has managed to double the share of renewables in final energy consumption, from 7.2% to 15.2%. For the power generation sector, the participation of RES is higher, as in 2016 it reached 30% for the European Union, while in Greece, the corresponding figure is 23.8% [15]. The concept of 'social acceptance' is used to assess the degree of readiness of citizens to accept renewable energy investments in their area [16,17]. According to another study, 'social acceptance' is a measure of the active or passive attitude of citizens towards different green technologies or products [18]. Under social acceptance, a body of literature also explores willingness to pay for greener electricity. Many studies from Greece reflect positive public attitudes towards various forms of renewable energy and social responsible actions [19–21], although only a few Greek studies access willingness to pay.

Within this research framework, the scope of this paper is to address the social and economic dimensions of renewable energy sources for an urban area of Greece, with two main research aims: (a) the examination of public perceptions about RES and (b) the estimation of willingness to pay (hereinafter WTP) for a greater expansion of RES into the Greek energy mix. The area of Nikaia was selected for the study, a densely populated municipality near the capital of Greece, Athens. The ambient air condition in Athens, is heavily burdened by traffic load, heating applications, and industrial facilities [22]. Furthermore, all previous Greek studies on WTP for renewables took place in semi-urban or rural areas of the Greek province, where severe environmental problems are less noticeable to the residents.

2. Literature Review

Social acceptance of green investments is monitored at both national and local levels, as it has been observed that citizens' attitudes may vary, not only between countries but also between regional entities of the same country [23–25]. The leaders in renewable energy production are Denmark and Germany. In the latter, more than 42% of electricity generation is produced by renewable sources [26]. In South Korea, active ecological awareness has been reported among citizens; most of them support policies which promote renewable forms of energy that remain state-owned [27]. In parallel, both the federal and the state US regulations are further motivating consumers through tax credits and discounts, so that the energy end-users can install solar energy systems [28]. In Portugal, there is a positive attitude towards innovative RES investments, and this social behavior is more pronounced for solar projects and new hydropower units [29]. The countries with the largest installed photovoltaic rated power are Germany, Italy, USA, China, Japan, Spain, France, Belgium, Australia, and the Czech Republic. These countries are mainly drawing their energy policies upon KWh guaranteed prices (FiT), low bank lending rates, national solar development goals, and tax reductions [30]. In the Netherlands, volunteers and local authorities play a very important role in the technological spur and large-scale applicability of photovoltaics [31].

Contrarily, an important factor that inhibits the wider adoption of RES-based energy systems, apart from the high cost of infrastructure, is the lack of publicly shared information and the behavior of citizens against RES technological advancements [14]. This social behavior has mainly been observed in economically developing regions or countries. Lack of information was reported in residents of rural, suburban, and urban areas in the Chinese context [32]. Another study was deployed in the Malaysian context, about views and perceptions of the local population towards solar energy and the installation of photovoltaics; it was concluded that the Malaysians hardly understood the incentives and the wider socio-economic benefits derived, thus they were reluctant to invest in photovoltaics [33]. Additionally, in the Middle East and North Africa, the attribute of social opposition was reported among interviewed citizens, since they expressed a biased behavior, significantly distorting anything that tends to become socially acceptable [34]. On the other hand, a study about the social acceptance of small hydropower plants (SHP) in India revealed that SHP projects are regionally challenging forms that can be directly utilized in the Indian energy mix of production [35].

Apart from social acceptance, many studies focus on the economic amount a consumer is prepared to pay for further expansion of RES in their area of residence, which is defined as willingness to pay (WTP). For estimation of the economic value that an individual hypothetically assigns to a non-market good, such as WTP for renewable energy, the contingent valuation method (CVM), is commonly used [36]. In this method, the respondent is directly asked usually through a questionnaire survey, to state his preference [36]. A positive relation has already been identified between WTP, income, and level of information [37,38]. In a study examining attitudes towards RES, Australian tourists were willing to pay 1–5% more for the existence of renewable energy systems within their accommodation units [39]. For the case of Sweden, by using binary logistic regression, it was found that people with increased environmental awareness are more likely to accept renewable energy [40]. Comparable results were proved, by a study on the factors influencing WTP for green electricity, noticing that a proactive attitude towards environmental issues can lead to a higher level of economic participation [41]. A study from China revealed that household income, knowledge of renewable energy, and education are positively associated with WTP, while age and perception of neighbors' non-participation have a negative impact on WTP [16].

Willingness to pay for renewable energy with the CVM method, has been estimated at 17 USD/household/month for Japan [42]; 4.24 USD/household/month for North Carolina, USA [43]; 14 USD/household/month for New Mexico, USA [44]; 2.7–3.3/household/month for Beijing, China [45]; 13–16 USD/household/month for Australia [46]; 2.3–4.3/household/month for Italy [47]; and 4.1 USD/household/month for Slovenia [48].

Under an economic view, RES expansion in Greece can lead to benefits estimated to be 4.9 euros/MWh and 4.4 euros/MWh for electricity produced from solar p/v's and biomass power stations, respectively. Economic benefits of 1.9 euros/MWh and 1.8 euros/MWh were also attributed to wind farms and hydropower plants electricity production [49]. Furthermore, in a study accessing Greek households WTP for greener electricity, evidence of positive association was found with income, level of information, and awareness on green investments [19]. In a relevant empirical study about tourists' WTP for renewable energy in the island of Crete, in Greece, positive association was found between the respondents' age, information status about RES, and previous experience by using a logit model [50]. In a study concerning WTP for biofuels expansion at the area of Thrace, in northern Greece, most of the car owners who took part in the survey, were willing to use biofuels and accept an increased cost of 0.079 €/L [51].

The significance of the energy sector in Greece, which is a developed country under a severe economic crisis since 2009, has also attracted a wider scientific interest covering:

- the applicability of enhanced exploitation of renewables [52];
- the spur of novel technologies upon electrical energy storage in electricity generation [53];
- the socio-cultural value of energy production as a common good that must be instilled in the secondary education systems [54];
- the energy autonomy and the forestry management of mountainous regions in satisfying the energy demand, either as energetically autonomous countries, or under bilateral national agreements of legal importing and exporting forestry biomass as supplementary sources of energy production [55,56];
- the public attitude that is generally positive towards RES and the high cost and lack of information as obstacles towards the wider support of RES-based energy investments [19,21,57];
- and the ongoing controversy between ecological, economic, and financial environment in Greece, since national policies should be oriented to 'green fiscal policies'. Greek national policy must be compatible with relevant European reports. National environmental policy should align with the restructuring of the community framework for the taxation of energy products, which draws the transition from income taxes towards a system based on the principle that payment should burden those who consume more energy and produce more air pollutants [58,59].

3. Materials and Methods

A questionnaire survey was conducted in Nikaia, which is situated in the western sector of Attica and has a population of 89,380 permanent residents, according to the 2011 census. Data was collected between November and December 2016. The method of random stratified sampling was used [60]. The stratification is carried out at municipality level, using the list of registered voters as the sampling frame. By using the electoral catalogues, we ensured that the sampling units are adults, residents of the municipality of Nikaia. The municipality of Nikaia was divided into 159 polling stations with 93,851 registered voters during the parliamentary elections of January 2015 [61]. We used the electoral lists of the elections of January 2015, containing voters per polling station, as retrieved from the Ministry of Interior [61]. For creation of the sample, we numbered all voters for each polling station in ascending order, according to our nominal voter list. Then we randomly selected voters, by separate draw per polling station, according to the percentage the polling station represents over the total voters. To ensure randomness, we used the random number generator of Microsoft Excel, version 2007. Random numbers are given by the 'RANDBETWEEN' function, which is compiled as follows: '=RANDBETWEEN (bottom, top)'. The 'RANDBETWEEN' function can produce as many integer numbers as desired by the user, between the bottom and top boundaries. Special emphasis has been put on calculating the correct sample size, since it affects not only the accuracy of the measurement but also the conclusions about the population [62]. For the calculation of the appropriate sample size, because the dispersion of the variables of our investigation is not available, we performed a preliminary survey in the area, by collecting 50 questionnaires. By using this pivot sample, we were able to calculate the variance, standard deviation and the ratio for each variable. Regarding sample size estimation, in the case of unknown population variance and for the case of a large sample, the following equation is used [62]

$$n = \frac{4s^2 \times (Z_{1-\alpha/2})}{D^2} \tag{1}$$

where n is the estimated sample size, s is the calculated standard deviation resulting from the control sample, $Z_{1-a/2}$ values derive from the confidence level selected by the investigator based on the normal distribution table and D is the overall width of the desired confidence level as determined by the researcher or as given by similar research. In our sample, the variable with the greater standard deviation 'age' (mean = 40.4, s = 14.24). By using Equation (1), sample size was estimated as can be seen in the calculation below

$$n = \frac{4 \times 203 \times 1.96}{2^2} = 397.88 \tag{2}$$

The appropriate sample size was rounded up at 400 persons. The proportion equation for variables expressed in percentages, gave lower sample size estimates. Face-to-face method was used to fill out the questionnaires. If the respondents were absent or refused, one more effort was made to capture their opinion. This mainly happened with senior respondents (>65), who were reluctant to respond to the questionnaire. In this case, we used the previous process to select new sampling units, thus our study better represents the views of citizens aged 18–65.

The questionnaire includes 16 multifaceted questions on RES, which form 73 variables, about respondents' viewpoint of use, information and acceptance of RES. Previous surveys on renewables for Greece were taken into consideration for appropriate questionnaire design [19,63–67].

The main research goal is to assess public opinion on renewable energy sources, perceived advantages and disadvantages, and willingness to pay for RES energy. A second research goal is to locate the main dimensions of public attitude towards WTP for RES expansion. Statistical methods include principal components analysis, one-way ANOVA, and binary logit regression. For the purposes of the analysis, Stata/MP 13.0 and SPSS v.17 were used.

4. Results

Most of the respondents are males (52.3%), found in the age category of 41–55 (35.5%), as presented in Table 1 and Figure 1. For verification of sampling accuracy, in Figures 1 and 2, age distributions for both the sample and the total population of Nikaia, according to the 2011 census, are provided [68]. The two bar-charts are comparable, with the exemption of the category of adults over 65, who were, in many cases, unwilling to take part in our survey. The mean sample age is 40.4.

Table 1. Respondents' age distribution.

		Age			
		Frequency	**Percent**	**Valid Percent**	**Cumulative Percent**
	18–30	113	28.3	28.3	28.3
	31–40	106	26.5	26.5	54.8
Valid	41–55	142	35.5	35.5	90.3
	56–65	34	8.5	8.5	98.8
	>65	5	1.3	1.3	100.0
	Total	400	100.0	100.0	

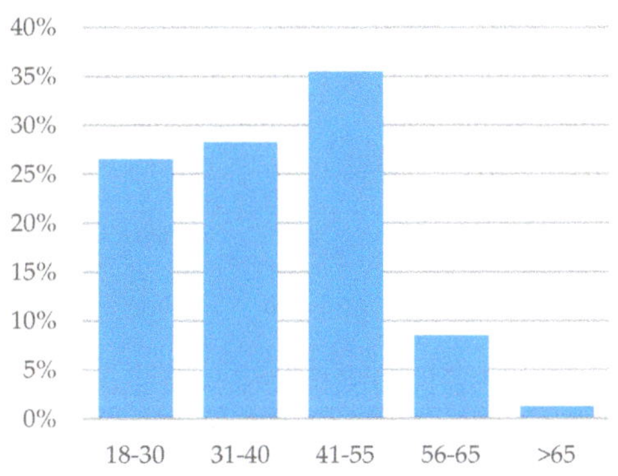

Figure 1. Bar chart depicting age distribution of sample.

Most of the respondents are high school graduates (38.0%), followed by university graduates (35.0%). Around 67% of the respondents have an annual family income of up to 20,000 euros, while one-third of the sample population stated that their annual income does not exceed 10,000 euros.

The respondents are many civil servants or private employees (57%), while around 25% of the sample are students, unemployed, or homemakers. Concerning RES use, 239 respondents, who correspond to 60% of the sample, reported that they use renewable energy technologies (active or passive) in their everyday life. In a multiple response question about ownership of different types of renewable energy technologies, 95% of RES users have installed solar water heaters, while another 13% of them are using photovoltaic systems (PVs). Only four RES users have invested in wind energy systems and two in geothermal energy systems (Table 2).

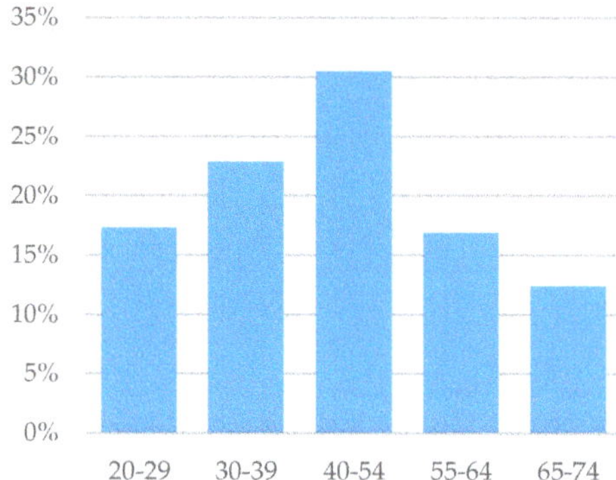

Figure 2. Bar chart depicting age distribution of population.

Table 2. Used renewable energy technologies (multiple response).

Technology	Frequency	% of RES Users
Solar water heater	227	95.0%
Solar P/V	31	13.0%
Wind turbines	4	1.7%
Geothermal	2	0.8%
Biofuels	7	2.9%
Passive solar systems	7	2.9%

In a question accessing the reasons for not investing in renewable energy systems (Table 3), 37.3% of non-RES users, indicated "high installation costs", followed by lack of information (30.4%).

Table 3. Main reasons for not using any kind of renewable energy technology.

Reasons for Not Using RES	Frequency	%
High installation cost	60	37.3%
I do not have the right information	49	30.4%
Low reliability	18	11.2%
Complex installation process	9	5.6%
Systems hazards	8	5.0%
Legislative environment difficulties	8	5.0%
High maintenance cost	5	3.1%
Difficulty of use	3	1.9%
Possible fire spread	1	0.6%

The internet is the preferred method for retrieving information about renewable energy (43.5%), followed by television (28.8%), as can be seen in Table 4.

Table 4. Main sources of retrieving information on renewable energy sources.

Sources of Information on RES	Frequency	%
Television	115	28.8
Radio	18	4.5
Newspapers/magazines	59	14.8
Internet	174	43.5
Environmental organizations	22	5.5
Friends	12	3.0

Regarding respondents' self-evaluation on their knowledge degree upon RES, 54% and 42% of the sample are adequately informed about solar power and wind power, respectively. On the contrary, respondents are inadequately informed about hydropower, geothermal, and biomass applications. The results are presented in Table 5.

Table 5. Degree of knowledge upon various RES technologies.

RES Type	Poor	Fair	Average	Good	Excellent
Wind	6.0	20.8	31.5	24.3	17.5
Solar	2.8	13.8	29.3	28.0	26.3
Hydrodynamic	14.3	32.8	28.5	14.5	10.0
Geothermal	30.5	33.3	18.8	11.3	6.3
Biomass	34.3	35.8	15.5	9.3	5.3

In response to perceived RES benefits (see Table 6), most of the respondents (51.5%) consider environmental protection to be an utmost importance parameter, followed by the increase of energy independence. In a question concerning respondents' opinion about "further RES expansion", most of them (83%) gave a positive answer.

Table 6. Perceived RES benefits.

RES Advantages	Strongly Disagree	Disagree	Neutral	Argee	Strongly Agree
Life quality	0.3	1.5	13.5	40.0	44.8
Environmental protection	0.3	1.5	11.5	35.3	51.5
Economic development	0.3	1.3	19.3	41.8	37.5
Green development	0.5	2.8	13.5	39.3	44.0
New labor positions	0.5	2.3	20.3	38.5	38.5
Reduced oil dependence	0.0	1.3	13.3	40.0	45.5
Energy independence	0.0	1.5	16.8	35.0	46.8

In a dichotomous type question about WTP for a further expansion of 10% of RES share, in the electricity generation mix, more than one-third of our sample gave a positive answer. Out of those respondents, a percentage of 28.9 are willing to accept an increase of 6–10 euro in their electricity bill, while a percentage of 52.4 are willing to pay more than 10 euros per quarter, as it can be seen in Table 7. By taking the mean of each class of Table 7, multiplying with frequency and dividing the total sum by the total number of cases, we estimated that the mean WTP for a 10% increase of RES penetration in electricity mix, is 26.5 euros per household, quarterly.

Our results led to higher estimation of WTP for greener electricity compared to other studies on WTP for RES electricity within the Greek context [19,51,69–71], as presented in Table 8. A possible explanation for this is that our survey took place in an urban area near Athens, where the need for environmental protection is a matter of top priority among the respondents. Ambient air condition in the Attica region is heavily burdened by traffic loads, heating applications, and industrial facilities [22].

Table 7. Willingness to pay for a further 10% RES expansion to the electricity mix.

How Much Money Would you be Willing/or Pay Every Quarter?		
Classes	**Frequency**	**%**
2€–5€	28	18.8
6€–10€	43	28.9
11€–30€	28	18.8
31€–50€	27	18.1
51€–100€	19	12.8
>100€	4	2.7
Total	149	100.0

Table 8. WTP for renewable energy expansion in Greece.

Authors	Year	Area	WTP for green energy
Zografakis et al. [19]	2010	Crete Island	16.33 euros quarterly
Koundouri et al. [69]	2009	Rhodes Island	11.60 euros quarterly *
Markantonis & Bithas [70]	2009	Country level	13.93 euros quarterly *
Kontogianni et al. [71]	2013	Lesvos Island	138–180 euros one-off payment
Savvanidou et al. [51]	2010	Thrace	0.079 €/L for car biofuel

* calculated according to data included in the research paper.

To access the drivers of WTP, we applied a factorial analysis under the PCA method, by inputting all variables concerning respondents' aspects on various RES issues. Under this methodology, each identified component interprets a percentage of the variance that has not been interpreted by previous components. In the context of social sciences, an explained percentage of 60% of the variance or less can be accepted [72]. Kaiser's criterion (eigenvalue > 1) was used for factor identification. KMO criterion and Bartlett's test of sphericity were applied prior to the PCA method to measure question and sampling adequacy. Regarding the number of observations in proportion to the number of variables, this ratio should be at least 5:1 and ideally 10:1 [72], a condition fulfilled by our sample. The result of this analysis via KMO index and Bartlett test of sphericity (Table 9), revealed satisfactory values (KMO = 0.86 and $p < 0.001$ on Bartlett). This conclusion implied that statistically significant correlations exist between questions, and that the sample size meets the criteria to be used for factor analysis [30].

Table 9. KMO and Barlett test for factor analysis appropriateness.

Kaiser-Meyer-Olkin Measure of Sampling Adequacy		0.867
	Approx. Chi-Square	9670.2
Bartlett's Test of Sphericity	df	780
	Sig.	0.000

Out of the initial number of variables, nine components were identified by the Kaiser criterion, explaining a total of 68% of the observed variance; a percentage which is considered satisfactory. A rotation of the initial factors was afterwards performed by the Varimax method. The rotation enabled the simplification of the initial factor table. Regarding the nature of the questions, respondents' attitudes towards RES fall into the components presented in Table 10.

By using the F1 component named "RES perceived benefits", a positive relationship with WTP exists, as verified by the one-way ANOVA method, presented inTable 11. When the score on perceived RES benefit rises, WTP becomes higher, as depicted in the means plot (Figure 3).

Table 10. Main components of public attitudes towards RES.

F1: Perceived benefits
F2: Perceived disadvantages
F3: RES energy subsidies
F4: Actions for expansion
F5: Institutional promotion barriers
F6: Economical obstacles
F7: Price compared to fossil fuels
F8: Motivation by the social-legal framework
F9: Purchase with interest free installments

Table 11. One-way ANOVA between the variables "willingness to pay for RES" and "perceived benefits from RES".

F1: Perceived Benefits					
	Sum of Squares	df	Mean Square	F	Sig.
Between groups	10.907	5	2.181	2.423	0.038
Within groups	128.767	143	0.900		
Total	139.674	148			

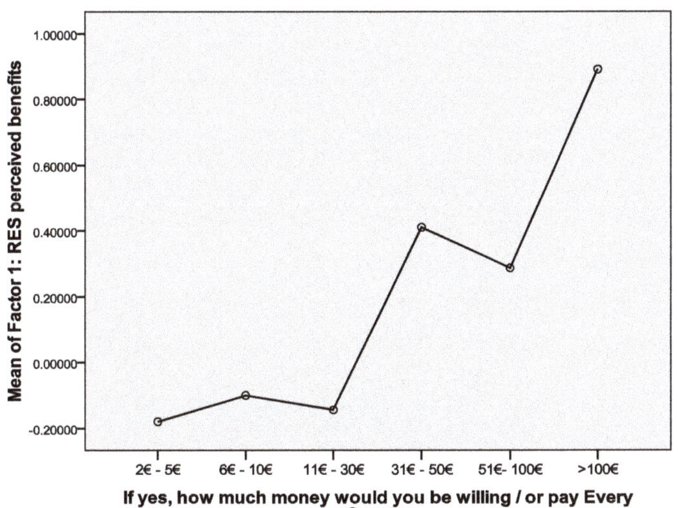

Figure 3. Depiction of a positive relationship between WTP and perceived RES advantages.

To further assess the desire for additional payment (WTP) for energy from renewable sources, we used a binary logit model in STATA/MP 13.0, setting the dichotomous variable WTP ('yes/no') as dependent. The independent variables we used are the socio-demographic characteristics of the sample (age, income, occupation, education, and gender) and the nine components of Table 10 (variables F1–F9), reflecting respondents' views on RES. The reference category for the dependent variable of WTP is 'yes'.

As seen in Table 12, McFadden's R squared, which is the default 'pseudo R^2' measure reported by Stata [73], equals 17.83%. The initial model, including all the variables, is presented in Table 13. Model coefficients are presented under column (B) and the odds ratio is given in the most-right column labeled "Exp(B)".

Table 12. Initial model, pseudo R square.

Logistic Regression	Number of Obs =	399
	LR chi2(13) =	90.57
	Prob > chi2 =	0
Log likelihood = −208.68486	Pseudo R^2 =	0.1783

Table 13. Variables included in the initial logit model for the estimation of WTP.

WTP	Coef. (B)	Std. Err.	Z	P > z (sig.)	(95% Conf.	Interval)	Odds Ratio Exp(B)
EDUCATION	0.242	0.101	2.390	0.017	0.043	0.440	1.273
OCCUPATION	0.038	0.075	0.500	0.618	−0.110	0.185	1.038
INCOME	0.003	0.004	0.920	0.356	−0.004	0.011	1.003
GENDER	−0.170	0.250	−0.680	0.495	−0.660	0.319	0.843
AGE	−0.086	0.139	−0.620	0.536	−0.358	0.186	0.918
F1	0.108	0.124	0.870	0.384	−0.136	0.352	1.114
F2	−0.506	0.126	−4.010	0.000	−0.753	−0.259	0.603
F3	0.425	0.132	3.230	0.001	0.167	0.682	1.529
F4	0.517	0.134	3.870	0.000	0.255	0.780	1.678
F5	0.444	0.128	3.460	0.001	0.192	0.696	1.559
F6	0.110	0.122	0.900	0.370	−0.130	0.349	1.116
F7	−0.020	0.120	−0.160	0.870	−0.255	0.216	0.981
F8	0.501	0.129	3.890	0.000	0.248	0.754	1.650
Constant	−1.868	0.894	−2.090	0.037	−3.620	−0.115	0.154

By looking at sig. values, in column (P > z) of Table 13, we observe that the variables of OCCUPATION, INCOME, GENDER, AGE, F1, F6, and F7 do not make a significant contribution to the model, having sig. > 0.05 at 95% confidence level. Therefore, we decided to drop those variables by using the stepwise method, provided by Stata/MP 13.0. The optimal solution was found after seven iterations, as presented in Table 14.

Table 14. Stepwise regression, variables removed according the sig. < 0.05 criteria.

. Stepwise, pr (.05): Logistic WTP EDUCATION OCCUPATION INCOME GENDER AGE F1 F2 F3 F4 F5 F6 F7 F8
Step 1: p = 0.8703 >= 0.0500 removing F7
Step 2: p = 0.6116 >= 0.0500 removing OCCUPATION
Step 3: p = 0.4735 >= 0.0500 removing GENDER
Step 4: p = 0.4370 >= 0.0500 removing AGE
Step 5: p = 0.4595 >= 0.0500 removing F6
Step 6: p = 0.2982 >= 0.0500 removing INCOME
Step 7: p = 0.3404 >= 0.0500 removing F1

By looking at Table 15, McFadden's R^2 of the final model, equals 17.05%, indicating a weak, although respectable capability to explain WTP variation. Hosmer & Lemeshow goodness-of-fit test statistic of the final iteration is greater than 0.05 (Table 16), indicating that the model fits the data at an acceptable level (p = 0.148 > 0.05).

Table 15. Final model, pseudo R square.

Logistic Regression	Number of Obs =	399
	LR chi2(13) =	86.62
	Prob > chi2 =	0
Log likelihood = −210.65756	Pseudo R^2 =	0.1705

Table 16. Goodness of fit test, Hosmer and Lemeshow.

Step	Chi-Square	df	Sig.
1	34.359	8	0.000
2	30.616	8	0.000
3	37.694	8	0.000
4	23.014	8	0.003
5	17.928	8	0.022
6	17.442	8	0.026
7	12.063	8	0.148

All variables included in the final model are presented in Table 17, along with their coefficient, the confidence intervals, and the odds ratio. Interpreting the results of Table 17, we clarify that column "B", includes the coefficients of the logit model while "Exp(B)" shows the odds ratios, or the marginal probabilities, for the predictors. The odds ratios are the exponentiation of the coefficients.

Table 17. Variables included in the final logit model for the estimation of WTP.

WTP	Coef.	Std. Err.	z	P > z	(95% Conf.	Interval)	Odds Ratio
	(B)			(sig.)			Exp(B)
EDUCATION	0.239	0.098	2.430	0.015	0.046	0.432	1.270
F8	0.487	0.126	3.860	0.000	0.240	0.734	1.627
F3	0.404	0.128	3.160	0.002	0.153	0.655	1.498
F4	0.534	0.129	4.130	0.000	0.280	0.788	1.706
F5	−0.469	0.116	−3.730	0.000	−0.716	−0.223	0.599
F2	−0.498	0.123	−4.040	0.000	−0.740	−0.256	0.608
Contant	−2.080	0.534	−3.890	0.000	−3.126	−1.033	0.125

Thus, the final form of the model is:

$$\text{logit}(p) = \log(p/(1-p)) = -2.080 - 0.498F2 + 0.40F3 + 0.534F4 - 0.469F5 + 0.487F8 + 0.239 \quad (3)$$

Out of the initial 15 explanatory variables, 6 were statistically significant. Negative coefficients mean that a one-unit increase in those variables, minimizes the odds that the user remains in the reference category of "WTP = yes", by 1-Exp(B). The variables negatively associated with WTP are F2 (Perceived disadvantages of RES) and F5 (Institutional promotion barriers). On the other hand, variables with a positive coefficient, F2 (Perceived benefits from RES), EDUCATION, F3 (Subsidies for RES), and F8 (Motivation by socio-political framework), are positively associated with "WTP = yes".

For example, the odds ratio coefficient, under column Exp(B) of variable "EDUCATION" says that, holding all other explanatory variables at a fixed value, we will see 27% increase in the odds of a respondent belonging in "WTP = yes", for a one unit increase in the educational level, since exp(0.239) = 1.270 [74]. The same explanation applies to variables F3, F4, and F8.

On the other hand, a one-unit increase in F2 (Perceived disadvantages of RES) holding all explanatory variables fixed, decreases the odds of a respondent belonging in the category "WTP = yes" by 39.2%, since 1-exp(B) = 1 − 0.608 = 0.392 [74]. The same applies to variable F5, which also has a negative coefficient.

The overall predictability of the model is depicted in Table 18. Overall, 76.2% of the respondents were correctly identified. Specifically, 93.7% of the respondents who are not willing to pay more for RES expansion (WTP = no) and 36.8% of those willing to pay the extra cost (WTP = yes), were classified in the correct category. We decided to check for differences in personal characteristics of the respondents because of the low predictability of the final model for the 'WTP = yes' category. For this purpose, we applied Pearson's chi-squared test of independence between all personality characteristics variables and WTP, noticing a relationship between WTP and the dichotomous variable

'RES user'. More specifically, RES users are more willing to pay than non-RES users. By applying a filter and selecting all respondents who are 'RES users', we re-run our final binary logit model and took the following table:

Table 18. Logit model overall prediction capabilities.

Observed		Predicted		
		WTP		Percentage Correct
		yes	no	
WTP	yes	39	67	36.8
	no	15	223	93.7
Overall percentage				76.2

As we see in Table 19, our final model has significant better prediction capabilities (Pseudo R^2 = 0.440) when it is applied only to RES users. Several interesting studies are also reporting that personality traits, like environmental awareness or eco-consciousness, are positively associated with environmentally responsible behavior [75–77].

Table 19. Logit model, prediction capabilities when applied to RES users.

Observed		Predicted		
		WTP		Percentage Correct
		yes	no	
WTP	yes	54	35	60.7
	no	21	125	85.6
Overall percentage				76.2

5. Discussion and Conclusions

Concerning policy implications, in previous Greek studies, a positive public attitude was denoted [19,21,57]. It is noteworthy that the social acceptability and the perceived advantages of RES diffusion in the national energy mix stems from the adverse environmental and healthcare implications from the ongoing air pollution and ecosystems' deterioration caused by the overexploitation of the carbon-based fossil fuels. Our survey results suggest that the prolonged economic recession in the Greek economy motivated citizens to undertake market research for cost-effective energy choices, especially regarding their household expenses. Most of the respondents have installed water heaters while more than half of them have in-depth knowledge of solar energy systems followed by wind energy technologies. Environmental protection is outlined as the most important reason for installing RES technologies, followed by reduced oil dependency. On the other hand, the respondents reported that high installation costs and lack of information are dominant reasons for not installing any kind of RES technology. Willingness to pay for an expansion of 10% of RES into the current electricity mix was estimated at 26.5 euros/quarterly on the electricity bill, higher than previous Greek studies which ranged between 11–16 euros/quarterly [19,51,69–71]. Our higher estimate of WTP is attributed to respondents' perception upon the role of RES in improving the environmental quality, which plays a decisive role to the wider acceptability of green investments [17,20,25]. The results of WTP are comparable to other European countries like Spain and Slovenia. On the other hand, Japan and Australia are amongst the countries with the highest estimated WTP. According to previous studies, WTP was found to be related to income, age, education, and environmental awareness. By using a logit model, we discovered that WTP is positively related to education status, subsidies provided by the state, actions for RES expansion undertaken by the state and motivation by the socio-political framework. On the other hand, the perceived disadvantages of RES and institutional barriers negatively affected WTP.

Moreover, since WTP was found to be associated with the role of the state, local stakeholders—such as municipality and local authorities, as well as private-owned investment companies—should take that into consideration for the implementation of a successful national environmental policy [28,31]. The state must also ensure access to environmental information for citizens to stimulate investment desire and participation [19,21,57]. Citizens' active participation promotes the implementation of the renewable energy targets agreed upon by the national governments.

Concerning the limitations of this study, it must be noted that it was undertaken at a densely populated urban area, in which citizens' prioritization is mainly determined by the cost-effective purchase for commercial goods and services and not by the environmental protection of the nearby provinces and the surrounded suburban region. Contrarily, future research orientations under the same methodological approach, can be implemented to purely rural and island areas, in which it is anticipated that, among the local population, environmental protection and the improvement of life quality criteria should take advantage of the economically convenient and less costly energy choices that the Greek countryside offers. Finally, the inclusion of personality trait variables [75–77] in logit models evaluating WTP may provide better fit and better forecasting accuracy.

Author Contributions: Stamatios Ntanos, Miltiadis Chalikias, and Garyfallos Arabatzis designed the experimental framework and analyzed the data. Stamatios Ntanos and Mitiadis Chalikias carried out the implementation, performed the calculations and the computer programming. Grigorios L. Kyriakopoulos and Skordoulis Michalis gathered and implemented all the theoretical background of the paper, having the input from the experimental development. Arabatzis Garyfallos, Grigorios L. Kyriakopoulos, and Stamatios Ntanos reviewed and discussed the results of the study.

Conflicts of Interest: The authors declare no conflicts of interest.

References

1. Mariani, F.; Perez-Barahona, A.; Raffin, N. Life expectancy and the environment. *J. Econ. Dyn. Control* **2010**, *34*, 798–815. [CrossRef]

2. Lam, C.W.; Lim, S.; Schoenung, J.M. Environmental and risk screening for prioritizing pollution prevention opportunities in the U.S. printed wiring board manufacturing industry. *J. Hazard. Mater.* **2011**, *189*, 315–322. [CrossRef] [PubMed]

3. Ockenden, C.M.; Deasy, C.; Quinton, J.N.; Surridge, B.; Stoate, C. Keeping agricultural soil out of rivers: Evidence of sediment and nutrient accumulation within field wetlands in the UK. *J. Environ. Manag.* **2014**, *135*, 54–62. [CrossRef] [PubMed]

4. Van Gent, H.A.; Rietveld, P. Road transport and the environment in Europe. *Sci. Total Environ.* **1993**, *129*, 205–218. [CrossRef]

5. International Energy Agency (IEA). CO_2 Emissions from Fuel Combustion Highlights 2017. Available online: https://www.iea.org/publications/freepublications/publication/co2-emissions-from-fuel-combustion-highlights-2017.html (accessed on 10 January 2018).

6. Bell, A.R.; Cook, B.I.; Anchukaitis, K.J.; Buckley, B.M.; Cook, E.R. Repurposing climate reconstructions for drought prediction in Southeast Asia: A letter. *Clim. Chang.* **2011**, *106*, 691–698. [CrossRef]

7. Canadell, J.G.; Kirschbaum, M.U.F.; Kurz, W.A.; Sanz, M.J.; Schlamadinger, B.; Yamagata, Y. Factoring out natural and indirect human effects on terrestrial carbon sources and sinks. *Environ. Sci. Policy* **2007**, *10*, 370–384. [CrossRef]

8. Boden, T.A.; Marland, G.; Andres, R.J. *Global, Regional, and National Fossil-Fuel CO_2 Emissions*; Carbon Dioxide Information Analysis Center, Oak Ridge National Laboratory, U.S. Department of Energy: Oak Ridge, TN, USA, 2017.

9. Dyson, F. Can we Control the Carbon Dioxide in the Atmosphere? *Energy* **1976**, *2*, 217–291. [CrossRef]

10. Bach, W. Impact of increasing atmospheric CO_2 concentrations on global climate: Potential consequences and corrective measures. *Environ. Int.* **1979**, *2*, 215–228. [CrossRef]

11. Garrett, C.W. On global climate change, carbon dioxide, and fossil fuel combustion. *Progress Energy Combust. Sci.* **1992**, *18*, 369–407. [CrossRef]

12. IPCC. Climate Change 2014: Synthesis Report. Contribution of Working Groups I, II and III to the Fifth Assessment Report of the Intergovernmental Panel on Climate Change, Geneva, Switzerland. Available online: https://www.ipcc.ch/report/ar5/wg3/ (accessed on 7 December 2017).

13. The Paris Agreement. Work Programme under the Paris Agreement. Available online: http://unfccc.int/paris_agreement/items/9485.php (accessed on 20 January 2018).

14. Vasseur, V.; Kemp, R. The adoption of PV in the Netherlands: A statistical analysis of adoption factors. *Renew. Sustain. Energy Rev.* **2015**, *41*, 483–494. [CrossRef]

15. Eurostat. Share of R.E.S. in Gross Final Energy Consumption for 2018. Available online: http://ec.europa.eu/eurostat/statistics-explained/images/d/dc/Renewable_energy_statistics-2018-v1.xlsx (accessed on 3 February 2018).

16. Liu, W.; Wang, C.; Mol, A. Rural public acceptance of renewable energy deployment: The case of Shandong in China. *Appl. Energy* **2013**, *102*, 1187–1196. [CrossRef]

17. Caporale, D.; De Lucia, C. Social acceptance of on-shore wind energy in Apulia Region (Southern Italy). *Renew. Sustain. Energy Rev.* **2015**, *52*, 1378–1390. [CrossRef]

18. Rosso-Cerón, A.M.; Kafarov, V. Barriers to social acceptance of renewable energy systems in Colombia. *Curr. Opin. Chem. Eng.* **2015**, *10*, 103–110. [CrossRef]

19. Zografakis, N.; Sifaki, E.; Pagalou, M.; Nikitaki, G.; Psarakis, V.; Tsagarakis, K.P. Assessment of public acceptance and willingness to pay for renewable energy sources in Crete. *Renew. Sustain. Energy Rev.* **2010**, *14*, 1088–1095. [CrossRef]

20. Arabatzis, G.; Myronidis, D. Contribution of SHP Stations to the development of an area and their social acceptance. *Renew. Sustain. Energy Rev.* **2011**, *15*, 3909–3917. [CrossRef]

21. Kaldellis, J.K.; Kapsali, M.; Kaldelli, E.; Katsanou, E. Comparing recent views of public attitude on wind energy, photovoltaic and small hydro applications. *Renew. Energy* **2013**, *52*, 197–208. [CrossRef]

22. Valavanidis, A.; Vlachogianni, T.; Loridas, S.; Fiotakis, C. Atmospheric Pollution in Urban Areas of Greece and Economic Crisis, University of Athens, Department of Chemistry. 2016. Available online: http://www.chem.uoa.gr/wp-content/uploads/epistimonika_themata/atmosph_pollut_greece.pdf (accessed on 8 January 2017).

23. Bertsch, V.; Hall, M.; Weinhardt, C.; Fichtner, W. Public acceptance and preferences related to renewable energy and grid expansion policy: Empirical insights for Germany. *Energy* **2016**, *114*, 465–477. [CrossRef]

24. Tabi, A.; Wüstenhagen, R. Keep it local and fish-friendly: Social acceptance of hydropower projects in Switzerland. *Renew. Sustain. Energy Rev.* **2017**, *68*, 763–773. [CrossRef]

25. Enevoldsen, P.; Sovacool, B. Examining the social acceptance of wind energy: Practical guidelines for onshore wind project development in France. *Renew. Sustain. Energy Rev.* **2016**, *53*, 178–184. [CrossRef]

26. Eleftherakis, D. Citizens towards Renewable Energy sources. Available online: http://www.skai.gr/news/environment/article/305772/stis-protovoulies-ton-politon-i-uperishusi-ton-ananeosimon-pigon-energeias/ (accessed on 25 November 2018). (In Greek)

27. Shin, J.; Woo, J.; Huh, S.Y.; Lee, J.; Jeong, G. Analyzing public preferences and increasing acceptability for the Renewable Portfolio Standard in Korea. *Energy Econ.* **2014**, *42*, 17–26. [CrossRef]

28. Zhai, P.; Williams, E. Analyzing consumer acceptance of photovoltaics (PV) using fuzzy logic model. *Renew. Energy* **2012**, *41*, 350–357. [CrossRef]

29. Ribeiro, F.; Ferreira, P.; Araújo, M.; Braga, A.C. Public opinion on renewable energy technologies in Portugal. *Energy* **2014**, *69*, 39–50. [CrossRef]

30. Sahu, B.K. A study on global solar PV energy developments and policies with special focus on the top ten solar PV power producing countries. *Renew. Sustain. Energy Rev.* **2015**, *43*, 621–634. [CrossRef]

31. Huijben, J.C.C.M.; Verbong, G.P.J. Breakthrough without subsidies? PV business model experiments in the Netherlands. *Energy Policy* **2013**, *56*, 362–370. [CrossRef]

32. Yuan, X.; Zuo, J.; Ma, C. Social acceptance of solar energy technologies in China—End users' perspective. *Energy Policy* **2011**, *39*, 1031–1036. [CrossRef]

33. Muhammad-Sukki, F.; Ramirez-Iniguez, R.; Abu-Bakar, S.H.; McMeekin, S.G.; Stewart, B.G. An evaluation of the installation of solar photovoltaic in residential houses in Malaysia: Past, present, and future. *Energy Policy* **2011**, *39*, 7975–7987. [CrossRef]

34. Hanger, S.; Komendantova, N.; Schinke, B.; Zejli, D.; Ihlal, A.; Patt, A. Community acceptance of large-scale solar energy installations in developing countries: Evidence from Morocco. *Energy Res. Soc. Sci.* **2016**, *14*, 80–89. [CrossRef]
35. Höffken, J.I. A closer look at small hydropower projects in India: Social acceptability of two storage-based projects in Karnataka. *Renew. Sustain. Energy Rev.* **2014**, *34*, 155–166. [CrossRef]
36. Breffle, W.S.; Morey, E.R.; Lodder, T.S. Using Contingent Valuation to Estimate a Neighborhood's Willingness to Pay to Preserve Undeveloped Urban Land. *Urban Stud.* **1998**, *35*, 715–727. [CrossRef]
37. Roe, B.; Teisl, F.M.; Levy, A.; Russell, M. US consumers' willingness to pay for green electricity. *Energy Policy* **2001**, *29*, 917–925. [CrossRef]
38. Zarnikau, J. Consumer demand for 'green power' and energy efficiency. *Energy Policy* **2003**, *31*, 1661–1672. [CrossRef]
39. Dalton, G.J.; Lockington, D.A.; Baldock, T.E. A survey of tourist attitudes to renewable energy supply in Australian hotel accommodation. *Renew. Energy* **2008**, *33*, 2174–2185. [CrossRef]
40. Ek, K. Public and private attitudes towards "green" electricity: The case of Swedish wind power. *Energy Policy* **2005**, *33*, 1677–1689. [CrossRef]
41. Hansla, A.; Gamble, A.; Juliusson, A.; Gärling, T. Psychological determinants of attitude towards and willingness to pay for green electricity. *Energy Policy* **2008**, *36*, 768–774. [CrossRef]
42. Nomura, N.; Akai, M. Willingness to pay for green electricity in Japan as estimated through contingent valuation method. *Appl. Energy* **2004**, *78*, 453–463. [CrossRef]
43. Whitehead, J.C.; Cherry, T.L. Willingness to pay for a Green Energy program: A comparison of ex-ante and ex-post hypothetical bias mitigation approaches. *Resour. Energy Econ.* **2007**, *29*, 247–261. [CrossRef]
44. Mozumder, P.; Vásquez, W.F.; Marathe, A. Consumers' preference for renewable energy in the southwest USA. *Energy Econ.* **2011**, *33*, 1119–1126. [CrossRef]
45. Guo, X.; Liu, H.; Mao, X.; Jin, J.; Chen, D.; Cheng, S. Willingness to pay for renewable electricity: A contingent valuation study in Beijing, China. *Energy Policy* **2014**, *68*, 340–347. [CrossRef]
46. Ivanova, G. Are consumers' willing to pay extra for the electricity from renewable energy sources? An example of Queensland, Australia. *Int. J. Renew. Energy* **2012**, *2*, 758–766.
47. Bigerna, S.; Polinori, P. Italian households' willingness to pay for green electricity. *Renew. Sustain. Energy Rev.* **2014**, *34*, 110–121. [CrossRef]
48. Zoric, J.; Hrovatin, N. Household willingness to pay for green electricity in Slovenia. *Energy Policy* **2012**, *47*, 180–187. [CrossRef]
49. Tourkolias, C.; Mirasgedis, S. Quantification and monetization of employment benefits associated with renewable energy technologies in Greece. *Renew. Sustain. Energy Rev.* **2011**, *15*, 2876–2886. [CrossRef]
50. Kostakis, I.; Sardianou, E. Which factors affect the willingness of tourists to pay for renewable energy? *Renew. Energy* **2012**, *38*, 169–172. [CrossRef]
51. Savvanidou, E.; Zervas, E.; Tsagarakis, K.P. Public acceptance of biofuels. *Energy Policy* **2010**, *38*, 3482–3488. [CrossRef]
52. Kyriakopoulos, G.L.; Arabatzis, G.; Chalikias, M. Renewables exploitation for energy production and biomass use for electricity generation. A multi-parametric literature-based review, In special issue: "Biomass Utilization Technology for Building of Recycling Society". *AIMS Energy* **2016**, *4*, 762–803. [CrossRef]
53. Kyriakopoulos, G.L.; Arabatzis, G. Electrical energy storage systems in electricity generation: Energy policies, innovative technologies, and regulatory regimes. *Renew. Sustain. Energy Rev.* **2016**, *56*, 1044–1067. [CrossRef]
54. Ntona, E.; Arabatzis, G.; Kyriakopoulos, G.L. Energy saving: Views and attitudes of students in secondary education. *Renew. Sustain. Energy Rev.* **2015**, *46*, 1–15. [CrossRef]
55. Kyriakopoulos, G.; Kolovos, K.; Chalikias, M. Environmental sustainability and financial feasibility evaluation of woodfuel biomass used for a potential replacement of conventional space heating sources. Part II: A Combined Greek and the nearby Balkan Countries Case Study. *Oper. Res. Int. J.* **2010**, *10*, 57–69. [CrossRef]
56. Chalikias, M.S.; Kyriakopoulos, G.; Kolovos, K.G. Environmental sustainability and financial feasibility evaluation of woodfuel biomass used for a potential replacement of conventional space heating sources. Part I: A Greek case study. *Oper. Res. Int. J.* **2010**, *10*, 43–56. [CrossRef]
57. Tsantopoulos, G.; Arabatzis, G.; Tampakis, S. Public attitudes towards photovoltaic developments: Case study from Greece. *Energy Policy* **2014**, *71*, 94–106. [CrossRef]

58. Barker, T.; Dagoumas, A.; Rubin, J. The macroeconomic rebound effect and the world economy. *Energy Effic.* **2009**, *2*, 411–427. [CrossRef]

59. Dagoumas, A. Energy and climatic policies in Greece and Europe. In *Energeia*; KEPE: Athens, Greece, 2014; pp. 7–10. Available online: https://energypress.gr/sites/default/files/media/KEPE%20ENERGY%202014.pdf (accessed on 10 December 2017). (In Greek)

60. Damianou, C. *Sampling Methodology: Techniques and Applications*, 3rd ed.; Aithra Publications: Athens, Greece, 1999. (In Greek)

61. Ministry of Interior. Election catalogues of the Greek Parliamentary Elections on January 2015 (Excel file). Available online: http://www.ypes.gr/el/Elections/ElectionCatalogues/ElectionDepartments/ (accessed on 2 February 2018). (In Greek)

62. Eng, J. Sample Size Estimation: How Many Individuals Should Be Studied? *Radiology* **2003**, *227*, 309–313. [CrossRef] [PubMed]

63. Kolovos, K.G.; Kyriakopoulos, G.; Chalikias, M.S. Co-evaluation of basic woodfuel types used as alternative heating sources to existing energy network. *J. Environ. Prot. Ecol.* **2011**, *12*, 733–742.

64. Kyriakopoulos, G.; Chalikias, M.S. The Investigation of Woodfuels' Involvement in Green Energy Supply Schemes at Northern Greece: The Model Case of the Thrace. *Procedia Technol.* **2013**, *8*, 445–452. [CrossRef]

65. Arabatzis, G.; Malesios, C. Pro-environmental attitudes of users and non-users of fuelwood in a rural area of Greece. *Renew. Sustain. Energy Rev.* **2013**, *22*, 621–630. [CrossRef]

66. Kaldellis, J.K. Social attitude towards wind energy applications in Greece. *Energy Policy* **2005**, *33*, 595–602. [CrossRef]

67. Ntanos, S.; Arabatzis, G.; Chalikias, M. The role of emotional intelligence as an underlying factor towards social acceptance of green investments. In Proceedings of the 8th International Conference on Information and Communication Technologies in Agriculture, Food and Environment, Chania, Greece, 21–24 September 2017; Volume 2030, pp. 341–351.

68. Hellenic Statistical Authority. Population census 2011, Table Γ01. Permanent population by sex and age groups. Settlements with a population of >50,000. Data for Nikaia. Available online: http://www.statistics.gr/el/statistics/-/publication/SAM03/2011 (accessed on 02 February 2018). (In Greek)

69. Koundouri, P.; Kountouris, Y.; Remoundou, K. Valuing a wind farm construction: A contingent valuation study in Greece. *Energy Policy* **2009**, *37*, 1939–1944. [CrossRef]

70. Markantonis, V.; Bithas, K. The application of the contingent valuation method in estimating the climate change mitigation and adaptation policies in Greece. An expert-based approach. *Environ. Dev. Sustain.* **2010**, *12*, 807–824. [CrossRef]

71. Kontogianni, A.; Tourkolias, C.; Skourtos, M. Renewables portfolio, individual preferences and social values towards RES technologies. *Energy Policy* **2013**, *55*, 467–476. [CrossRef]

72. Hair, J.F.; Black, W.C.; Babin, B.J.; Anderson, R.E. *Multivariate Data Analysis*, 7th ed.; Chapter "Explanatory Factor analysis"; Prentice Hall International: Englewood Cliffs, NJ, USA, 2010.

73. IDRE 2008. Multinomial Logistic Regression, Stata Annotated Output. Available online: https://stats.idre.ucla.edu/stata/output/multinomial-logistic-regression-2/ (accessed on 15 February 2018).

74. Kim, H.-Y. Statistical notes for clinical researchers: Logistic regression. *Restor. Dent. Endod.* **2017**, *42*, 342–348. [CrossRef] [PubMed]

75. Huijts, N.M.A.; Molin, E.J.E.; Steg, L. Psychological factors influencing sustainable energy technology acceptance: A review-based comprehensive framework. *Renew. Sustain. Energy Rev.* **2012**, *16*, 525–531. [CrossRef]

76. Parant, A.; Pascual, A.; Jugel, M.; Kerroume, M.; Felonneau, M.; Guéguen, N. Raising Students Awareness to Climate Change: An Illustration with Binding Communication. *Environ. Behav.* **2017**, *49*, 339–353. [CrossRef]

77. Stigka, E.; Paravantis, J.; Mihalakakou, G. Social acceptance of renewable energy sources: A review of contingent valuation applications. *Renew. Sustain. Energy Rev.* **2014**, *32*, 100–106. [CrossRef]

Article

Optimal Investment Planning of Bulk Energy Storage Systems

Dina Khastieva *,†, **Ilias Dimoulkas** † and **Mikael Amelin** †

Royal Institute of Technology (KTH), Teknikringen 33, 114 28 Stockholm, Sweden; iliasd@kth.se (I.D.); amelin@kth.se (M.A.)
* Correspondence: dinak@kth.com; Tel.: +46-760-744-605
† These authors contributed equally to this work.

Received: 30 January 2018; Accepted: 21 February 2018; Published: 27 February 2018

Abstract: Many countries have the ambition to increase the share of renewable sources in electricity generation. However, continuously varying renewable sources, such as wind power or solar energy, require that the power system can manage the variability and uncertainty of the power generation. One solution to increase flexibility of the system is to use various forms of energy storage, which can provide flexibility to the system at different time ranges and smooth the effect of variability of the renewable generation. In this paper, we investigate three questions connected to investment planning of energy storage systems. First, how the existing flexibility in the system will affect the need for energy storage investments. Second, how presence of energy storage will affect renewable generation expansion and affect electricity prices. Third, who should be responsible for energy storage investments planning. This paper proposes to assess these questions through two different mathematical models. The first model is designed for centralized investment planning and the second model deals with a decentralized investment approach where a single independent profit maximizing utility is responsible for energy storage investments. The models have been applied in various case studies with different generation mixes and flexibility levels. The results show that energy storage system is beneficial for power system operation. However, additional regulation should be considered to achieve optimal investment and allocation of energy storage.

Keywords: energy storage; power system planning; wind power generation; stochastic processes

---

1. Introduction

1.1. Motivation

The flexibility of a power system is defined by how well it can cope with variability and uncertainty and balance the production and consumption. Variability and uncertainty come from various sources such as time-varying demand and generation based on variable renewable sources as well as different contingencies such as line and generation outages.

Power systems are designed to handle demand variability and uncertainty as well as the majority of the contingencies. However, the increasing interest in variable renewable generation such as wind-based generation raises concerns on the need to increase the flexibility of the systems to accommodate large scale varying renewable energy sources. The capacity of wind energy installations is constantly increasing. For example, in Europe, the share of wind-based energy increased from 2.5 to 15.6% just over 15 years [1]. Current percentage of wind-based electricity generation in the European generation mix is now even greater than hydro based electricity generation which is 15.5%. Such a share of variable wind energy is still considered relatively low. In addition, the current state of a flexibility of the majority European power systems is proved to be sufficient to handle variability and uncertainty of the present wind based generation. However, if the trend will continue, power systems might have to improve the flexibility of the system. Based on current European targets, 20% consumption of energy should come from renewable generation by the year 2020. The target has been set by 2020 Climate and Energy package and will require even higher installed capacity of renewable generation due to variability and uncertainty of the renewable sources. Thus, wind power penetration is expected to grow substantially just over next few years. In addition, more ambitious targets are expected to be set for 2030 by Winter Package which is still under development. Increase in large scale renewable generation will contribute to higher volatility of wholesale electricity prices, higher balancing costs and system maintenance costs as well as large curtailments of renewable generation output. Thus, additional flexibility will be required [2,3].

The flexibility of the power system is provided mainly through flexible generation units with fast response time and flexible demand. One of the most flexible and least expensive generation units is hydro. The presence of hydropower in a power system clearly has a positive impact on the flexibility of the system. Hydrothermal power systems generally have good ramping capability and energy storage possibility in the form of hydro reservoirs. Thus, power system operators can use the flexibility of the hydropower generation to balance variable renewable electricity generation and load. However, for a large-scale expansion of wind power (or other variable generation such as solar) existing hydro flexibility might not be sufficient. More importantly, expansion of hydropower generation is difficult and in some cases is even impossible due to limited natural resources. In addition to hydropower, the flexibility of the system can then be improved by increasing the capacity of existing power plants, adding additional fast-ramping thermal generation capacity, demand response or energy storage capability. In this paper, we address the possibility to provide additional flexibility by adding energy storage capacity considering different storage technologies.

1.2. Knowledge Gap

Energy storage is not a new concept and was used for decades in power system, however predominantly pumped-hydro energy storage was in operation. Almost 99% of installed bulk energy storage capacity comes from pumped hydro and new installation of such energy storage is limited due to the same reasons as hydropower. However, other technologies such as compressed air energy storage (CAES) and various types of batteries are mature and available for applications on transmission level. In addition, other technologies for energy storage systems (ESS) are also under development and will be commercially available in foreseeable future. A database with a list of existing energy storage projects around the world is available in [4]. Energy storage systems are capable of providing additional flexibility on different time frames to power system operation by charging at peak hours and discharging when additional electricity is required. Such flexibility is very desirable for systems with high share of variable renewable generation. In addition, energy storage technologies are very fast and can be deployed at different capacities and power capabilities depending on the needs of the system. According to [5] the need in additional storage capacity in Europe alone is expected to double by 2050 mostly due to renewable generation capacity increase and additional balancing needs connected to that growth.

Energy storage systems (ESS) have multiple applications and can be beneficial at different levels of the electricity system. Various literature provide an overview on possible applications and assessment of energy storage benefits. In [6] a comprehensive analysis of possible energy storage applications and suitable energy storage technologies is presented. Applications may vary from energy arbitrage to grid upgrade investments deferral. The most promising applications for energy storage include energy arbitrage, balancing services and renewable generation support. Different ways how energy storage systems could be used for balancing applications, especially in presence of a large amount of variable renewable generation, were studied in [7,8], while [9] includes benefits of energy storage as a flexibility source. In addition, [10,11] analyze how energy storage can be beneficial for supporting variable wind power generation and [12] presents benefits of energy storage from a technical point of view and its effect on maximum wind power penetration. A review of modeling techniques of energy storage given different objectives is provided in [13] and includes more than 150 papers on the energy storage assessment subject. The literature provides evidence that energy storage is beneficial for renewable generation support and can be profitable under certain assumptions, however high capital cost is seen as the main obstacle in energy storage market development. Cost evaluation and calculation of different energy storage technologies is presented in [14,15].

The aforementioned papers have shown that additional capacity of flexibility sources such as energy storage will be required to reach future renewable targets and energy storage might be profitable in the systems with a high share of renewables but the financial profitability of the energy storage is still strongly dependent on the size and location of the deployed energy storage system. Optimal planning of energy storage under different conditions and objectives have been studied in [10,16–22]. In addition, [23–27] investigated joint optimal allocation and sizing of energy storage. In [28] the authors also show that energy storage is beneficial for renewable generation expansion and that joint optimization of renewable generation and flexibility sources including energy storage results in much higher cost savings than when investment planning is procured separately. However, these papers consider centralized investments planning which does not ensure profitability of the energy storage system itself and does not consider profit maximizing behaviour of the energy storage investor. Should flexibility sources such as energy storage be a market asset or system asset is an open question in power systems. Under current European regulation energy storage cannot be used to obtain profit if it is owned by system operators. Thus, current development of energy storage will mostly depend on independent investors which have profit maximizing objectives and other constraints on expected profit. A profit maximizing bilevel approach for investment planning of energy storage systems which will ensure that the owner of the energy storage will maximize its benefits has been proposed in [23,29,30]. However, neither of the proposed models include other sources of flexibility such as hydro and flexible demand which are currently the main competitors of emerging energy storage systems. Moreover, these models do not take into account possible growth of renewable generation.

1.3. Modeling Methodology

As in [23] this paper proposes a bilevel investment planning of strategic energy storage investor following the modeling approach proposed in [31] for generation investment planning. The approach allows to model behaviour of the strategic investor considering power system operation and locational marginal prices as an output of the operation. The modeling approach proposed in [31] allows to simulate operation of power system close to realistic operation and including many details such as dynamics of energy limited resources including energy storage, hydro power and flexible demand. Thus, the prices obtained to calculate energy storage profit are more realistic than using other mathematical models. In addition, the paper uses a technique to reformulate bilevel problem into single level linear program. Thus the obtained optimization problem can be solved with standard solvers such as CPLEX.

1.4. Contribution

This paper proposes two different mathematical models for joint energy storage sizing and allocation along with renewable generation expansion. The renewable generation expansion is ensured by expected renewable generation target constraint which sets the lower bound on renewable generation as a percentage of total consumption. The first model is for a centralized operation and investment planning while the second model is designed for an independent energy storage owner who is responsible for energy storage investments while the operation is still on a centralized planner. The proposed models can manage different generation sources (including thermal, hydro and variable renewable generation) along with flexible demand. The energy storage investment decisions are made over a portfolio of different energy storage technologies with varying properties for efficiency, self-discharge, etc. The owner of energy storage systems can decide which energy storage units to invest in and where to allocate them. The model has been applied to a case study under different cost parameters and various levels of installed flexibility. The proposed models and case study in this paper differ from the ones existing in the literature on four main points:

- First, the investment planning includes other sources of the flexibility of the system (hydro power and consumption flexibility) which can create competition for energy storage systems and affect the revenue stream.
- Second, energy storage investments are made along with renewable generation expansion and takes into account renewable generation targets present nowadays in Europe and USA.
- Third, the decentralized planning model in addition to investment return constraints includes payback period constraint which make the simulation of investment decision on energy storage closer to real life investment planning. Moreover, a solution to the bilevel problem has been suggested.
- Fourth, the paper presents a comparative analysis based on several case studies of systems with different generation mix and different levels of congestion. The results contribute to an understanding of the benefits of energy storage under different planning strategies and dependency of existing flexibility and type of flexibility on the profitability of the energy storage and possible effect on system congestion.

The models in this paper will be effective not only to help independent investment planning of energy storage owner considering expected growth of renewable generation but also to analyze the influence of existing flexibility in the system on energy storage investments.

1.5. Structure of the Paper

The paper is organized as follows. Section 2 introduces a centralized planning model followed by a bilevel mathematical formulation of a decentralized approach and a brief description of a one-level reformulation and linearization techniques. Section 3 presents case studies, results and an analysis of results. Finally, Section 4 provides conclusions and a discussion on a future work.

2. Energy Storage Investment Decision and Allocation Problem

The models in this paper consider two investment decisions: wind power generation expansion and energy storage investments. The first model (which is referred to as the centralized model) assumes perfect competition which could effectively be modeled by assuming that all operation and investment decisions are made by a single cost minimizing entity. The second model (which is referred to as the decentralized model) it is assumed that a separate entity makes decisions about energy storage investments, while the rest of the system remains a perfectly competitive market environment. The model assumes that the energy storage investor is a leader while the centralized planner is a follower meaning that first the decision on energy storage is made and afterwards based on that decision the centralized planner can expand renewable generation capacity and decide on operation dispatch. Thus, the energy storage entity can benefit from taking decision beforehand and strategically place energy storage while the centralized player can react to the investments and update its renewable generation capacity based on new flexibility in the system.

Both of the approaches have to take into account power system operation decisions, however the objectives are different. Thus, two different mathematical models were created in order to address the energy storage investment decision problem from two different ownership prospective. The following assumptions on energy storage investment decisions are taken for both models:

1. The energy storage investor can choose between energy storage modules of different technologies, where each module has fixed energy capacity, power capability and other technical parameters such as self-discharge and efficiency.
2. The energy storage charge and discharge efficiency as well as the self-discharge rate are fixed parameters and do not vary based on the charge/discharge output level or the energy level of the storage.

Additional assumptions are taken for each player.

1. The centralized player is responsible for generation expansion investments into renewable energy.
2. An independent investor will only make investments which will reach break-even within a given time. For example, typical expected payback period in long term investment planning is five years.
3. An independent investor has a lower limit on minimum investments returns.
4. The financial benefit of the energy storage is obtained through energy arbitrage.
5. The energy storage utility can exchange information with the centralized player which is in charge of the optimal dispatch of the generation, flexible load and energy storage in the system. The centralized player receives information from energy storage owner about invested and available energy storage energy capacity and power capability of each unit. On the other hand the centralized player provides information about dispatch of each energy storage unit and electricity prices.
6. The power system is represented by a DC load flow model.

2.1. Centralized Energy Storage Investment Decision and Allocation Problem

The investment decision problem for a centralized player is described in this section. The centralized player is responsible for the short-term operation of the power system and for the decisions on renewable generation expansion and investments and allocation of energy storage units. The problem is formulated as a mixed integer linear problem. Integer variables are used to allocate energy storage units while wind power generation investments are assumed to be continuous variables.

The objective function is described through Equation (1) and reflects the cost minimization of the whole system. The total cost consists of three main parts. The objective function is to minimize the cost of the scheduled day ahead generation dispatch $A_{p,s}$ based on the marginal generation cost of thermal units mc_j, energy storage charge and discharge cost mc_e and charges to activate flexible load Δd_d. The marginal cost of the energy storage charging or discharging is usually equal to zero (except for compressed air energy storage which uses natural gas or fuel in the discharging process). However, to take into account fast degradation connected to the cycling of some energy storage technologies such as batteries, an additional variable cost is assigned to each charge and discharge. In order to evaluate operation of energy limited resources with storage capability we also consider future value of stored energy in hydro reservoirs and energy storage systems calculated in $B_{k,t,s}$. FC_k is expected future price of electricity at the end of each operational period k. In addition to the variable costs, the system operator also minimize the investment cost into generation expansion and energy storage C_t. In this model we do not simulate all hours of operation of power system. Instead, we use selected days (for example number of seasons $k = 4$, number of selected days $d = 1$ and number of operation periods of each day $l = 24$ h). Thus, in order to match the simulated short-term operation costs of each year with and investment costs we use scale factor ψ which can be calculated through simple formula $\frac{8760}{k*l*d}$ (for the given example it will be equal to 91.25).

$$\text{Minimize}: f_{LL} = \sum_{\Omega_s}(\psi \sum_t \sum_s \pi_s(\sum_{k,l} A_{p,s} - \sum_k B_{k,t,s}) + C_t) \tag{1}$$

where:

$$A_{p,s} = \sum_j mc_j g_{j,p,s} + \sum_d mc_d(\uparrow \Delta d_{d,p,s} + \downarrow \Delta d_{d,p,s}) + \sum_e mc_e(g^{ch}_{e,p,s} + g^{dch}_{e,p,s}) \tag{2}$$

$$B_{k,t,s} = FC_{k,t}(\sum_e SOC_{e,t,k,L,s} + \sum_h \sigma_h v_{h,t,k,L,s}) \tag{3}$$

$$C_t = \sum_w C_{w,t}(G^{max}_{w,t} - G^{max}_{w,t-1}) + \sum_e C_{e,t}(y_{e,t} - y_{e,t-1}) \tag{4}$$

The minimization problem is a subject to various constraints.

Energy storage investment constraint (5) ensures that invested energy storage unit is available at the later time periods after the investment was made.

$$y_{e,t} \geq y_{e,t-1} \forall e, t \tag{5}$$

Equation (6) represents the power balance for each node of the system. Total generation and net injection should be equal to the total demand.

$$\sum_j In_{,j}g_{j,p,s} - \sum_e In_{,e}(g^{ch}_{e,p,s} - g^{dch}_{e,p,s}) + \sum_h In_{,h}g_{h,p,s} + \sum_w In_{,w}g_{w,p,s} - \sum_d In_{,d}D_{d,p}$$
$$- \sum_d In_{,d} \uparrow \Delta d_{d,p,s} + \sum_d In_{,d} \downarrow \Delta d_{d,p,s} + \sum_m In_{,m}f_{n,m,p,s} = 0 \; : (\lambda_{n,p,s}) \quad \forall p, s, n. \tag{6}$$

The energy balance constraint (7) represents the state of the charge of the energy storage unit. The dynamics of energy storage are very similar to hydropower. The main difference is that energy storage will convert surplus of electricity and store it in a different form of energy or in the form of an electromagnetic field and then convert it back when it is demanded [32]. The conversion of electricity into another form of energy induces some losses. These losses could be represented through efficiency coefficient ϵ_e of the energy storage. Also, the energy storage have a self-discharge rate ϕ_e which also cause the losses of energy. The use of efficiencies in the modeling also prevents energy storage to charge and discharge at the same time, therefore binary variables are not required.

$$SOC_{e,p,s} = \phi_e SOC_{e,t,k,l-1,s} + \epsilon_e g^{ch}_{e,p,s} - 1/\epsilon_e g^{dch}_{e,p,s} \; : (\lambda^{SOC}_{e,p,s}) \; \forall p, s, e. \tag{7}$$

Renewable generation operation constraints (8). In this model the wind-based generation could be spilled when there is an excess of generation or not enough ramping capability. Thus, (8) is used to determine actually utilized wind power $g_w(p,s)$.

$$0 \leq g_{w,p,s} \leq wp_{p,s}G_w + wp_{p,s}G^{max}_{w,t} \; : (\underline{v}_{w,p,s}, \overline{v}_{w,p,s}) \quad \forall p, s, w. \tag{8}$$

Equation (9) reflects the renewable generation penetration target which is set by the system (regulator). The equation ensures that expected wind generation at target year (RTY) and further on will be greater or equal to target values (RG^{min}).

$$\sum_s \pi_s \sum_{w,k,l} g_{w,p,s} \geq RG^{min} \sum_{d,k,l} D_{d,p} \; : (\beta_t) \quad \forall t \geq RTY. \tag{9}$$

Wind power generation investment constraints (10) which ensures that invested generation capacity stays on in the further periods.

$$0 \leq G^{max}_{w,t-1} \leq G^{max}_{w,t} \; : (\tau_{w,t}) \quad \forall t, w. \tag{10}$$

Hydro power operation constraints. The hydrological balance constraint (11) represents the hourly reservoir water level including previous content, direct inflow fl_h, spillage s_h and hydro discharge u_h used for power generation. The power generated by hydro units is determined through a linear function (12). This means that the efficiency of the hydro unit is assumed to be constant. Another approach is to use a piecewise linear function. This method is described in detail in [33].

$$v_{h,p,s} = v_{h,t,k,l-1,s} - u_{h,p,s} - s_{h,p,s} + fl_{h,k,l} + \sum_{h*} u_{h*,t,k,l-\tau_{h*},s} + \sum_{h*} s_{h*,t,k,l-\tau_{h*},s} \; : (\lambda^{res}_{h,p,s})$$
$$\forall t, k, l, s, h. \tag{11}$$

$$g_{h,p,s} = \sigma_h u_{h,p,s} \; : (\lambda^{Gen}_{h,p,s}) \; \forall p, s, h. \tag{12}$$

Ramping constraints of thermal generation units and hydro power units (13) and (14).

$$- RD_j^{max} \leq g_{j,t,k,l,s} - g_{j,t,k,l-1,s} \leq RU_j^{max} \; : (\underline{\kappa}_{j,p,s}, \overline{\kappa}_{j,p,s}) \quad \forall t,k,l,s,j. \tag{13}$$

$$- RD_h^{max} \leq g_{h,t,k,l,s} - g_{h,t,k,l-1,s} \leq RU_h^{max} \; : (\underline{\kappa}_{h,p,s}, \overline{\kappa}_{h,p,s}) \quad \forall t,k,l,s,h. \tag{14}$$

The problem considers DC power flows. Power flows $f_{n,m}$ are calculated using Equation (15) and are subject to power flow limits (21).

$$f_{n,m,p,s} - \frac{100}{X_{n,m}}(\theta_{n,p,s} - \theta_{m,p,s}) = 0 \; : (\lambda_{n,m,p,s}^{Line}) \quad \forall n,m,p,s. \tag{15}$$

The flexible demand can be increased or decreased. Thus, two different variables are used for upward demand change $\uparrow \Delta d_{d,p,s}$ and for downwards $\downarrow \Delta d_{d,p,s}$ for each hour. However, the total energy should be maintained for each operational period. Equation (16) is enforced to ensure that the total energy demand for each operational period is equal to the initial value.

$$\sum_l \uparrow \Delta d_{d,t,k,l,s} = \sum_l \downarrow \Delta d_{d,t,k,l,s} \; : (\lambda_{d,t,k,s}^D) \quad \forall t,k,s,d. \tag{16}$$

$$0 \leq\, \uparrow \Delta d_{d,p,s} \leq D_d^{max} \; : (\uparrow \underline{\omega}_{d,p,s}, \uparrow \overline{\omega}_{d,p,s}) \quad \forall p,s,d. \tag{17}$$

Upper and lower limit constraints (18)–(29) of decision variables.

$$SOC_e^{min} y_{e,t} \leq SOC_{e,p,s} \leq SOC_e^{max} y_{e,t} \; : (\overline{\gamma}_{e,p,s} \underline{\gamma}_{e,p,s}) \quad \forall p,s,e. \tag{18}$$

$$- \Theta \leq \theta_{n,p,s} \leq \Theta \; : (\underline{\rho}_{n,p,s}, \overline{\rho}_{n,p,s}) \quad \forall n,p,s. \tag{19}$$

$$\theta_{n=1,p,s} = 0 \; : (\rho 0_{p,s}) \; \forall p,s \tag{20}$$

$$- T_{n,m}^{max} \leq f_{n,m,p,s} \leq T_{n,m}^{max} \; : (\underline{\mu}_{n,m,p,s}, \overline{\mu}_{n,m,p,s}) \quad \forall n,m,p,s. \tag{21}$$

$$G_h^{min} \leq g_{h,p,s} \leq G_h^{max} \; : (\underline{\nu}_{h,p,s}, \overline{\nu}_{h,p,s}) \quad \forall p,s,h. \tag{22}$$

$$G_j^{min} \leq g_{j,p,s} \leq G_j^{max} \; : (\underline{\nu}_{j,p,s}, \overline{\nu}_{j,p,s}) \quad \forall p,s,j. \tag{23}$$

$$0 \leq\, \downarrow \Delta d_{d,p,s} \leq D_d^{max} \; : (\downarrow \underline{\omega}_{d,p,s}, \downarrow \overline{\omega}_{d,p,s}) \quad \forall p,s,d. \tag{24}$$

$$V_h^{min} \leq v_{h,p,s} \leq V_h^{max} \; : (\overline{\sigma}_{h,p,s} \underline{\sigma}_{h,p,s}) \quad \forall p,s,h. \tag{25}$$

$$0 \leq u_{h,p,s} \leq U_h^{max} \; : (\overline{\vartheta}_{h,p,s} \underline{\vartheta}_{h,p,s}) \quad \forall p,s,h. \tag{26}$$

$$0 \leq s_{h,p,s} \leq S_h^{max} \; : (\overline{\theta}_{h,p,s} \underline{\theta}_{h,p,s}) \quad \forall p,s,h. \tag{27}$$

$$0 \leq g_{e,p,s}^{ch} \leq G_e^{max} y_{e,t} \; : (\overline{\xi}_{e,p,s}^{ch}, \underline{\xi}_{e,p,s}^{ch}) \quad \forall p,s,e. \tag{28}$$

$$0 \leq g_{e,p,s}^{dch} \leq G_e^{max} y_{e,t} \; : (\overline{\xi}_{e,p,s}^{dch}, \underline{\xi}_{e,p,s}^{dch}) \quad \forall p,s,e. \tag{29}$$

It should be noted, that set p is used to simplify the notation. It contains all time period indices (year, t, season, k, and hour, l). The index p is used in the equation when all these sets are indexed together. If the equation is used just for one of the subsets, the set p is not used and the original three sets are written.

The decisions of the system operator include short term operation of the power system and investments into expansion of wind-based generation for each candidate node. $\Omega_s = \{g_j, g_w, g_h, g_e^{ch}, g_e^{dch}, \uparrow \Delta d_d, \downarrow \Delta d_d, f_{n,m}, SOC_e, u_h, v_h, s_h, G_w^{max}, y_{e,t}\}$ is the set of decision variables of the problem.

2.2. Independent Investment Planning. MPEC Model

This section describes the model when energy storage investment decisions are taken by an independent, profit-maximizing player while generation expansion decisions are in the hands of a centralized player. The problem can be described as a mathematical problem with an equilibrium constraint (MPEC) trough a bilevel program. In the upper level the energy storage owner can decide in which energy storage units to invest and where to put them while obtaining the prices and charge/discharge dispatches from the centralized player. Therefore, the optimal operation and generation investment planning model is included as a lower level problem.

$$\begin{aligned} \underset{y_{e,t}}{Maximize} \\ f_{UL} = \textstyle\sum_{t,e} Pw_t(\psi \sum_{k,l,s} \pi_s(\lambda_{n,p,s} In_e(g_{e,p,s}^{dch} - g_{e,p,s}^{ch}) - mc_e(g_{e,p,s}^{ch} + g_{e,p,s}^{dch}) + FC_{k,t} SOC_{e,p,s} Y(l=L)) \\ - C_{e,t}(y_{e,t} - y_{e,t-1})) \end{aligned} \quad (30)$$

S.t:

$$\psi \textstyle\sum_{t^*=t,k,l,e,s}^{t+PBP} \pi_s(\lambda_{n,t^*,k,l,s} In_e(g_{e,t^*,k,l,s}^{dch} - g_{e,t^*,k,l,s}^{ch}) - mc_e(g_{e,t^*,k,l,s}^{ch} + g_{e,t^*,k,l,s}^{dch})) \geq \textstyle\sum_e C_{e,t}(y_{e,t} - y_{e,t-1}) \ \forall t \quad (31)$$

$$\psi \sum_{p,s,e} \pi_s(\lambda_{n,p,s} In_e(g_{e,p,s}^{dch} - g_{e,p,s}^{ch}) - mc_e(g_{e,p,s}^{ch} + g_{e,p,s}^{dch})) \geq IR \sum_{t,s,e} C_{e,t}(y_{e,t} - y_{e,t-1}) \quad (32)$$

$$y_{e,t} \geq y_{e,t-1} \ \forall e,t \quad (33)$$

where

$$\left\{ \lambda_{n,p,s}, g_{e,p,s}^{ch}, g_{e,p,s}^{dch} \right\} \in \underset{\Omega_s/y_{e,t}}{\arg Min} \quad f_{LL}^* = \textstyle\sum_t(\psi \sum_s \pi_s(\sum_{k,l} A_{p,s} - \sum_k B_{k,t,s}) \\ + \sum_w C_{w,t}(G_{w,t}^{max} - G_{w,t-1}^{max})) \quad (34)$$

S.t:

$$(6) - (29) \quad (35)$$

The described problem is a stochastic, mixed integer problem which is optimized over t investment planning periods where each of the them consists of k seasons and l operation hours. The objective function (30) is to maximize the profit from energy storage operation which consists of a revenue stream from selling energy at price $\lambda_{n,p,s}$ while the energy storage discharges $g_{e,p,s}^{dch}$ minus the costs of buying energy for charging $g_{e,p,s}^{ch}$, the operational costs and the investments costs. Short term operation revenue and cost are multiplied by a discount factor ψ to scale the operation and investment costs and make them comparable. Equation (31) enforces the break-even constraint for energy storage investment, i.e., that the overall investments for each period should payback in a given amount of years, whereas (32) ensures that returns on the investments will be sufficiently large. Charge, discharge and price variables are obtained through the lower level problem (34).

The proposed model is a bilevel mixed integer problem. Bilevel programming models are a powerful tool for problems with multiple-criteria decision-making models. Such models can be solved in various ways and one of the them is by reformulating the given bilevel model into a one-level model. The reformulation is illustrated in Figure 1. Step 1 shows the original bilevel formulation. The lower level is a linear problem and therefore could be equivalently represented by the Karush-Kuhn-Taker (KKT) optimality conditions. KKT optimality conditions consist of primal feasibility equations, stationary conditions and complementary slackness conditions. This reformulation will not affect the optimality since the KKT conditions are both necessary and sufficient [34]. The result of replacing the lower level problem by its KKT conditions is shown Step 2. However, the optimization problem in Step 2 is non-linear and therefore the complementary slackness conditions are replaced by the strong duality condition which implies that $f_{UL}^* = f_{UL}^{\star dual}$. The final reformulated problem is shown in Step 3.

The reformulation steps described above were applied on the independent investment planing problem (30)–(35). The model was transformed from a bilevel mixed integer non-linear model into an equivalent one-level mixed integer non-linear problem. Stationary conditions and complementary slackness conditions for lower level problem (34) are derived in the Appendix of this paper. Two set of non-linearities were identified. The following sections explain how these non-linearities can be reformulated.

Step 1	*Step 2*	*Step 3*
$\text{Minimize}: f_{UL}$ $\quad x,y$	$\text{Minimize}: f_{UL}$ $\quad x,y$	$\text{Minimize}: f_{UL}$ $\quad x,y$
S.t:	S.t:	S.t:
$h_{UL}(x,y) = 0$	$h_{UL}(x,y) = 0$	$h_{UL}(x,y) = 0$
$g_{UL}(x,y) \leq 0$	$g_{UL}(x,y) \leq 0$	$g_{UL}(x,y) \leq 0$
$\text{Where } \{y\} \in \quad\equiv$	$KKT conditions:$	$\quad\equiv\quad$ $KKT conditions:$
$\arg \underset{y}{\text{Min}} f_{LL}$	$h_{LL}(y) = 0$	$h_{LL}(y) = 0$
S.t:	$g_{LL}(y) \leq 0$	$g_{LL}(y) \leq 0$
$h_{LL}(y) = 0 : (\lambda)$	$\{Stationary\ conditions\}:$	$\{Stationary\ conditions\}$
$g_{LL}(y) \leq 0 : (\mu)$	$\nabla f_{LL}(y) + \lambda \nabla g_{LL}(y) + \mu \nabla h_{LL}(y) = 0$	$\nabla f_{LL}(y) + \lambda \nabla g_{LL}(y) + \mu \nabla h_{LL}(y) = 0$
	$\{Complimentary\ slackness\ conditions\}$	$\{Strong\ duality\ condition\}$
	$\mu g_{LL}(y) = 0$	$f_{LL} = f_{LL}^{dual}$
	$\mu \geq 0$	$\mu \geq 0$

Figure 1. One-level equivalent reformulation steps.

2.3. Strong Duality Condition

One set of non-linearities appears in the strong duality conditions of the reformulated one-level problem and are reformulated using the big-M approach.

$$\sum_{s,p} \Big[\sum_d (D_d^{max}(\uparrow \overline{\omega}_{d,p,s} + \downarrow \overline{\omega}_{d,p,s}) + \sum_n I_{n,d} D_{d,p} \lambda_{n,p,s}) + \sum_w w p_{p,s} G_w \overline{v}_{w,p,s} + \sum_j (G_j^{max} \overline{v}_{j,p,s} +$$

$$RU_j^{max} \overline{\kappa}_{j,p,s} + RD_j^{max} \underline{\kappa}_{j,p,s} - \sum_j G_j^{min} \underline{v}_{j,p,s}) + \sum_h (G_h^{max} \overline{v}_{h,p,s} + RU_h^{max} \overline{\kappa}_{h,p,s} + RD_h^{max} \underline{\kappa}_{h,p,s} - G_h^{min} \underline{v}_{h,p,s} +$$

$$V_h^{max} \overline{\sigma}_{h,p,s} - V_h^{min} \underline{\sigma}_{h,p,s} + U_h^{max} \overline{\vartheta}_{h,p,s} + S_h^{max} \overline{\theta}_{h,p,s} + f l_{h,k,l} \lambda_{h,p,s}^{res}) + \sum_n \Theta(\overline{\rho}_{n,p,s} + \underline{\rho}_{n,p,s}) +$$

$$\sum_{n,m} T_{n,m}^{max} (\underline{\mu}_{n,m,p,s} + \overline{\mu}_{n,m,p,s}) + \underbrace{\sum_e G_e^{max} \overline{\varsigma}_{e,p,s}^{ch} y_{e,t}}_{\hat{\varsigma}_{e,p,s}^{ch}} + \underbrace{\sum_e G_e^{max} \overline{\varsigma}_{e,p,s}^{dch} y_{e,t}}_{\hat{\varsigma}_{e,p,s}^{dch}} + \underbrace{\sum_e SOC_e^{max} \cdot \overline{\gamma}_{e,p,s} y_{e,t}}_{\hat{\gamma}_{e,p,s}^{u}} -$$

$$\quad\quad\quad\quad\quad\quad\quad L1 \quad\quad\quad\quad\quad\quad L2 \quad\quad\quad\quad\quad\quad L3 \quad\quad\quad\quad\quad (36)$$

$$\underbrace{\sum_e SOC_e^{min} \cdot \underline{\gamma}_{e,p,s} y_{e,t}}_{\hat{\gamma}_{e,p,s}^{l}} \Big] - \sum_t RG^{min} \sum_{d,k,l} D_{d,p} \beta_t = -\sum_t (\psi \sum_s \pi_s (\sum_{k,l} A_{p,s} - \sum_k B_{k,t,s}))$$

$$\quad\quad L4$$

$$+ \sum_w C_{w,t} (G_{w,t}^{max} - G_{w,t-1}^{max}))$$

The non-linear terms L1, L2, L3 and L4 can be reformulated by introducing new variables $\hat{\varsigma}_{e,p,s}^{ch}$, $\hat{\varsigma}_{e,p,s}^{dch}$, $\hat{\gamma}_{e,p,s}^{u}$, $\hat{\gamma}_{e,p,s}^{l}$ and using big-M reformulation technique. Stationary conditions and complementary slackness conditions for lower level problem (34) used in the linearizion process are derived as in Equations (A1)–(A16) and (A17)–(A49) respectively in the Appendix of this paper.

$$\left. \begin{array}{l} \hat{\varsigma}_{e,p,s}^{ch} - \overline{\varsigma}_{e,p,s}^{ch} \leq M(1 - y_{e,t}) \ \forall e, p, s \\ \hat{\varsigma}_{e,p,s}^{ch} \leq M y_{e,t} \ \forall e, p, s \\ \hat{\varsigma}_{e,p,s}^{ch} \geq 0 \ \forall e, p, s \end{array} \right\} L1 \quad\quad (37)$$

$$\begin{aligned}
\widehat{\underline{\zeta}}_{e,p,s}^{dch} - \overline{\zeta}_{e,p,s}^{dch} &\leq M(1 - y_{e,t}) \quad \forall e,p,s \\
\widehat{\underline{\zeta}}_{e,p,s}^{dch} &\leq M y_{e,t} \quad \forall e,p,s \\
\widehat{\underline{\zeta}}_{e,p,s}^{dch} &\geq 0 \quad \forall e,p,s
\end{aligned} \quad \Bigg\} \quad L2 \tag{38}$$

$$\begin{aligned}
\widehat{\gamma}_{e,p,s}^{u} - \overline{\gamma}_{e,p,s} &\leq M(1 - y_{e,t}) \quad \forall e,p,s \\
\widehat{\gamma}_{e,p,s}^{u} &\leq M y_{e,t} \quad \forall e,p,s \\
\widehat{\gamma}_{e,p,s}^{u} &\geq 0 \quad \forall e,p,s
\end{aligned} \quad \Bigg\} \quad L3 \tag{39}$$

$$\begin{aligned}
\widehat{\gamma}_{e,p,s}^{l} - \underline{\gamma}_{e,p,s} &\leq M(1 - y_{e,t}) \quad \forall e,p,s \\
\widehat{\gamma}_{e,p,s}^{l} &\leq M y_{e,t} \quad \forall e,p,s \\
\widehat{\gamma}_{e,p,s}^{l} &\geq 0 \quad \forall e,p,s
\end{aligned} \quad \Bigg\} \quad L4 \tag{40}$$

Big-M reformulation technique is used to convert a logical constraint into a set of linear constraints corresponding to the same feasible set. If the disjunctive parameter is chosen carefully then the reformulated problem will be equivalent to the original one. The big-M reformulation does not affect the size of the problem. However, the disjunctive parameters involved in the reformulation create computational issues for the solver. A disjunctive parameter that is not tuned affects the convergence of the problem [35]. The literature provides several methods to tune the big-M parameter. The methodologies for tuning big-M can be found in [35,36]. The methods are proved to provide good approximations of the big-M parameters under certain conditions but additional large scale optimization problems should be solved for each case and the optimality still cannot be guaranteed. The problem of the disjunctive parameter tuning becomes especially hard when the reformulation involves variables without physical upper or lower limits which is the case in our proposed model. In this paper we use a simple iterative method to tune big-M parameters. We iteratively solve the proposed model while increasing the big-M parameter till it does not affect the solution of the problem.

2.4. Reformulation of the Objective Function

Another set of non-linearities $\lambda_{n,p,s} I_{n,e}(g_{e,p,s}^{dch} - g_{e,p,s}^{ch})$ is found in objective function (30) and can be linearized following algebraic manipulation steps as shown below.

First step is to express $\lambda_{n,p,s}$ as a linear combination of other decision variables using stationary conditions of the lower level problem (A13) and (A14)

$$\begin{aligned}
\sum_{p,e} \lambda_{n,p,s} I_{n,e}(g_{e,p,s}^{dch} - g_{e,p,s}^{ch}) &\overset{(A13),(A14)}{=} \sum_{p,e}\big((-\psi\pi_s mc_e + 1/\epsilon_e \lambda_{e,p,s}^{SOC} - \underline{\zeta}_{e,p,s}^{dch} \\
&+ \overline{\zeta}_{e,p,s}^{dch})g_{e,p,s}^{dch} + (-\psi\pi_s mc_e - \epsilon_e \lambda_{e,p,s}^{SOC} - \underline{\zeta}_{e,p,s}^{ch} + \overline{\zeta}_{e,p,s}^{ch})g_{e,p,s}^{ch}\big) \forall s
\end{aligned} \tag{41}$$

Using complementary slackness conditions (A32)–(A35) for Equations (28) and (29) respectively we can further simplify the previous algebraic expression (41) as in (42):

$$\begin{aligned}
&\sum_{p,e}\big((-\psi\pi_s mc_e + 1/\epsilon_e \lambda_{e,p,s}^{SOC})g_{e,p,s}^{dch} + G_e^{max}\overline{\zeta}_{e,p,s}^{dch} + (-\psi\pi_s mc_e - \epsilon_e \lambda_{e,p,s}^{SOC})g_{e,p,s}^{ch} + G_e^{max}\overline{\zeta}_{e,p,s}^{ch}\big) = \\
&\sum_{p,e}\big(G_e^{max}(\overline{\zeta}_{e,p,s}^{dch} + \overline{\zeta}_{e,p,s}^{ch}) - \psi\pi_s mc_e(g_{e,p,s}^{dch} + g_{e,p,s}^{ch}) - \lambda_{e,p,s}^{SOC}(\epsilon_e g_{e,p,s}^{ch} - 1/\epsilon_e g_{e,p,s}^{dch})\big)
\end{aligned} \tag{42}$$

Equation (42) still contains non-linear term $\lambda_{e,p,s}^{SOC}(\epsilon_e g_{e,p,s}^{ch} - 1/\epsilon_e g_{e,p,s}^{dch})$. Thus, we apply additional algebraic manipulations. We first use energy balance constraint of energy storage (7) and express the charge and discharge variables $g_{e,p,s}^{ch}$ and $g_{e,p,s}^{dch}$ through state of charge variables $SOC_{e,t,k,l,s}$ and then we use stationary condition (A15) to represent the primary state of charge variables $SOC_{e,t,k,l,s}$ through linear combination of Lagrange multipliers.

$$\sum_{p,e} \lambda_{e,p,s}^{SOC}(\epsilon_e g_{e,p,s}^{ch} - 1/\epsilon_e g_{e,p,s}^{dch}) \overset{(7)}{=} \sum_{p,e}(\lambda_{e,p,s}^{SOC}(SOC_{e,p,s} - \phi_e SOC_{e,t,k,l-1,s})) =$$

$$\sum_{p,e}(SOC_{e,p,s}(\lambda_p^{SOC} - \phi_e \lambda_{e,t,k,l+1,s}^{SOC})) \overset{(A15)}{=} \sum_{p,e}(SOC_{e,p,s}(-\overline{\gamma}_{e,p,s} + \underline{\gamma}_{e,p,s} - \psi \pi_s FC_{k,t} Y(l=L)))\forall s \tag{43}$$

Using complementary slackness conditions (A36) and (A37) for Equation (18) we can simplify Equation (43) and replace the rest of the non-linear terms through linear combination of linear terms as in (44).

$$\sum_{p,e}(-SOC_e^{max}\hat{\gamma}_{e,p,s}^u + SOC_e^{min}\hat{\gamma}_{e,p,s}^l - \psi \pi_s FC_{k,t} SOC_{e,L,k,t,s} Y(l=L))\forall s \tag{44}$$

By combining the algebraic expressions obtained in (42) and (44) we can now present the non-linear term $\lambda_{n,p,s} I_{n,e}(g_{e,p,s}^{dch} - g_{e,p,s}^{ch})$ through linear combination of linear terms as in (45)

$$\sum_{p,e}(\lambda_{n,p,s} I_{n,e}(g_{e,p,s}^{dch} - g_{e,p,s}^{ch})) \overset{(42),(44)}{=} \sum_{p,e}(G_e^{max}(\hat{\xi}_{e,p,s}^{dch} + \hat{\xi}_{e,p,s}^{ch}) - \psi \pi_s mc_e(g_{e,p,s}^{dch} + g_{e,p,s}^{ch})$$

$$+SOC_e^{max}\hat{\gamma}_{e,p,s}^u - SOC_e^{min}\hat{\gamma}_{e,p,s}^l + \psi \pi_s FC_{k,t} SOC_{e,l,k,t,s} Y(l=L))\forall s \tag{45}$$

2.5. One-Level Problem Formulation

The initial bilevel problem is now transformed into a one-level mixed integer linear problem, which is repeated here for the sake of clarity.

$$\underset{\Omega_s \cup \Omega_p}{Maximize:} \tag{46}$$

$$f_{UL} = \sum_{e,t} Pw_t(\psi \sum_{l,k,s} \pi_s W(p,e,s) - C_{e,t}(y_{e,t} - y_{e,t-1}))$$

S.t:

$$W(p,e,s) = G_e^{max}(\hat{\xi}_{e,p,s}^{dch} + \hat{\xi}_{e,p,s}^{ch}) - (\psi\pi_s + 1)mc_e(g_{e,p,s}^{dch} + g_{e,p,s}^{ch}) + SOC_e^{max}\hat{\gamma}_{e,p,s}^u - SOC_e^{min}\hat{\gamma}_{e,p,s}^l$$

$$+(\psi\pi_s + 1)FC_{k,t} SOC_{e,L,k,t,s} \tag{47}$$

$$y_{e,t} \geq y_{e,t-1} \forall e,t \tag{48}$$

$$\psi \sum_{t^*=t+1,k,l,e}^{t+PBP} \pi_s W(p,e,s) \geq \sum_e C_{e,t}(y_{e,t} - y_{e,t-1}) \ \ \forall t \tag{49}$$

$$\psi \sum_{p,e,s} \pi_s W(p,e,s) \geq IR \sum_{t,e} C_{e,t}(y_{e,t} - y_{e,t-1}) \tag{50}$$

$$(6) - (29) \tag{51}$$

{Stationary condition}

$$(A1) - (A16) \tag{52}$$

{Strong duality condition}

$$\sum_{s,p} \big[\sum_d (D_d^{max}(\uparrow \overline{\omega}_{d,p,s} + \downarrow \overline{\omega}_{d,p,s}) + \sum_n I_{n,d} D_{d,p} \lambda_{n,p,s}) + \sum_w wp_{p,s} G_w \overline{v}_{w,p,s} + \sum_j (G_j^{max}\overline{v}_{j,p,s} +$$

$$RU_j^{max}\overline{\kappa}_{j,p,s} + RD_j^{max}\underline{\kappa}_{j,p,s} - \sum_j G_j^{min}\underline{v}_{j,p,s}) + \sum_h (G_h^{max}\overline{v}_{h,p,s} + RU_h^{max}\overline{\kappa}_{h,p,s} + RD_h^{max}\underline{\kappa}_{h,p,s} - G_h^{min}\underline{v}_{h,p,s} +$$

$$V_h^{max}\overline{\sigma}_{h,p,s} - V_h^{min}\underline{\sigma}_{h,p,s} + U_h^{max}\overline{\vartheta}_{h,p,s} + S_h^{max}\underline{\vartheta}_{h,p,s} + fl_{h,k,l}\lambda_{h,p,s}^{res}) + \sum_n \Theta(\overline{\rho}_{n,p,s} + \underline{\rho}_{n,p,s}) + \tag{53}$$

$$\sum_{n,m} T_{n,m}^{max}(\underline{\mu}_{n,m,p,s} + \overline{\mu}_{n,m,p,s}) + \sum_e (SOC_e^{max}\hat{\gamma}_{e,p,s}^u - SOC_e^{min}\hat{\gamma}_{e,p,s}^l + G_e^{max}(\hat{\xi}_{e,p,s}^{ch} + \hat{\xi}_{e,p,s}^{dch}))\big] -$$

$$\sum_t RG^{min} \sum_{d,k,l} D_{d,p}\beta_t = -\sum_t(\psi \sum_s \pi_s(\sum_{k,l} A_{p,s} - \sum_k B_{k,t,s}) + \sum_w C_{w,t}(G_{w,t}^{max} - G_{w,t-1}^{max}))$$

{Big-M reformulation constraints}

$$(37) - (40) \tag{54}$$

where:

$$\Omega_p = \{\lambda_n, \lambda_e^{SOC}, \lambda_h^{res}, \tau 0_w, \tau_w, \phi_w, \underline{\xi}_e^{ch}, \overline{\xi}_e^{ch}, \underline{\xi}_e^{dch}, \overline{\xi}_e^{dch}, \underline{\mu}_{n,m}, \overline{\mu}_{n,m}, \overline{\sigma}_h, \underline{\sigma}_h, \underline{\nu}_j, \overline{\nu}_j, \underline{\kappa}_h, \overline{\kappa}_h, \underline{\kappa}_j, \overline{\kappa}_j,$$
$$\underline{\nu}_w, \overline{\nu}_w, \underline{\omega}_d, \overline{\omega}_d, \underline{\gamma}_e, \overline{\gamma}_e, \underline{\vartheta}_h, \overline{\vartheta}_h, \underline{\theta}_h, \overline{\theta}_h, \widehat{\gamma}_e, \widehat{\xi}_e^{ch}, \widehat{\xi}_e^{dch}\}$$

3. Case Study

The case study tries to answer the following questions. First, how will the presence of energy storage investment option affect system operation cost, electricity prices and wind-based generation expansion? Second, how will the planning approach (centralized and decentralized) affect the investment decisions on energy storage and will the results be different for systems with congested transmission capacity? Third, can energy storage benefit from congestion in the system under decentralized planning? Therefore the following simulation steps were performed. First, the centralized planning model without energy storage investment possibility was simulated, Case 1. Second, energy storage investment option was added and centralized and decentralized planning were simulated under different flexibility set-ups, Case 2 and Case 3 respectively. Third, the second step was repeated for the systems with and without transmission congestion. In addition, a case study without renewable generation expansion was performed, Case 4. In this case study we fix renewable generation capacity in the level to satisfy renewable penetration target and simulate energy storage investment planning under both centralized and decentralized planning models.

The models from section II have been tested on the IEEE 30-node test system. The generation mix has been varied to obtain different flexibility levels of the system and compare optimal energy storage investments. In the first and second set-ups, which is referred to as the thermal system (T) and thermal system with demand response (T+D), the generation consists of thermal units and wind power in set-up T and thermal units, wind power and flexible demand in system T-D. In the third and fourth set-ups, which is referred to as the hydro-thermal (H-T) system and hydro-thermal system with flexible demand (H-T+D), some of the thermal units of T and T+D system respectively are replaced by hydro units. The total installed capacity of generation units remains the same; however, the total expected ramping capability of the system is changed based on the thermal unit characteristics. The generation mix and total expected ramping capability can be found in Table 1.

The total expected ramping capability (*R_Total*) of the system is measured in MW per minute and calculated as an expected maximum reserves which could be provided by each plant, energy storage and flexible demand. A formula is provided to calculate total expected ramping capability of the system:

$$R\_Total = \sum_s \pi_s \frac{1}{T*l} \left(\sum_p \left(\sum_j min\{RU_j^{max}, G_j^{max} - g_{j,p,s}\} + \sum_h min\{RU_h^{max}, G_h^{max} - g_{h,p,s}\} \right. \right.$$
$$\left. \left. + \sum_e (SOC_{e,p,s} - SOC_e^{min}) + \sum_d (D_d^{max} - \downarrow \Delta d_{d,p,s}) \right) \right) \tag{55}$$

The total expected ramping capability is used to compare the flexibility levels of different case study set-ups. It is calculated based on hourly available energy capacity of the thermal and hydro generation considering ramping limits and available energy which can be obtained through energy storage and demand response. Energy storage and demand response are considered to have very fast ramping capability and therefore no ramping limits are imposed on these sources of flexibility. However, it should be noted that this approach will not capture the full dynamics of energy limited resources such as hydro power, flexible demand and energy storage, but can be used to approximate the flexibility of the system. A more exact measurement of flexibility of the system is outside of the scope of this paper. The measurement is calculated based on the up-ramping capability of the system and a similar index can be calculated based on the down-ramping capability of the system. However, in this system, the down-ramping capability is always larger than the up-ramping capability (especially if consider possibility to curtail wind power) and is therefore not analyzed any further. The transmission capacity connecting wind-based generation with load were reduced in order to create congestion in the system and analyze the impact of additional flexibility and behaviour of both planning strategies. In addition, the systems with initially congested transmission capacity were compared to the cases where transmission capacity was increased and congestion was eliminated.

Table 1. Test system input data.

	Thermal		Hydro-Thermal	
	Capacity	**Node**	**Capacity**	**Node**
Thermal, (MW)	600	1, 2, 22, 27, 23, 13	300	1, 2, 23
Hydro, (MW)	-	-	300	22, 27
Wind, (MW)	100	5, 6	100	5, 6
Max. demand, (MW)	600	-	600	-
Flexible demand	10%	-	10%	-
Transmission limits, (MW)	100	-	100	-
Congested transmission limits, (MW)	70	-	70	-
Ramping capability, (MW)	300	-	420	-
Renewable generation target	20%	-	20%	-

3.1. System Description

The IEEE 30-node test system was chosen to test and analyze presented investment models. The initial input data is presented in Table 1. The installed capacity is chosen to be almost the same as the peak demand in order to force additional investments into renewable generation. Practically, this situation reflects the decision to close large power plants in the system such as nuclear or coal and replace the required generation by investments into a wind-based generation and an additional flexibility source such as energy storage unit. In addition, the investments in renewable generation is ensured through lower limit constraint on expected generation from renewable generation.

A moment matching technique is used to generate the wind power generation scenarios [37]. The technique provides various advantages. The main one is that the technique allows to use a relatively few numbers of distinct scenarios and therefore reduces the computational difficulty for solving the stochastic program. The investment decisions in energy storage are made considering two different energy storage technologies available: compressed air energy storage (CAES) and batteries. Both of these technologies could be used for bulk energy storage, mature and commercially available. Each technology represented through a set of energy storage units of fixed energy capacity and power capability which could be invested in. The technical characteristics of each unit of each technology as well as the energy capacity and the power capability are presented in Table 3.

Case studies presented in this paper consider different levels of capital costs of energy storage. Initially, capital costs were assumed to be high to represent current state of the energy storage market. We expect a cost reduction each year of up to 5 %, i.e., $C_{e,t} = 0.95C_{e,t-1}$, to take into account predicted reduction of the capital costs in the future and development of new technologies. The initial capital costs for the first year were taken from [15] for energy storage and from [38] for wind power. The costs were updated using the present worth factor based on (56) and parameters presented in Table 2 and in Table 4.

$$Pw_t = \frac{(1+inf)^t}{(1+dis)^t}$$

(56)

Table 2. Investment cost assumptions.

Parameter	Value
Annual inflation rate, (*inf*)	2%
Discount rate, (*dis*)	10%

Thus, the investment decisions could be delayed for later periods when the conditions will be more financially favourable. Investment decision planning includes 10 consecutive periods which represent years. Each year consists of four consecutive operational periods which represent each season of that year. Figure 2 shows the time line for the operation planning and investments decisions.

Table 3. Energy storage characteristics.

Technology	CAES	Battery
Energy storage capacity, SOC_e^{max}, (MWh)	100	15
Power limit, G_e^{max}, (MW)	20	6
Energy conversion efficiency, (ε)	0.75	0.85
Self discharge of energy storage, (η)	0.78	0.99
Initial state of charge	50%	50%
Capital cost, Energy, ($/kWh)	5	400
Capital cost, Power, ($/kW)	700	400
Maximum number of units	5	10

For the hydro-thermal generation mixes the limits for hydro reservoirs at the end of the each operational period were set based on the outputs of the weekly schedule and are allowed to be deviated up to 10% of the scheduled amount for each operational period of each year. This was done to simulate long term hydro power scheduling and avoid overuse of hydropower.

Table 4. Investment parameters.

Parameter	Value
Planning period, (T)	10 years
Investments return parameter, (IR)	1.2
Payback period limit, (PBP)	5
Short term operation period, (l)	74 h
Renewable penetration target year, (RTY)	5th year

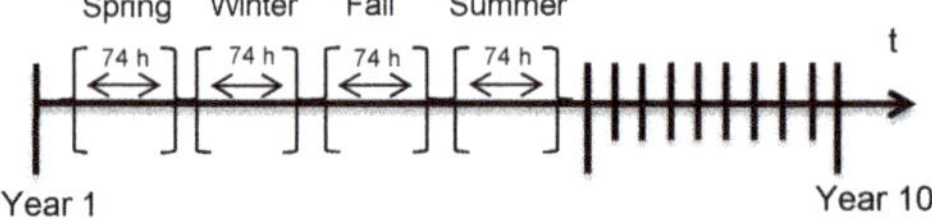

Figure 2. Investment planning time line.

3.2. Results and Discussions

Tables 5 and 6 present the results for the case studies listed above. Table 5 presents the summarized investment results and the influence of these investments on operation cost, renewable generation spillage and presence of congestion in the system. Table 6 presents more detailed data on energy storage investment, such as the time period when the investment was made, which technology was chosen and the node it was placed in.

First of all, the results show that energy storage could be a financially beneficial investment both, under centralized and decentralized planning. However, in hydro-thermal systems energy storage investments are not profitable for independent investors. No investments were made under decentralized planning in hydro-thermal systems while under centralized planning 45 MW of energy storage were installed which is 8% of the total installed generation capacity of the system. Considerable investments were made under decentralized planning in thermal only systems. 90 MW of energy storage consisting of batteries were installed compared to 115 MW of storage capacity consisting of CAES and a batteries under centralized planning. While decentralised planning ensures that the owner of energy storage system earns sufficient benefits in a fixed payback period time, the investments made are still beneficial for the whole system by reducing electricity prices and relieving transmission congestion. However, the benefits are much lower than under centralized planning and mainly due to lower investments in energy storage and as a consequence in renewable generation expansion. For example in the thermal only system the investments made under decentralized planning resulted in 12% price reduction compared to 20% price reduction under centralized planning while the standard deviation of the price was reduced from 12.6 to 6.2 in decentralized planning and till 4.3 under centralized planning. Curtailed wind remained relatively high (5%) under decentralized planning while centralized planning allowed to almost completely eliminate wind curtailment. Moreover energy storage investments reduced substantially installed capacity of the wind power generation which was required by renewable generation target set by the system. In centralized planning 115 MW of energy storage investment in thermal system helped to reduce required installed capacity of wind generation from 288 MW to 241 MW.

Another interesting observation was that in the system with congested transmission capacity, an independent profit maximizing energy storage owner will choose the placement of large energy storage units in such a way to keep the congestion in the system. On the other hand, under centralized planning more investments will be made just to relieve the congestion and reduce the renewable generation spillage and total operation cost. The difference is noticeable when comparing the system with three congested lines to the system without congestion. The average price and variability of the price is reduced under both centralized and decentralized planning with all generation mix set-ups while energy storage investments are also lower than in the case studies with congested lines with the same generation mix set-ups. For example the results in Table 6 show that similar quantities of energy storage were deployed under decentralized planning for congested and not congested systems but the nodes of placement were different. In the congested system under decentralized planning energy storage was placed at nodes 4, 6 and 8. In addition, the average price difference between centralized and decentralized planning was 9 % while in the non-congested system the price difference was equal to zero and the energy storage units were placed at nodes 4, 8 and 25. In centralized planning energy storage investments contribute to substantial reduction of wind power spillage and wind power generation investments which were forced by renewable generation target. Moreover, average price and price variability also were significantly lowered by additional energy storage investments. Decentralized planning also resulted in energy storage investments however the overall benefits of the system from these investments was lower.

In addition decentralized planning of energy storage especially in already congested systems can have a negative impact on system operation and further congested power system. This could be the case when variable renewable generation is planned beforehand and the flexibility requirements are set afterword. Thus, the flexibility providers can benefit from strategic placement and internalize existing system congestion. However, the results of the case studies show that independent energy storage investments will still contribute to congestion relief but in much lower volumes than in centralized planning. Congestion relief under decentralized planning is mainly due to the assumption that system operator follows energy storage investment decision and expands renewable generation according to the decision made on upper level. Thus, energy storage owner can not congest system further and considerably influence on locational marginal prices at the congested nodes. The case study Case 4 also proves this. The investments on energy storage when renewable generation capacity was fixed considerably increased price volatility and did not relieve congestion at all.

Table 5. Case study results.

	Congested				Not Congested	
	T	**T+D**	**H-T**	**H-T+D**	**T**	**H-T**
Case 1. Base case (no ES investments)						
Available capacity, (MW)	600	600	600	600	600	600
Wind power investments, (MW)	288	274	273	271	281	271
Ramping capability, (MWh)	300	360	420	480	300	420
Number of congested lines	3	1	3	1	0	0
Curtailed wind, (%)	10	8	11	6	7	5
Std of price	14.2	13.1	5.2	5.7	13.3	4.5
Average price, ($)	52	45	32	32	45	34
Case 2. Centralized planning						
Wind power investments, (MW)	241	235	232	238	248	236
Energy storage investments, (MW)	115	45	45	45	100	30
Ramping capability, (MWh)	390	390	455	510	384	447
Std of price	4.3	3.9	3.4	3.1	4.3	3.1
Average price, ($)	40	40	30	30	40	30
Number of congested lines	1	0	1	0	0	0
Curtailed wind, (%)	≥ 1	≥ 1	≥ 1	≥ 1	≥ 1	≥ 1
Case 3. Decentralized planning						
Wind power investments, (MW)	257	241	273	271	270	271
Energy storage investments, (MW)	90	45	0	0	45	0
Ramping capability, (MWh)	330	390	420	480	390	420
Std of price	6.2	5.4	5.2	5.7	5.2	4.5
Average price, ($)	44	40	35	35	40	34
Number of congested lines	2	1	2	1	0	0
Curtailed wind, (%)	6	6	9	6	7	4
Case 4. Decentralized planning with fixed wind investments						
Energy storage investments, (MW)	90	45	45	45	100	30
Ramping capability, (MWh)	330	390	420	480	384	447
Std of price	6.9	6.1	5.8	5.9	4.3	3.1
Average price, ($)	46	41	38	37	40	30
Number of congested lines	3	1	3	1	0	0
Curtailed wind, (%)	6.5	6.8	9	6	≥ 1	≥ 1

In addition, the investments in energy storage under independent investment planning were done predominately in batteries while in centralized planning the investments also include compressed air energy storage (CAES). Under centralized planning both congested and non-congested systems deployed one CAES unit while none of the case studies under decentralized planning invested in CAES. The large size of CAES and associated high total capital cost makes it harder for independent investors to get the payback of the investments in reasonable time. Thus, the payback limit constraints introduced in independent planning model restricts these investments.

Based on the case study results we can suggest that centralized ownership model provides more benefits to the power systems and ensures an effective short-term operation. However, current regulations existing in Europe prohibit the use of energy storage technologies for energy arbitrage if they are owned by the system operator. Thus, decentralized ownership model is the only valid option to provide energy arbitrage. The decentralized ownership of energy storage without proper regulation may potentially congest the system and result in inefficient development of power system and reduced deployments of renewable generation such as wind. Additional research on various regulations on energy storage investments should be performed in order to fully answer the question who should be responsible for energy storage investments.

Table 6. Energy storage investment results. Bat: Battery.

| | Congested | | | | Not Congested | | | |
| | Centralized | | Decentralized | | Centralized | | Decentralized | |
	CAES	Bat.	CAES	Bat.	CAES	Bat.	CAES	Bat
Thermal system								
Time period, (t)	t1	t1	-	t5	t1	t1	-	t5
Node, (n)	3	25	-	4, 6, 8	4	25	-	4, 8, 25
Number of units	1	6	-	1	1	3	-	
Hydro-Thermal system								
Time period, (t)	-	t1	-	-	-	t1	-	-
Node, (n)	-	3, 25, 15	-	-	-	3, 25	-	-
Number of units	-	3	-	-	-	2	-	-

4. Conclusions

This paper presents two mathematical models for centralized and decentralized investment planning of energy storage and wind power generation expansion. The decentralized investment planning is formulated as MPEC model, where a single energy storage investor is interacting with a centralized operator representing a perfect market environment. Both models are useful to investigate the interactions between variability of renewable generation and the flexibility provided by energy storage. The models include a wide range of generation mixes which allows to model different types of the system with different flexibility levels just by varying the input parameters.The proposed models allow to evaluate the differences between centralized and decentralized planning. Additional constraint on investments return and payback periods express the constraints of a profit maximizing company when it faces investment planning decisions. The models were applied on a case studies with various levels of flexibility and different levels of congestion in transmission capacity.

The following main conclusions were obtained from the case studies:

- First, energy storage can be beneficial to the whole system by reducing spillage of renewable generation and relieving congestion of transmission capacity under both centralized and decentralized planning approaches. However, there are still a big gap between centralized and decentralized planning approaches. More investments are made under centralized planning and the cost and the average price reduction under centralized planning is much higher.
- Second, if treated as a market asset (decentralized planning) energy storage can profit from strategically placing energy storage units and contribute on increase to transmission congestion of power system and additional wind spillage.
- Third, negative impact of strategic behavior of energy storage can be reduced if renewable generation decisions are taken simultaneously.
- Fourth, the case studies demonstrate that decentralized unregulated allocation planning for energy storage potentially may cause congestion in the system. Thus, additional studies on proper regulation for energy storage is necessary.

The gap between centralized and decentralized planning could be reduced if independent energy storage owner were able to have additional profits apart from energy arbitrage. An additional profit stream could increase the investments in energy storage under decentralized planning. This is the case especially in hydro-thermal system where investments in energy storage are generally less profitable under both centralized and decentralized investment planning. The revenue streams can include participation in balancing markets and provision of reserves. On the other hand increasing penetration of variable renewable generation will also increase the potential profitability and need in energy storage systems. Based on case study results energy storage considerably reduces wind spillage and therefore coordinated investment planning with renewable generation might increase investments in energy storage even under decentralized planning. Otherwise, in order to ensure sufficient flexibility in the system grid, the owner should be eligible and responsible for investments in energy storage. For example such a strategy was chosen in California and Oregon by passing energy storage mandates on energy storage installations.

Future research steps could be identified as the following.

- First, proposed decentralized model considers monopoly on energy storage investments and does not take into account additional competition from investments made on other flexibility sources such as hydro, flexible demand or flexible generators. Thus, an EPEC model could be developed to coordinate the investment and evaluate the dependency.

- Second, the models consider only one revenue stream which comes from providing energy arbitrage, however additional revenue streams such as provision of balancing services should be also considered to further evaluate the profitability of energy storage.

- Third, the initial formulation of the decentralized planning model is presented as a mixed integer non-linear bilevel model and later reformulated as a mixed integer linear one-level problem. The suggested technique for reformulation and linearization reduces the complexity of the model and makes it possible to find an optimal solution with reasonable computational time. However, the linearized model is still complex and a higher number of nodes and decision variables will increase the computational time. In order to apply the models to larger systems, it could be beneficial to investigate decomposition techniques (ex. Benders decomposition).

- Fourth, the choice of the number of days and operational hours also affects the computational time and the energy storage evaluation require rather large operational period to observe the charge and discharge cycles. Thus, the selection of the critical operational periods for energy storage evaluation is also a subject for future research.

Acknowledgments: Dina Khastieva has been awarded an Erasmus Mundus Ph.D. Fellowship in Sustainable Energy Technologies and Strategies (SETS) program. The authors would like to express their gratitude towards all partner institutions within the program as well as the European Commission for their support.

Author Contributions: Authors contributed equally to this work

Conflicts of Interest: The authors declare no conflict of interest.

Nomenclature

Indices

d	Demand;
e	Energy storage systems;
h	Hydro based generation;
j	Thermal generation;
k	Operation period (seasons);
l	Operation period (hours);
n, m	Nodes of the system;
p	Superset for l, k, t;
s	Scenarios;
t, t^*	Planning period (years);
w	Wind based generation;

Binary Variables

y_e	Energy storage investment decision variable

Continuous Variables

$\uparrow \Delta d_d, \downarrow \Delta d_d$	Up and down regulated flexible load, (MW);
$f_{n,m}$	Power flow between node n and m, (MW);
g_e^{ch}, g_e^{dch}	Charge and discharge of energy storage, (MW);
g_j, g_h, g_w	Output of thermal, hydro and wind generation units, (MW);
G_w^{max}	Expanded wind generation capacity, (MW);
s_h	Spillage of a hydro unit h;

SOC_e	State of charge of an energy storage, (MWh);
u_h	Hydro discharge of a hydro unit h
v_h	Reservoir level of a hydro unit h;
θ_n	Voltage angles at node n, $(p.u)$;
λ_n	Price at node n, $(\$/MW)$;
λ_e^{SOC}	Lagrange multipliers, $(\$/MW)$, for energy balance constraint for ES;
$\lambda_{n,m}^{Line}$	Lagrange multipliers, $(\$/MW)$, for power flow constraints;
λ_d^{D}	Lagrange multipliers, $(\$/MW)$, for demand response constraints;
$\lambda_{h,p,s}^{Gen}$	red Lagrange multipliers,$(\$/MW)$, for hydro power generation constraint
λ_h^{res}	Lagrange multipliers, $(\$/MW)$, for hydrological balance constraints;
$\tau 0_w, \tau_w$	Lagrange multipliers $(\$/MW)$ for generation investment constraints;
$\underline{\zeta}_e^{ch}, \overline{\zeta}_e^{ch}$	Lagrange multipliers, $(\$/MW)$, for energy storage charge constraints;
$\underline{\zeta}_e^{dch}, \overline{\zeta}_e^{dch}$	Lagrange multipliers, $(\$/MW)$, for energy storage discharge constraints;
$\underline{\mu}_{n,m}, \overline{\mu}_{n,m}$	Lagrange multipliers, $(\$/MW)$, for line constraints;
$\overline{\sigma}_h, \underline{\sigma}_h$	Lagrange multipliers, $(\$/MW)$, for water reservoir volume constraints;
$\underline{\nu}_j, \overline{\nu}_j$	Lagrange multipliers, $(\$/MW)$, for generator j constraints;
$\underline{\nu}_h, \overline{\nu}_h$	Lagrange multipliers, $(\$/MW)$, for generator h constraints;
$\underline{\nu}_w, \overline{\nu}_w$	Lagrange multipliers, $(\$/MW)$, for generator w constraints;
$\underline{\kappa}_j, \overline{\kappa}_j$	Lagrange multipliers, $(\$/MW)$, for generator j constraints;
$\underline{\kappa}_h, \overline{\kappa}_h$	Lagrange multipliers, $(\$/MW)$, for generator h constraints;
$\underline{\rho}_n, \overline{\rho}_n, \rho$	Lagrange multipliers, $(\$/MW)$, for voltage angle constraints;
$\uparrow \underline{\omega}_d, \uparrow \overline{\omega}_d$	Lagrange multipliers, $(\$/MW)$, for demand d constraints;
$\downarrow \underline{\omega}_d, \downarrow \overline{\omega}_d$	Lagrange multipliers, $(\$/MW)$, for demand d constraints;
$\underline{\gamma}_e, \overline{\gamma}_e$	Lagrange multipliers, $(\$/MW)$ for energy storage SOC constraints;
$\underline{\vartheta}_h, \overline{\vartheta}_h$	Lagrange multipliers, $(\$/MW)$, for spillage constraints;
$\underline{\theta}_h, \overline{\theta}_h$	Lagrange multipliers, $(\$/MW)$, for water flow constraints;
β_t	Lagrange multipliers for renewable target constraint;
Parameters	
C_w	Capital cost of wind gen. expansion, $(\$/MW)$;
C_e	Capital cost of energy storage block, $(\$/block)$;
D_d	Non-dispatchable load, (MW);
D_d^{max}, D_d^{min}	Limits of flexible load, (MW);
dis	Discount rate;
FC	Expected future cost of electricity for period k ;
fl_h	Inflow of hydro unit h;
σ_h	Efficiency of hydro unit h;
τ_{h*}	Hydro discharge time delay;
$G_j^{max}, G_h^{max}, G_e^{max}$	Upper generation limits, (MW);
$G_j^{min}, G_h^{min}, G_e^{min}$	Lower generation limits, (MW);
G_w	Existing capacity of wind power generation (MW);
inf	Annual inflation rate;
$I_{n,j}, I_{n,h}, I_{n,w}$	Incidence matrix for thermal, hydro and wind generation units;
$I_{n,e}$	Incidence matrix for energy storage units;
$I_{n,d}$	Incidence matrix for flexible demand units;
$I_{n,m}$	Incidence matrix for transmission;
IR	Investments return coefficient;
L	Operation time period;
M	Big-M parameter,sufficiently large number;
PBP	Payback period, (years) ;
Pw_t	Present worth factor ;
RG^{min}	Renewable generation penetration target ;
R_Total	Total expected ramping capability of a system ;
RU_j^{max}, RU_h^{max}	Ramp-up hourly limits, (MW);
RD_j^{max}, RD_h^{max}	Ramp-down hourly limits, (MWh);

S_h^{max}	Maximum spillage of hydro units;
SOC_e^{max}	Storage capacity, (MWh);
$T_{n,m}^{max}$	Transmission line capacity, (MW);
T	Investment planning period;
RTY	Target year for renewable generation penetration;
U_h^{max}	Maximum flow of hydro units;
V_h^{max}	Maximum reservoir of hydro units;
$wp_{p,s}$	Wind power output for each scenario as percentage of capacity;
ϵ_e	Energy conversion efficiency;
ϕ_e	Self discharge of energy storage;
π_s	Scenario probability;
mc_e, mc_j, mc_d	marginal costs of energy storage units,thermal units and flexible demand units
ψ	Scaling factor for operation and investment values
$Y(*)$	1 if $*$ is true and 0 otherwise;

Appendix A. Stationary Conditions

$$\psi\pi_s mc_d + I_{n,d}\lambda_{n,p,s} + \uparrow \underline{\omega}_{d,p,s} - \uparrow \overline{\omega}_{d,p,s} + \lambda_{d,t,k,s}^D = 0 \quad \forall d,p,s \tag{A1}$$

$$\psi\pi_s mc_d - I_{n,d}\lambda_{n,p,s} + \downarrow \underline{\omega}_{d,p,s} - \downarrow \overline{\omega}_{d,p,s} - \lambda_{d,t,k,s}^D = 0 \quad \forall d,p,s \tag{A2}$$

$$\psi\pi_s mc_j + I_{n,j}\lambda_{n,p,s} + \underline{v}_{j,p,s} - \overline{v}_{j,p,s} + \underline{\kappa}_{j,p,s} - \overline{\kappa}_{j,p,s} - \underline{\kappa}_{j,t,k,l+1,s} + \overline{\kappa}_{j,t,k,l+1,s} = 0 \quad \forall j,p,s \tag{A3}$$

$$I_{n,w}\lambda_{n,p,s} + \beta_t Y(t \geq RTY) + \underline{v}_{w,p,s} - \overline{v}_{w,p,s} = 0 \quad \forall w,p,s \tag{A4}$$

$$I_{n,h}\lambda_{n,p,s} + \underline{v}_{h,p,s} - \overline{v}_{h,p,s} + \lambda_{h,p,s}^{res} + \underline{\kappa}_{h,p,s} - \overline{\kappa}_{h,p,s} - \underline{\kappa}_{h,t,k,l+1,s} + \overline{\kappa}_{h,t,k,l+1,s} = 0 \quad \forall h,p,s \tag{A5}$$

$$\underline{\vartheta}_{h,p,s} - \overline{\vartheta}_{h,p,s} + \lambda_{h,p,s}^{res} = 0 \quad \forall h,p,s \tag{A6}$$

$$\underline{\theta}_{h,p,s} - \overline{\theta}_{h,p,s} + \lambda_{h,p,s}^{res} = 0 \quad \forall h,p,s \tag{A7}$$

$$\underline{\sigma}_{h,p,s} - \overline{\sigma}_{h,p,s} - \lambda_{h,p,s}^{res} + \lambda_{h,t,k,l+1,s}^{res} - \pi_s\sigma_h FC_{k,t}Y(l=L) = 0 \quad \forall h,p,s \tag{A8}$$

$$\underline{C}_{w,t} + \tau 0_w - \tau_{w,t=2} + \sum_{s,p} wp_{p,s}\overline{v}_{w,p,s} = 0 \quad \forall w \; (t=1) \tag{A9}$$

$$\tau_{w,t} \quad \tau_{w,t-1} + \sum_{s,p} wp_{p,s}\overline{v}_{w,p,s} - 0 \quad \forall w,l \; (1 < t < T) \tag{A10}$$

$$-\underline{C}_{w,t} + \tau_{w,T} - \phi_{w,T} + \sum_{s,p} wp_{p,s}\overline{v}_{w,p,s} = 0 \quad \forall w \; (t=T) \tag{A11}$$

$$-\lambda_{n,p,s} + \underline{\mu}_{n,m,p,s} - \overline{\mu}_{n,m,p,s} + \lambda_{n,m,p,s}^{Line} = 0 \quad \forall n,m,p,s \tag{A12}$$

$$\psi\pi_s mc_e - I_{n,e}\lambda_{n,p,s} + \epsilon_e\lambda_{e,p,s}^{SOC} + \underline{\zeta}_{e,p,s}^{ch} - \overline{\zeta}_{e,p,s}^{ch} = 0 \quad \forall e,p,s \tag{A13}$$

$$\psi\pi_s mc_e + I_{n,e}\lambda_{n,p,s} - 1/\epsilon_e\lambda_{e,t,l,s}^{SOC} + \underline{\zeta}_{e,p,s}^{dch} - \overline{\zeta}_{e,p,s}^{dch} = 0 \quad \forall e,p,s \tag{A14}$$

$$-\overline{\gamma}_{e,p,s} + \underline{\gamma}_{e,p,s} - \lambda_{e,p,s}^{SOC} + \phi_e\lambda_{e,t,k,l+1,s}^{SOC} - \psi\pi_s FC_{k,t}Y(l=L) = 0 \quad \forall e,p,s \tag{A15}$$

$$-\sum_m \frac{100}{X_{n,m}}\lambda_{n,m,t,p}^{Line} + \sum_m \frac{100}{X_{m,n}}\lambda_{m,n,t,p}^{Line} + \underline{\rho}_{n,t,p} - \overline{\rho}_{n,t,p} + \rho 0_{p,s}Y(n=1) = 0 \quad \forall n,p,s \tag{A16}$$

Appendix B. Complementary Slackness Conditions for Lower Level Problem

$$(g_{w,p,s} - wp_{p,s}G_w + wp_{p,s}G_{w,t}^{max})\bar{v}_{w,p,s} = 0 \ : \ (\underline{v}_{w,p,s'}) \quad \forall p,s,w. \tag{A17}$$

$$g_{w,p,s}\underline{v}_{w,p,s} = 0 \quad \forall p,s,w. \tag{A18}$$

$$(RG^{min}\sum_{d,k,l} D_{d,p} - \sum_s \pi_s \sum_{w,k,l} g_{w,p,s})\beta_t = 0 \quad \forall t \geq RTY. \tag{A19}$$

$$(G_{w,t-1}^{max} - G_{w,t}^{max})\tau_{w,t} = 0 \quad \forall t,w. \tag{A20}$$

$$G_{w,t1}^{max}\tau 0_w = 0 \quad \forall w. \tag{A21}$$

$$((g_{j,t,k,l,s} - g_{j,t,k,l-1,s}) + RD_j^{max})\underline{\kappa}_{j,p,s} = 0 \quad \forall t,k,l,s,j. \tag{A22}$$

$$((g_{j,t,k,l,s} - g_{j,t,k,l-1,s}) - RU_j^{max})\bar{\kappa}_{j,p,s} = 0 \quad \forall t,k,l,s,j. \tag{A23}$$

$$((g_{h,t,k,l,s} - g_{h,t,k,l-1,s}) + RD_h^{max})\underline{\kappa}_{h,p,s} = 0 \quad \forall t,k,l,s,h. \tag{A24}$$

$$((g_{h,t,k,l,s} - g_{h,t,k,l-1,s}) - RU_h^{max})\bar{\kappa}_{h,p,s} = 0 \quad \forall t,k,l,s,h. \tag{A25}$$

$$(g_{h,p,s} - G_h^{min})\underline{v}_{h,p,s} = 0 \quad \forall p,s,h. \tag{A26}$$

$$(g_{h,p,s} - G_h^{max})\bar{v}_{h,p,s} = 0 \quad \forall p,s,h. \tag{A27}$$

$$(g_{j,p,s} - G_j^{min})\underline{v}_{j,p,s} = 0 \quad \forall p,s,j. \tag{A28}$$

$$(g_{j,p,s} - G_j^{max})\bar{v}_{j,p,s} = 0 \quad \forall p,s,j. \tag{A29}$$

$$(s_{h,p,s} - S_h^{max})\bar{\theta}_{h,p,s} = 0 \quad \forall p,s,h. \tag{A30}$$

$$s_{h,p,s}\underline{\theta}_{h,p,s} = 0 \quad \forall p,s,h. \tag{A31}$$

$$(g_{e,p,s}^{ch} - G_e^{max}y_{e,t})\bar{\xi}_{e,p,s}^{ch} = 0 \quad \forall p,s,e. \tag{A32}$$

$$g_{e,p,s}^{ch}\underline{\xi}_{e,p,s}^{ch} = 0 \quad \forall p,s,e. \tag{A33}$$

$$(g_{e,p,s}^{dch} - G_e^{max}y_{e,t})\bar{\xi}_{e,p,s}^{dch} = 0 \quad \forall p,s,e. \tag{A34}$$

$$g_{e,p,s}^{dch}\underline{\xi}_{e,p,s}^{dch} = 0 \quad \forall p,s,e. \tag{A35}$$

$$(SOC_{e,p,s} - SOC_e^{max}y_{e,t})\bar{\gamma}_{e,p,s} = 0 \quad \forall p,s,e. \tag{A36}$$

$$(SOC_{e,p,s} - SOC_e^{min}y_{e,t})\underline{\gamma}_{e,p,s} \quad \forall p,s,e. \tag{A37}$$

$$(\theta_{n,p,s} + \Theta)\underline{\rho}_{n,p,s} = 0 \quad \forall n,p,s. \tag{A38}$$

$$(\theta_{n,p,s} - \Theta)\bar{\rho}_{n,p,s} = 0 \quad \forall n,p,s. \tag{A39}$$

$$(f_{n,m,p,s} + T_{n,m}^{max})\underline{\mu}_{n,m,p,s} = 0 \quad \forall n,m,p,s. \tag{A40}$$

$$(f_{n,m,p,s} - T_{n,m}^{max})\bar{\mu}_{n,m,p,s} \quad \forall n,m,p,s. \tag{A41}$$

$$(\uparrow \Delta d_{d,p,s} - D_d^{max}) \uparrow \overline{\omega}_{d,p,s} = 0 \quad \forall p,s,d. \tag{A42}$$

$$\uparrow \Delta d_{d,p,s} \uparrow \underline{\omega}_{d,p,s} = 0 \quad \forall p,s,d. \tag{A43}$$

$$(\downarrow \Delta d_{d,p,s} - D_d^{max}) \downarrow \overline{\omega}_{d,p,s} = 0 \quad \forall p,s,d. \tag{A44}$$

$$\downarrow \Delta d_{d,p,s} \downarrow \underline{\omega}_{d,p,s} = 0 \quad \forall p,s,d. \tag{A45}$$

$$(v_{h,p,s} - V_h^{max})\overline{\sigma}_{h,p,s} = 0 \quad \forall p,s,h. \tag{A46}$$

$$(v_{h,p,s} - V_h^{min})\underline{\sigma}_{h,p,s} = 0 \quad \forall p,s,h. \tag{A47}$$

$$u_{h,p,s}\underline{\vartheta}_{h,p,s} = 0 \quad \forall p,s,h. \tag{A48}$$

$$(u_{h,p,s} - U_h^{max})\overline{\vartheta}_{h,p,s} = 0 \quad \forall p,s,h. \tag{A49}$$

References

1. REN21. *Renewables 2016 Global Status Report*; REN21 Secretariat: Paris, France, 2016.
2. Holttinen, H.; Milligan, M.; Ela, E.; Menemenlis, N.; Dobschinski, J.; Rawn, B.; Bessa, R.; Flynn, D.; Gomez-Lazaro, E.; Detlefsen, N. Methodologies to Determine Operating Reserves Due to Increased Wind Power. *IEEE Trans. Sustain. Energy* **2012**, *3*, 713–723.
3. Hirth, L.; Ueckerdt, F.; Edenhofer, O. Integration costs revisited—An economic framework for wind and solar variability. *Renew. Energy* **2015**, *74*, 925–939.
4. Department of Energy. Global Energy Storage Database. Available online: https://www.energystorageexchange.org/ (accessed on 15 January 2018).
5. International Energy Agency. *Technology Roadmap—Solar Photovoltaic Energy*; International Energy Agency: Paris, France, 2014.
6. Eyer, J.; Corey, G. *Energy Storage for the Electricity Grid: Benefits and Market Potential Assessment Guide*; Sandia National Laboratories: Albuquerque, NM, USA, 2010.
7. Akhil, A.A.; Huff, G.; Currier, A.B.; Kaun, B.C.; Rastler, D.M.; Chen, S.B.; Cotter, A.L.; Bradshaw, D.T.; Gauntlett, W.D. *DOE/EPRI 2013 Electricity Storage Handbook in Collaboration with NRECA*; Sandia National Laboratories: Albuquerque, NM, USA, 2013.
8. Das, T.; Krishnan, V.; McCalley, J.D. Assessing the benefits and economics of bulk energy storage technologies in the power grid. *Appl. Energy* **2015**, *139*, 104–118.
9. Heydarian-Forushani, E.; Golshan, M.; Siano, P. Evaluating the benefits of coordinated emerging flexible resources in electricity markets. *Appl. Energy* **2017**, *199*, 142–154.
10. Qin, M.; Chan, K.W.; Chung, C.Y.; Luo, X.; Wu, T. Optimal planning and operation of energy storage systems in radial networks for wind power integration with reserve support. *IET Gener. Transm. Distrib.* **2016**, *10*, 2019–2025.
11. Liu, J.; Wen, J.; Yao, W.; Long, Y. Solution to short-term frequency response of wind farms by using energy storage systems. *IET Renew.e Power Gener.* **2016**, *10*, 669–678.
12. Edmunds, R.; Cockerill, T.; Foxon, T.; Ingham, D.; Pourkashanian, M. Technical benefits of energy storage and electricity interconnections in future British power systems. *Energy* **2014**, *70*, 577–587.
13. Zucker, A.; Hinchliffe, T.; Spisto, A. *Assessing Storage Value in Electricity Markets*; JRC Scientific and Policy Report; Joint Research Centre: Brussels, Belgium, 2013.
14. Schoenung, S. *Energy Storage Systems Cost Update*; Sandia National Laboratories: Albuquerque, NM, USA, 2011.
15. Zakeri, B.; Syri, S. Electrical energy storage systems: A comparative life cycle cost analysis. *Renew. Sustain. Energy Rev.* **2015**, *42*, 569–596.
16. Chen, H.; Zhang, R.; Li, G.; Bai, L.; Li, F. Economic dispatch of wind integrated power systems with energy storage considering composite operating costs. *IET Gener. Transm. Distrib.* **2016**, *10*, 1294–1303.

17. Hu, Z.; Jewell, W.T. Optimal generation expansion planning with integration of variable renewables and bulk energy storage systems. In Proceedings of the 2013 1st IEEE Conference on Technologies for Sustainability (SusTech), Portland, OR, USA, 1–2 August 2013; pp. 1–8.

18. Qiu, T.; Xu, B.; Wang, Y.; Dvorkin, Y.; Kirschen, D. Stochastic Multi-Stage Co-Planning of Transmission Expansion and Energy Storage. *IEEE Trans. Power Syst.* **2016**, *32*, 643–651.

19. Dehghan, S.; Amjady, N. Robust Transmission and Energy Storage Expansion Planning in Wind Farm-Integrated Power Systems Considering Transmission Switching. *IEEE Trans. Sustain. Energy* **2016**, *7*, 765–774.

20. Solomon, A.; Kammen, D.M.; Callaway, D. The role of large-scale energy storage design and dispatch in the power grid: A study of very high grid penetration of variable renewable resources. *Appl. Energy* **2014**, *134*, 75–89.

21. Makarov, Y.V.; Du, P.; Kintner-Meyer, M.C.; Jin, C.; Illian, H.F. Sizing energy storage to accommodate high penetration of variable energy resources. *IEEE Trans. Sustain. Energy* **2012**, *3*, 34–40.

22. Alnaser, S.W.; Ochoa, L.F. Optimal Sizing and Control of Energy Storage in Wind Power-Rich Distribution Networks. *IEEE Trans. Power Syst.* **2016**, *31*, 2004–2013.

23. Dvorkin, Y.; Fernandez-Blanco, R.; Kirschen, D.S.; Pandzic, H.; Watson, J.P.; Silva-Monroy, C.A. Ensuring Profitability of Energy Storage. *IEEE Trans. Power Syst.* **2016**, *31*, 611–623.

24. Fern, R.; Dvorkin, Y.; Xu, B.; Wang, Y.; Kirschen, D.S. *Energy Storage Siting and Sizing in the WECC Area and the CAISO System*; University of Washington: Seattle, DC, USA; 2016; pp. 1–31.

25. Dvijotham, K.; Chertkov, M.; Backhaus, S. Storage sizing and placement through operational and uncertainty-aware simulations. In Proceedings of the Annual Hawaii International Conference on System Sciences, Waikoloa, HI, USA, 6–9 January 2014; pp. 2408–2416.

26. Zhang, F.; Hu, Z.; Song, Y. Mixed-integer linear model for transmission expansion planning with line losses and energy storage systems. *IET Gener. Transm. Distrib.* **2013**, *7*, 919–928.

27. Wogrin, S.; Gayme, D.F. Optimizing Storage Siting, Sizing, and Technology Portfolios in Transmission-Constrained Networks. *IEEE Trans. Power Syst.* **2015**, *30*, 3304–3313.

28. Go, R.S.; Munoz, F.D.; Watson, J.P. Assessing the economic value of co-optimized grid-scale energy storage investments in supporting high renewable portfolio standards. *Appl. Energy* **2016**, *183*, 902–913.

29. Nasrolahpour, E.; Kazempour, S.J.; Zareipour, H.; Rosehart, W.D. Strategic Sizing of Energy Storage Facilities in Electricity Markets. *IEEE Trans. Sustain. Energy* **2016**, *7*, 1462–1472.

30. Cui, H.; Li, F.; Fang, X.; Chen, H.; Wang, H. Bi-Level Arbitrage Potential Evaluation for Grid-Scale Energy Storage Considering Wind Power and LMP Smoothing Effect. *IEEE Trans. Sustain. Energy* **2017**, doi:10.1109/TSTE.2017.2758378.

31. Kazempour, S.J.; Conejo, A.J.; Ruiz, C. Generation Investment Equilibria With Strategic Producers; Part I: Formulation. *IEEE Trans. Power Syst.* **2013**, *28*, 2613–2622.

32. Grbovic, P. Energy Storage Technologies and Devices. In *Ultra-Capacitors in Power Conversion Systems: Analysis, Modeling and Design in Theory and Practice*; Wiley-IEEE Press: Hoboken, NJ, USA, 2014; p. 336.

33. Ramos, T.P.; Marcato, A.L.M.; da Silva Brandi, R.B.; Dias, B.H.; da Silva Junior, I.C. Comparison between piecewise linear and non-linear approximations applied to the disaggregation of hydraulic generation in long-term operation planning. *Int. J. Electr. Power Energy Syst.* **2015**, *71*, 364–372.

34. Bertsekas, D.P. *Nonlinear Programming*; Athena Scientific Belmont: Belmont, MA, USA, 1999.

35. Hooker, J. *Logic-Based Methods for Optimization: Combining Optimization and Constraint Satisfaction*; John Wiley & Sons: Hoboken, NJ, USA, 2011; Volume 2.

36. Trespalacios, F.; Grossmann, I.E. Improved Big-M reformulation for generalized disjunctive programs. *Comput. Chem. Eng.* **2015**, *76*, 98–103.

37. Rubasheuski, U.; Oppen, J.; Woodruff, D.L. Multi-stage scenario generation by the combined moment matching and scenario reduction method. *Oper. Res. Lett.* **2014**, *42*, 374–377.

38. Breeze, P. Wind Power. *Power Gener. Technol.* **2014**, *1*, 223–242.

 *sustainability*

Article

A Spatial Decision Support System Framework for the Evaluation of Biomass Energy Production Locations: Case Study in the Regional Unit of Drama, Greece

Konstantinos Ioannou [1,*], Georgios Tsantopoulos [2], Garyfallos Arabatzis [2], Zacharoula Andreopoulou [3] and Eleni Zafeiriou [4]

[1] Hellenic Agricultural Organization "DEMETER", Forest Research Institute, Vasilika, Thessaloniki 57006, Greece

[2] Department of Forestry and Management of the Environment and Natural Resources, Democritus University of Thrace, Pantazidou 193, Orestiada 68200, Greece; tsantopo@fmenr.duth.gr (G.T.); garamp@fmenr.duth.gr (G.A.)

[3] Department of Forestry and Natural Environment, Lab. Of Forest Informatics, Aristotle University of Thessaloniki, POBOX 247, Thessaloniki 54124, Greece; randreop@for.auth.gr

[4] Department of Agricultural Development, Democritus University of Thrace, Pantazidou 193, Orestiada 68200, Greece; ezafeir@agro.duth.gr

* Correspondence: ioanko@fri.gr or ioannou.konstantinos@gmail.com; Tel.: +30-23-1046-1171 (ext. 225)

Received: 23 January 2018; Accepted: 12 February 2018; Published: 16 February 2018

Abstract: Renewable Energy Sources are expected to play a very important role in energy production in the following years. They constitute an energy production methodology which, if properly enabled, can ensure energy sufficiency as well as the protection of the environment. Energy production from biomass in particular is a very common method, which exploits a variety of resources (wood and wood waste, agricultural crops and their by-products after cultivation, animal wastes, Municipal Solid Waste (MSW) and food processing wastes) for the production of energy. This paper presents a Spatial Decision Support System, which enables managers to locate the most suitable areas for biomass power plant installation. For doing this, fuzzy logic and fuzzy membership functions are used for the creation of criteria layers and suitability maps. In this paper, we use a Multicriteria Decision Analysis methodology (Analytical Hierarchy Process) combined with fuzzy system elements for the determination of the weight coefficients of the participating criteria. Then, based on the combination of fuzzy logic and theAnalytic Hierarchy Process (AHP), a final proposal is created thatdivides the area into four categories regarding their suitability forsupporting a biomass energy production power plant. For the two optimal locations, the biomass is also calculated.The framework is applied to theRegional Unit of Drama, which is situated in Northern Greece and is very well known for the area's forest and agricultural production.

Keywords: fuzzy logic; AHP; biomass; Greece; land suitability

---

1. Introduction

Energy is a very important part of human life and the economy.Many researchers consider it as a key element forimproving quality of life [1,2], although there are differing opinions [3]. In any case, energy plays a vital role,whethernegative or positive, in our way of life.

Also, many researchers have proven that overthe last decades we are witnessing a constant increase in energy usage, which is followed by a constant need toincrease energy production [4–6]. The International Energy Outlook for the year 2017, published by the U.S. Energy Information

Administration [6], suggests that the total energy consumption will rise from 575 quadrillion Btus in 2015 to 735 Btus in 2040. Most of this increase will come from countries that are not part of the Organization for Economic Co-operation and Development (OECD). These countries are characterized by strong economic growth, increased access to marketed energy and quickly growing populations, which lead to rising energy demands. It is estimated that the energy consumption in non-OECD countries will increase by 41% between 2015 and 2040 in contrast to a 9% increase in OECD countries [6]. One of the most important drivers that leads to increased energy demand is population. Studies showed that in OECD countries the estimation of population growth is only 0.2% per year, whereas in non-OECD countries in Africa, Middle East and India are expected to experience among the highest rates of population growth in the world [6].

Apart from the energy usage increase caused from population growth there are numerous other reasons causing similar phenomena. The average household consumes more energy mainly because there are new appliances introduced and used every day [7]. Furthermore, it is estimated that the increase in the number of appliances like TV sets per household, gaming consoles, computers as well as the diffusion of usage of others (vacuum cleaners, dish washers etc.) also increases the average energy usage [8,9]. The same researchers have shown that between the years 2005 and 2011 an increase of 102 kWh/year has been made only from the usage of high definition consoles.

Other researchers, by analysing micro-level data from the National Survey of Family and Expenditure, revealed that the family and income structure of households affects appliance usage [7]. Additionally, it is proven that the need for larger houses also leads to a subsequent increase in energy consumption which is caused by the residents need to utilize the entire structure [9–11], Another increase factor in energy usage is caused by the increment of the number of cooling days. It is proven that since the year 1990 there is a constant increase in the number of cooling days and heating days per year, which causes a subsequent increase in energy demand [12–14].

Currently the most popular methods of energy production are mainly based on the usage of fossil fuels, which will represent 77% of energy production by 2040 according to IEO [6]. However, these methods have an enormous environmental impact mainly because they produce contaminants which play a very important role in climate change (fossil fuels) or their bi-products are difficult to be processed and stored (nuclear fuels) [15–20].

Furthermore, the usage of non-renewable energy sources has a limited time frame. Studies have proven that oil, coal and gas reserves will be extinct in 35, 107 and 37 years respectively, based on the current usage rate and energy demands prediction [21–23].

The most promising solution to these problems is the usage of renewable energy sources (RES) which can be exploited without the subsequent side effects caused by the typical energy production forms.

Under this scope the aim of this paper is the presentation of a methodology in the form of a spatial decision support system framework, which is based on the combination of fuzzy systems and multicriteria decision analysis and will enable researchers, managers and local or regional authorities to identify the locations where potentially a RES power plant can be installed in order to maximise its usage and minimise environmental annoyance and installation cost. Additionally, the methodology will provide sufficient data in order to make a first estimation regarding the overall energy potential of the proposed area.

Finally, in order to validate the results of the system, the methodology is applied to the Regional unit of Drama. The regional unit was selected because it combined all the types of biomass energy sources. In detail, Drama has the largest forest and transitional forest areas in Greece as well as significant agricultural areas, pastures. Additionally, in the regional unit there is the largest number of livestock units in the prefecture of Eastern Macedonia and Thrace and it is in the top 10 regional units in Greece [24].

2. Literature Review

2.1. Renewable Energy Sources

It is already stated that there is a rapidly increasing need for producing energy using cleaner more environmental friendly methods like Renewable Energy Sources (RES).

The most common forms of RES include the usage of hydroelectric power, solar and wind energy, geothermal energy and biomass for the production of electricity generation [25]. RES constitutes for many countries (developed and developing), an important, alternative energy source, which reduces energy dependency and strengthens the security of their energy supply [26].

Global electricity energy production from renewable energy is expected to grow 2.7 times until 2035 [27]. Renewables are the fastest-growing source of energy for electricity generation (Figure 1). The average increase is estimated to be 2.8%/year from 2015 to 2040. From all RES technologies, non-hydropower renewable resources are the fastest-growing energy sources for new generation capacity in both OECD and non-OECD regions. Non-hydropower renewables accounted for 7% of total world generation in 2015; their share in 2040 is 15% in the IEO2017 Reference case, with more than half of the growth coming from wind power [6].

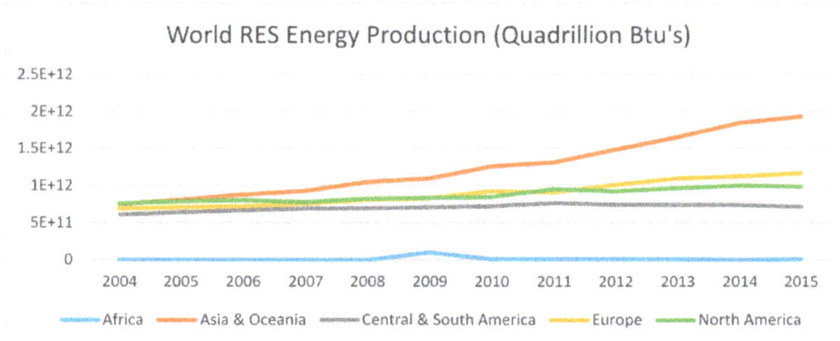

Figure 1. World energy production from RES in quadrillion Btusfor the years 2004–2015 [28].

In Europe, renewable energy production within the 28 member states in 2015 was 204 Mtoe (Figure 2) (8.08 quadrillion Btus). The quantity of renewable energy produced within the member states has increased overall by 70.2% from 2005 until 2015 [29]. This increase in RES usage is triggered by the EU legislation framework. The EU Directive 2009/28/EC set the obligatory level of contribution of RES to energy mixture as 20% by 2020. In order to achieve this level EU member must double the share of electricity from RES from 16% to over 30%. In 2012 the penetration of RES in all 28 member states was 14.1%. Sweden is the leading country with 51% of energy production covered by RES, followed by Finland, Austria and Latvia with 30% [29].

In Greece, due to its geographic position, there is an enormous potential for energy production by RES. The geo-morphologic profile and the weather conditions favour energy production from the exploitation of wind and solar energy [30,31]. In 2015 the percentage of electricity generated by RES was 22.1% [29].

From Figure 3 it is evident that Hydro Power, Solar and Biofuels are the dominant forms of RES production. However, there can be vast improvements in order for the country to reach the EU average.

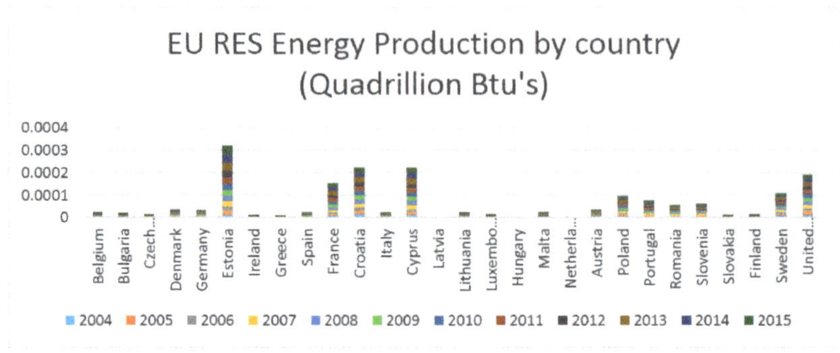

Figure 2. EU RES Energy Production in quadrillion Btus for the years 2004–2015 [29].

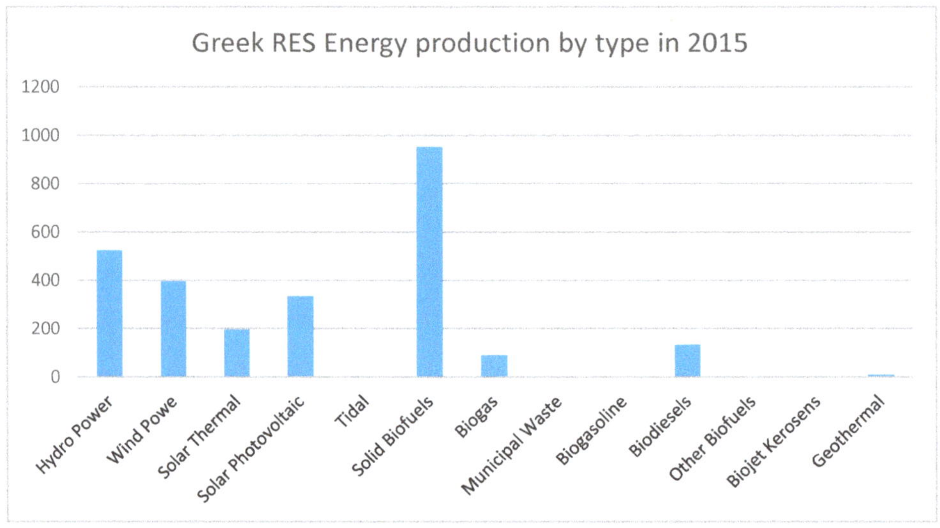

Figure 3. Energy production by RES in Greece for the year 2015 in Mtoes [29].

2.2. Biomass Energy Production

Biomass is a very important form of RES, mainly because its usage can be done in raw form without any process. Among other factors the growing interest in biomass usage is based on the following facts [32]:

- Its contribution to poverty reduction in developing countries
- Its ability to constantly meet energy demand
- Its capability on delivering energy in all forms people need (liquid, gas, heat and electricity)
- Its carbon dioxide neutrality
- It helps on the restoration of unproductive and degraded lands, increases biodiversity, soil fertility and water retention

The most common biomass energy sources are wood and wood waste, agricultural crops and their by-products after cultivation, animal wastes, Municipal Solid Waste (MSW) and food processing wastes. Currently the available biomass resources can provide approximately 6×10^{15} Btus of energy [33].

Wood and wood waste produce 64% of the total Biomass energy, followed by MSW (24%), agricultural wastes and by-products (5%) and landfill gases (5%) [33–35].

There are many economic benefits deriving from the usage of biomass as a source for energy production [36]:

- Inexpensive resources
- Locally distributed
- Price stability
- Generation of employment opportunities
- Biomass export opportunities
- Potentially inexhaustible fuel resource

In Greece, it is estimated that the total of the available biomass consists of 7,500,000 tones of Agricultural crop residues (cereals, maize, cotton, tobacco, sunflower, strawberries, olive kernels, vines etc.) and 2,700,000 tones from forestry residues (branches, roots, barks etc.). The vast percentage of this biomass remains unused and is causing the development of forest fires or is acting as a mean for disease spreading. Additionally, in Greece, the available agricultural and forest residues are estimated to be equivalent to 3–4 million tons of oil, which constitute approximately 30–40% of the total oil consumption in Greece. At the same time the potential of energy cultivations can surpass the energy produced by these residues. Until now, biomass in Greece is mainly used on a limited scale for heating purposes (households, greenhouses etc.) and in industry (cotton ginners, timber products, lime kilns etc.). In these cases, the main biomass products used are residues from wood industry, almond shells, cereal straws, ginning residues etc. From the aforementioned it is evident that the prospects of biomass usage in Greece are very promising. This is due to the fact that there is considerable potential, much of which is readily available. At the same time, the energy that can be produced is, in many cases, economically [37].

However, there are also significant disadvantages in biomass usage. Biomass fuels gave low energy density, collection and transportation can be cost prohibitive, Biomass usage for energy production requires a constant supply of fuels which are bulk and must be stored near the biomass plant. Additionally, there is also a constant requirement of other inputs which include water, crops and fossil energy which also have cost [38].

Therefore, although biomass usage for energy production is technologically well established, the price paid for electricity seldom offsets the full cost of the biomass fuel. It is clear that for the optimization of biomass exploitation for energy production a Decision Support System (DSS) is needed. This system must take under consideration biomass production as well as biomass accessibility, land uses etc. and provide managers with the optimal solution which will help them determine the best site for installing a biomass power plant.

2.3. MCDA and AHP

Multi Criteria Decision Analysis (MCDA) combined with Geographical Information Systems are valuable tools for the determination of the optimal solution for a problem. In general, MCDA is a sub-field of operational research which focuses on the development of decision support systems which can help in the resolution of complex problems which depend in a variety of criteria and parameters [39]. MCDA and GIS were used by many researchers for the solution of environmental problems. For the determination of the optimal solution to post fire regeneration [40], for the creation of an algorithm which will help on the determination of the optimal design of biomass supply chain networks [41], for determining the optimal location of a log yard [42] and the optimal usage of biomass for electricity generation [43].

Analytical Hierarchy Process (AHP) is an MCDA methodology introduced by Thomas L. Saaty. AHP is based on the determination of a goal as well as the criteria and alternatives which affect the achievement of the preselected goal. The pairwise comparisons among the criteria and the alternatives help the researchers to determine the weight coefficients of each of them. By determining the weight coefficients, the researchers know the level of effect to the final solution [44].

AHP in combination with GIS has been used for the selection of the optimal location of offshore wind farms [45], for performing solar farms feasibility analysis in India [46–48], for selecting the optimal location for solar PV power plants [49] and large wind farms [50,51], for reinforcing the hydropower strategy in Nepal [52] and for exploring geothermal resources and locations [53,54].

Fuzzy AHP is an enhanced version of the classic AHP methodology. The usage of fuzzy AHP for multiple criteria decision making requires scientific weight derivation from fuzzy pairwise comparison matrices [55]. Fuzzy AHP has been used in water loss managements in developing countries [56], for the evaluation of solar farms locations in Iran [57], for energy planning in Istanbul [58], for the evaluation of the capability of renewable energy sources to generate electricity [59–62], for the selection of solar-thermal plants investment projects [63], for creating spatial decision support systems for solar farm location planning [64], for selecting the optimal renewable energy type in Indonesia [65] and for analysing the assessment factors for renewable energy dissemination [66].

In this paper, we present a spatial DSS which based on the usage of Multi Criteria Decision Analysis (MCDA), fuzzy Analytical Hierarchy Process (AHP), Python programming will analyse a series of parameters and criteria affecting the selection of optimal installation sites and propose the optimal location. For the visualization of the installation site Geographical Information Systems (GIS) are used.

Multi criteria decision making using fuzzy Analytical Hierarchy Process is a powerful tool which enables researchers to explore various possible problem solutions and at the same time accepting the fact that the solution to a problem cannot only expressed as a binary result (yes or no). The usage of fuzzy techniques allows researchers to map the vagueness of an answer, by allowing the interviewed person to express an uncertainty to his answer. Furthermore, the usage of AHP, a well-established multi criteria decision making method can easily identify the weight coefficients of the problems criteria and thus allow the researchers to map the optimal solution more efficiently.

The combination of the aforementioned methodologies (MCDA and fuzzy system elements) with a spatial analysis tool like Geographic Information Systems can enhance the produced results by allowing researchers to determine more efficiently the locations were biomass energy production plants can be installed by determining the overall biomass potential.

3. Methodology

Energy produced from a biomass plant is directly related to the amount of biomass available as well as to the amount of accessible biomass in the wider area. The greater the amount of biomass available in a given area the greater the potential of electricity generation [67].

The information on biomass availability in Regional unit of Drama was provided from two sources, the biomass potential from shape files created from the Greek Centre for Renewable Energy Sources (CRES) and data regarding farms which were provided from the department of environment of Regional unit of Drama.

3.1. Criteria Selection

The installation of a biomass energy production plant is affected by some factors which can be classified in three main categories: Technical, Economic and Environmental. These factors depend on the geographical location, the socio-economical structure of the studied area and the biophysical attributes.

All of the aforementioned factors were expressed as GIS layers. For the creation of these layers' fuzzy logic and fuzzy membership functions were used. Each layer has a value ranging from zero to one. The value of one represents the most suitable site for the installation of a biomass energy production plant while the value of zero represents the least suitable site for installation of the plant. The Analytical Hierarchy Process was finally used to combine the three layers as well as the layer of human and environmental limitations (exclusion zones). The resulting layer provided the most suitable areas for the installation of biomass energy production plants in Regional unit of Drama.

3.2. Human and Environmental Considerations

Prior to selecting the most suitable locations for the installation site we must exclude the areas were due to legislation, human or environmental considerations we cannot propose the installation. These locations include cities, villages, settlements, lakes, roads, protected areas, rivers, wetlands, agricultural lands, wetlands etc. Greek legislation under law 2742/1999 regarding the National spatial planning and sustainable development clearly defines the areas were the installation of RES is forbidden (Table 1). These limitations were ultimately combined in a separate layer (Constraint Layer).

Table 1. Legislation, human and environmental considerations.

Criterion	Limitations	References
Slope	Between 0 and 10% for constructions and accessibility from vehicles	[68]
Heavily forested areas and Protected areas	Restriction: these zones are completely off limits	N2742/1999 (Greek Legislation)
Water Bodies, wetlands, rivers	500 m buffer zone for wetlands 400 m buffer zone for wetlands 300 m buffer zone for wetlands 500 m buffer zone for rivers 400 m buffer zone for rivers 1000 m from coast 200 m from rivers	[69] [70] [71] [69] [70] N2742/1999 (Greek Legislation)
Agricultural land, pastures, Orchards	Agricultural land, pastures and orchards are unsuitable	[68,72]
Urban areas	2000 m from urban areas 1–5 km from urban areas	N2742/1999 (Greek Legislation), [72]
Road and railroad network	500 m buffer zone 150 m buffer zone railroad area restriction	[73] [74] [75] N2742/1999 (Greek Legislation)

Constraint Areas Fuzzy Datasets

Current land use is one of the main criteria which affect the installation location. Water bodies, protected areas, urban and residential areas, roads, railroads and steep slopes cannot be used for the installation of the proposed power plant. Additionally, agricultural land, land covered by orchards or other high productivity areas cannot be used [76].

As a result, only areas with poor vegetation, logged areas or barren land can be considered as ideal locations.

For this reason, a buffer zone must be created around these areas. The buffer zone acceptability regarding the installation increases from zero to one from the border of the arable land to a distance of 400 m from that border [72].

Figure 4 shows the fuzzy membership value (μ) for land. Values between zero and one represent the land suitability inside the buffer zone. Clearly the suitability equals one outside the buffer zone.

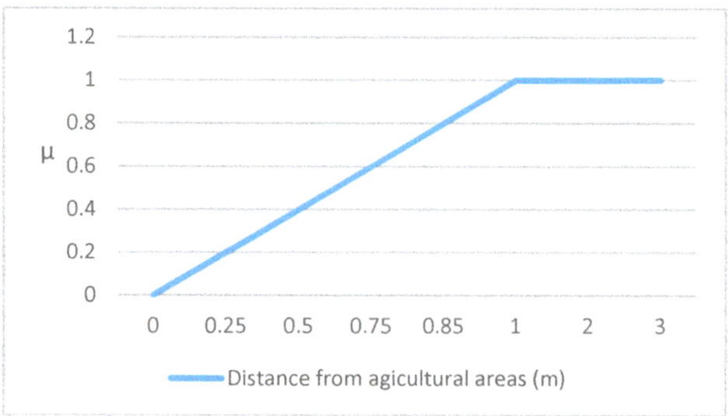

Figure 4. Agricultural land fuzzy membership function.

Data for the land uses in Regional unit of Drama where obtained from the CORINE programme. The fuzzy membership function considering that restriction is:

$$\mu_{PA} = \begin{cases} 0, \ x \leq 100 \\ \frac{x-100}{300}, \ 100 < x \leq 400 \\ 1, \ x > 400 \end{cases} \tag{1}$$

where: x, is the distance from the arable land, orchards etc. and μ_{PA} is the fuzzy membership function.

The power plants cannot be installed inside heavily forested areas or protected areas, therefore the buffer zone created around these types of areas is expressed through the following membership function:

$$\mu_F = \begin{cases} 0, \ x \leq 100 \\ \frac{x-100}{400}, \ 100 < x \leq 500 \\ 1, \ x > 500 \end{cases} \tag{2}$$

where: x is the distance from the protected or heavily forested areas and μ_F is the fuzzy membership function.

Similarly, we cannot propose as a potential installation site area inside or in close proximity to urban areas, water bodies and over road or railroad network. The membership function for the exclusion of this areas are:

$$\mu_u = \begin{cases} 0, \ x \leq 1000 \\ \frac{x-1000}{4000}, \ 1000 < x \leq 5000 \\ 1, \ x > 5000 \end{cases} \tag{3}$$

$$\mu_{wb} = \begin{cases} 0, \ x \leq 200 \\ \frac{x-200}{200}, \ 200 < x \leq 400 \\ 1, \ x > 400 \end{cases} \tag{4}$$

$$\mu_{rr} = \begin{cases} 0, \ x \leq 100 \\ \frac{x-100}{300}, \ 100 < x \leq 400 \\ 1, \ x > 400 \end{cases} \tag{5}$$

where μ_u, μ_{wb} and μ_{rr} are the fuzzy membership functions for urban areas, water bodies and rail and road network.

The last fuzzy membership function is created for determining the slope value. Slope is a topographic feature which is strongly related to the overall project cost. In order to select the optimal site, the slope value must be low. In any other case the accessibility of the plant will be reduced because trucks cannot access steep slopes and furthermore the overall construction cost will rise due to excavations or embankments. For the determination of slope, the Digital Elevation Model (DEM) of Regional unit of Drama was created based on satellite data provided by the Copernicus programme. From the DEM, the slope map was created, each cell of the raster slope map contains a slope value. The fuzzy membership function for the slope is:

$$\mu_s = \begin{cases} 1, x \leq 3 \\ \frac{10-x}{7}, 3 < x \leq 10 \\ 0, x > 10 \end{cases} \tag{6}$$

whereas x is the slope value expressed in percentage.

One of the constraint layers was modelled as a raster layer in GIS with a 50 m spatial resolution. Finally, all the constraint layers were multiplied together and formed the final constraint layer. Areas with value of zero will also be zero in the final layer. For example, if a location is inside a river buffer zone the location's value will be zero (fuzzy membership function is zero), the final constraint layer will be also zero even if the values of all the other layers are one. Areas with value of one in all sub layers will also have a value of one in the final constraint layers and are the most prominent areas because they lack restrictions. Finally, locations with values between zero and one will receive a final score based on the multiplication of each layer's values.

3.3. Techno-Economic Criteria

The amount of available biomass plays a key role in the selection of the proper installation site. For this reason, the biomass potential database provided by the Centre for Renewable Energy Sources (CRES). The database the entire regional unit and includes the following data:

- Point sources of biomass
- Arable crops
- Greenhouse crops
- Tree crops
- Vineyards
- Forests

The aforementioned data provide the residues from these types of cultivations for the various municipalities inside Regional unit of Drama and they are estimated in tones.

Figure 5 represents the biomass potential for the municipalities of Drama (in light blue). The biomass potential was calculated by adding the individual data provided by the database of CRES. Higher values are depicted in yellow, orange and red, while lower values are depicted in shades of green.

Another essential factor is the availability of transport links (railroads and road network). It is easier to move supplies through the existing transport network. Furthermore, the existence of transportation network reduces the cost and the probable damages to the environment which can be caused by the creation of new network. The maximum acceptable distance varies depending on the study. Our approach values potential locations near roads and railroads to be of better value compared to other potential sites which are further from the transportation network. Similar approaches have been made by researchers trying to allocate the optimal installation sites for other types of RES like wind farms and P/V farms [77–79].

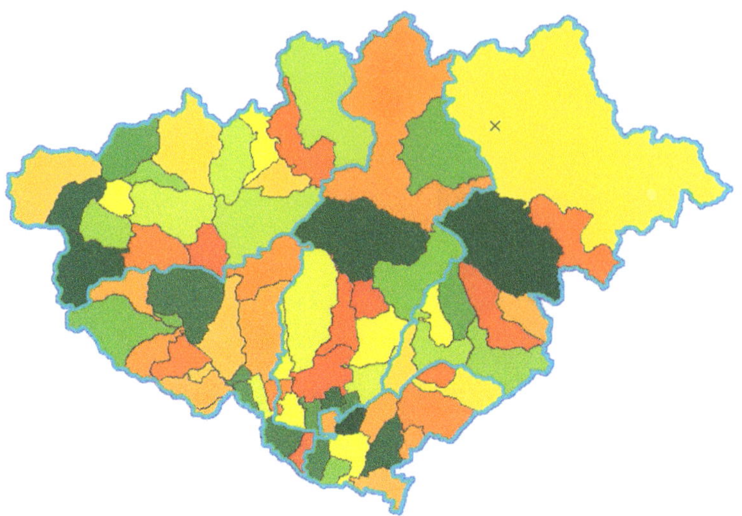

Figure 5. Regional unit of Drama Biomass Potential.

The membership function describing the distance from the transportation network is:

$$\mu_t = \begin{cases} 1, \ x \leq 500 \\ \frac{1000-x}{500}, \ 500 < x \leq 1000 \\ 0, \ x \geq 1000 \end{cases} \tag{7}$$

whereas x is the distance from the transportation network and μ_t is the fuzzy membership function.

Additional techno-economic criteria that must be taken under consideration is the proximity to farms and to the energy transfer grid. Proximity to farms is important because they constitute a continuing mean of biomass production from animal wastes. Studies have shown that animal waste can efficiently be used for energy production [80–83].

The membership function describing the distance from farms is:

$$\mu_f = \begin{cases} 1, \ x \leq 500 \\ \frac{3000-x}{2500}, \ 500 < x \leq 3000 \\ 0, \ x > 3000 \end{cases} \tag{8}$$

whereas x is the distance of the potential installation site and the farm location and μ_f is the membership function. From Equation (8) it is evident that the optimal installation locations are the ones which are 500 m or less from farms, less preferred location are inside a zone ranging from 500 m up to 3000 m and the least preferred locations are the ones which are located 3000 m or more from farms.

The final techno-economic criteria which must be taken under consideration is the distance from the existing energy transfer network. This criterion is similar to the proximity to the existing transportation network. Locations that are closer to the existing transfer network are preferred compared to locations which are further. The main reason is that the connection of the biomass power plant to the existing energy grid will be more economical and the environmental disturbance will be minimal. The member transfer function describing the distance from the transportation network is:

$$\mu_{te} = \begin{cases} 1, \ x \leq 100 \\ \frac{1000-x}{900}, \ 100 < x \leq 1000 \\ 0, \ x \geq 1000 \end{cases} \tag{9}$$

whereas x is the distance of the potential installation site and the energy transfer network and μ_{fe} is the membership function.

Membership function described from Equations (7)–(9) can be depicted in the following diagram (Figure 6).

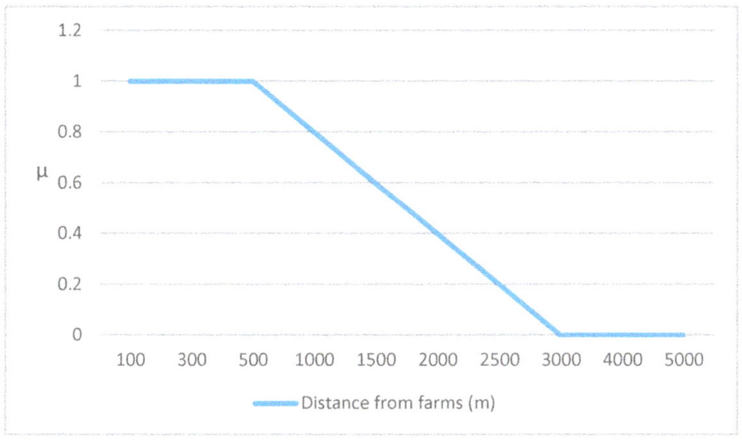

Figure 6. Distance from farms fuzzy membership function.

3.4. AHP

Analytical Hierarchy Process (AHP) is a well-established Multi Criteria Decision Methodology which is widely used in optimizing decision making [44]. It provides a comprehensive framework for structuring a decision problem, representing and quantifying the comprising elements, relating those elements to goals and evaluating possible solutions.AHP is based on pairwise comparisons among the criteria and the parameters affecting the achievement of a goal [44].

AHP is widely used in the field of sustainable energy [45,49–51].

The implementation of AHP involves the following steps [44]:

1. Determination of the goal, the alternatives involved in reaching this goal and the criteria affecting the alternatives.
2. Definition of the priorities among the alternatives by performing pairwise comparisons (Table 2).
3. Synthetization of the priorities to yield a set of overall priorities for the hierarchy.
4. Consistency check.
5. Final decision.

Table 2. Pairwise comparison values.

Definition	Index	Definition	Index
Equally Important	1	Equally important	1/1
Equally or slightly more important	2	Equally or slightly less important	1/2
Slightly more important	3	Slightly less important	1/3
Slightly to much more important	4	Slightly-to-way less important	1/4
Much more important	5	Way less important	1/5
Much-to-far more important	6	Way to far less important	1/6
Far more important	7	Far less important	1/7
Far more important to extremely more important	8	Far less important to extremely less important	1/8
Extremely more important	9	Extremely less important	1/9

In this study, we used PYTHON code for performing the AHP methodology in GIS. The code allows the determination of matrices for each criterion and alternative and performs the comparisons based on the user inputs. At the end, it provides the weight coefficients and performs a consistency check of the results.

For implementing the analysis, we defined a tree, with nodes for the criteria and the alternatives:

```
       +–(alt. a)
            +–[node 1]–+–(alt. b)
            |          +–(alt. c)
[root]–+
            |                      +–(alt. a)
            |          +–[node 2.1]–+–(alt. b)
            |          |            +–(alt. c)
            +–[node 2]–+
            |          |          +–(alt. a)
            +–[node 2.2]–+–(alt. b)
```

The example represents a two level AHP tree, having 4 nodes and 3 alternatives. there are three 'edge nodes,' known as "node 1,""node 2.1" and "node 2.2."

An example of the code used to calculate weight coefficients inside a node is shown below:

```
def __wc(n):
if not isinstance(n, AHPTreeNode):
return;

ret = {};
        # first, we process childrens
forch in n.iter_childrens():
ifisinstance(ch, AHPTreeNode):
        w = ch.calculate_weights();
        d = __dive(ch);
if d != None:
ret[ch.get_name()] = (w, d);
else: # leaf
ret[ch.get_name()] = w;
        ##

if ret != {}:
if n == sclf__TreeRoot:
        w = n.calculate_weights();
return {'Root': (w, ret)};
else:
return ret;
else:
return None;
```

4. Results

4.1. Constraint Layer

The Constraint layer has been produced by the multiplication of the individual constraint layers (slope, heavily forested and protected areas, rivers lakes and coasts, Agricultural land, pastures Orchards, Urban areas (towns, villages, settlements) and distance from Road and railroad network).

Figure 7, presents the constraint layers after the implementation of the fuzzy membership functions for each layer. Areas in shades of green are more suitable for the installation of the biomass energy production plant. Areas in shades of yellow are less suited and finally areas in red are not suitable for installation. The final constraint layer is presented in Figure 7. From this layer, we have excluded protected areas (Natura 2000, Ramsar sites etc.) regardless of their capability to support the installation, as Greek legislation prohibits the exploitation of these areas.

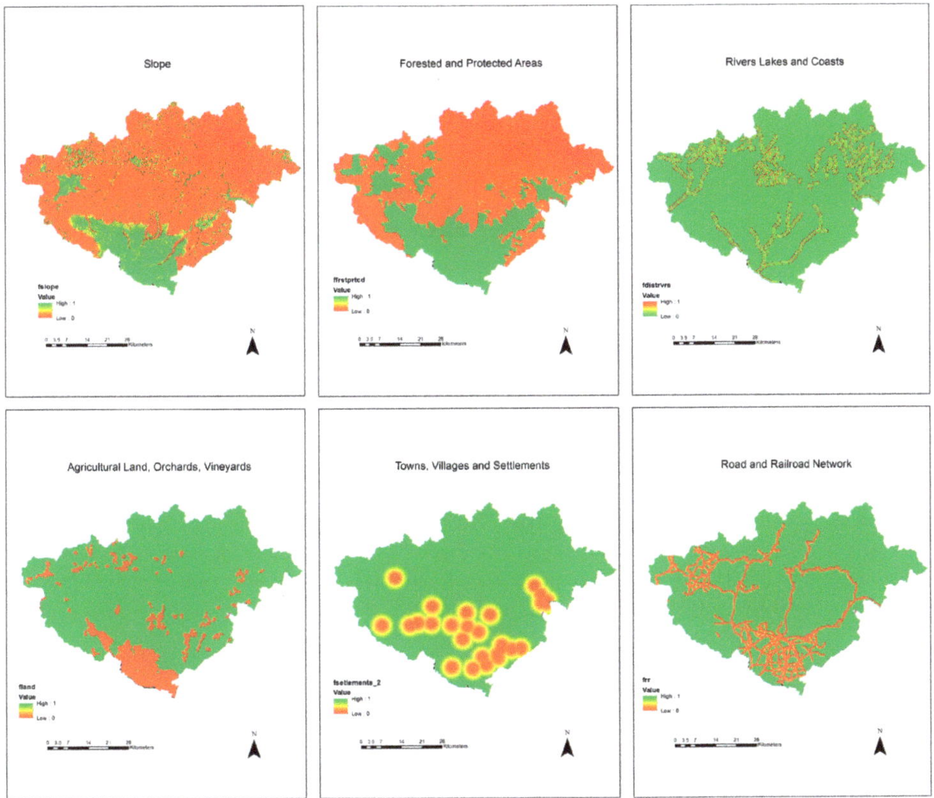

Figure 7. The individual constraint layers.

Figure 8 presents the result from the multiplication of the individual layers. The areas in purple are excluded as possible installation sites because they are protected. Areas in red are the least suitable for installing the power plant, whereas areas in green and yellow are the most suitable.

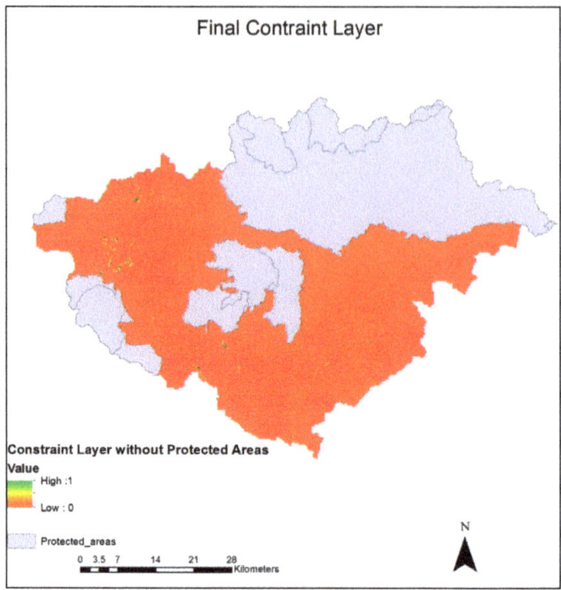

Figure 8. The final constraint layer.

4.2. Biomass Potential

Figure 9 shows the total biomass potential of Regional unit of Drama. As it is apparent the amount of biomass varies in the regional unit. Areas depicted in red colour have low biomass potential whereas areas in orange and yellow are more suitable for exploitation. Finally, areas in shades of green are the most suitable for the installation of power plant that exploit biomass.

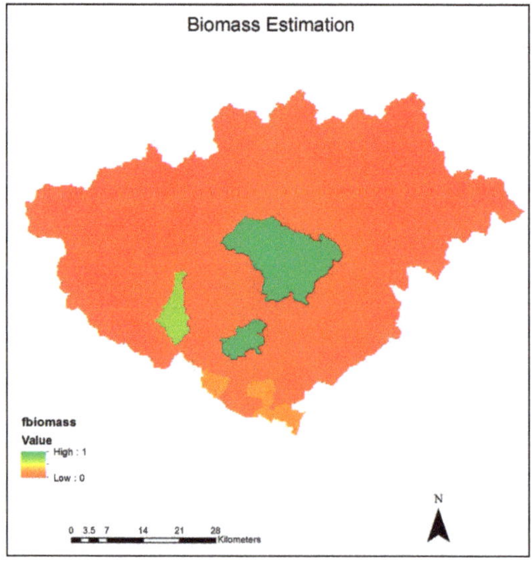

Figure 9. Biomass estimation in Regional unit of Drama.

4.3. Access to Transportation Network and Energy Grid

The access of the proposed installation site to the current transportation network as well as to the current energy grid plays a vital role. Almost all areas of the regional unit outside the protected areas have good access to the road network. The energy transfer grid depicts only medium and high voltage power lines which are suitable for long distances transfer (Figure 10).

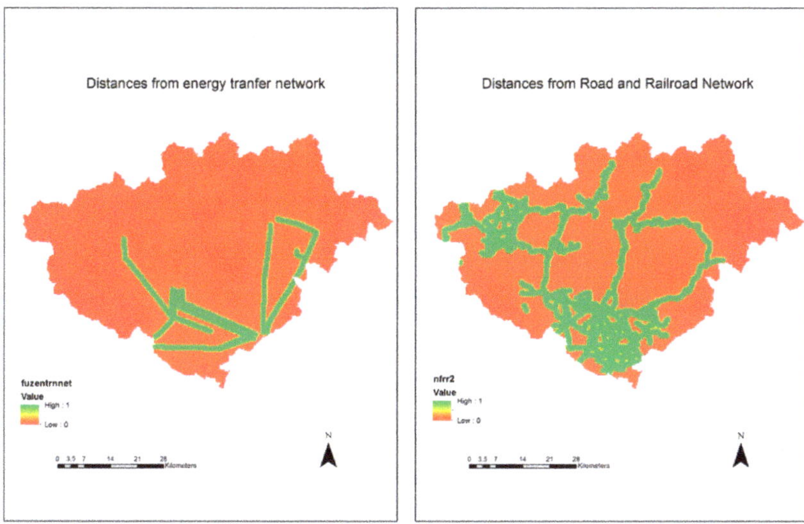

Figure 10. Distances from energy transfer network and transportation network.

4.4. Distances from Farms

Data for the location of farms in Regional unit of Drama were provided from the Department of Environment of the regional unit. The data include the locations of currently running farms and in a shape file and were transformed to raster using the GIS software (Figure 11).

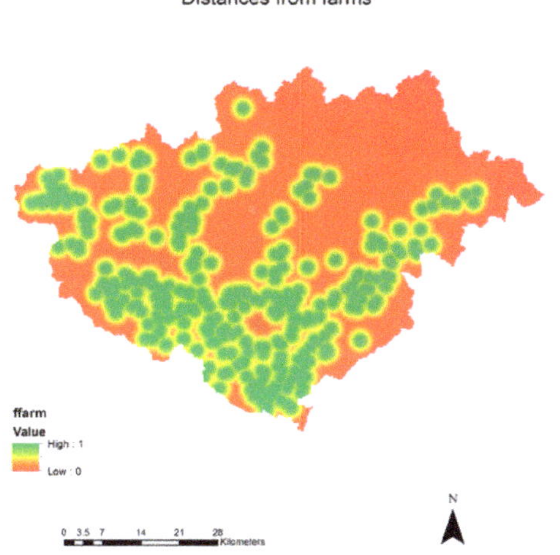

Figure 11. Distances from farms.

4.5. Determining Suitable Land

Finally, the aforementioned criteria were compared using the AHP methodology implemented in PYTHON. The importance of each criterion was set based on the opinions of experts as these were depicted in similar studies. Therefore, the greatest weight was given to the constraint layer, which determines the land suitability at the first level of the research. The second most important weight coefficient was the biomass potential of the area, followed by farm proximity, proximity to the energy grid and finally proximity to the current transportation network (Table 3).

Table 3. Weight coefficients after the implementation of AHP.

Criteria	Weight Coefficient
Constraint Layer	0.458
Transportation network	0.033
Proximity to farms	0.142
Energy grid	0.07
Biomass potential	0.297

For the determination of the suitable installation locations, areas which have a fuzzy value of zero in all layers were excluded at the beginning. Then the layers were combined based on the coefficients presented in Table 3. At the end, the final layer was created based on AHP (Figure 9).

Four suitability levels were created based on the fuzzy membership functions of the location, unsuitable (0, 0.3), poor (0.3, 0.5), good (0.5, 0.7) and excellent (0.7, 1). Figure 9 presents the result for Regional unit of Drama. It should be pointed out again that the areas on the North and West of the regional unit (in purple) are protected areas, which are not available for exploitation according to Greek legislation.

From Figure 12 it is also clear that only a small portion of the entire regional unit area is available for exploitation. However, these areas if managed properly are sufficient for providing sufficient energy.

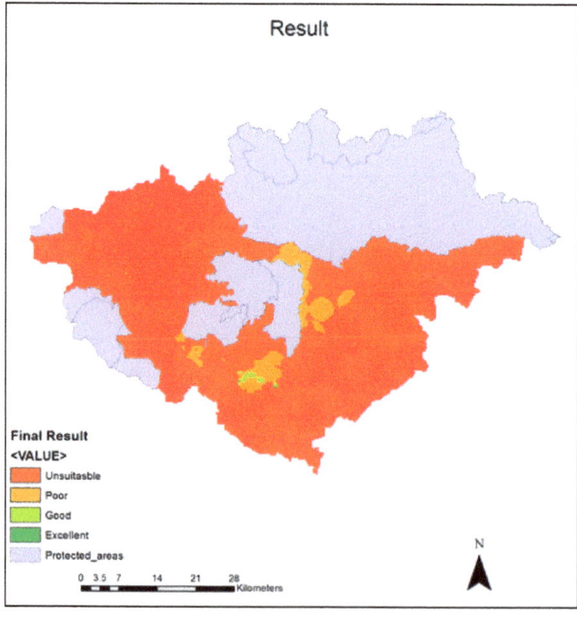

Figure 12. Land suitability for biomass power plant installation.

4.6. Results Validation

For the areas with suitability levels good or excellent (Figure 9) an effort was made to estimate the biomass potential based on the maps provided by CRES. These maps include arable crops, greenhouse residuals, tree crops, vineyards and forest residuals. The results of the estimation showed that although the selected areas represent only the 8% of the total area of the regional unit of Drama the biomass potential is estimated to be 25.5% of the total biomass potential. This corresponds to approximately 586.067 tones of biomass available for energy production.

This result must be enhanced by incorporating in these areas the number farm situated in these regions. The spatial analysis showed that in these areas there are a total of 44 farm units which can further enhance biomass production.

These results also show the efficiency of the proposed methodology by evaluating an entire regional unit and localizing the proposed installation locations to a small fraction of the total regional unit's area.

5. Discussion

Renewable energy sources are expected to play a vital role in energy production in the future. Both the environmental degradation caused by the usage of fossil fuels as well as the legislation framework developed, aim toward the dissemination and increase of their usage. Although there are a variety of RES forms, it is up to the managers to select the most appropriate type, in each case, based on the environmental constraints and production characteristics of the region under investigation, as well as the type that causes the least reactions from citizens [84,85]. The exploitation of national RES potential contributes both to the diversification of the national energy mix and to the security of energy supply, while at the same time strengthening the development of the local and national economy [31].

During the last years, the adoption of common European policies on energy sector, mainly in relation with the necessities underlined by international contracts to limit greenhouse gases, has affected both the European and National energy systems. In particular, an increasing penetration of RES is observed both in power generation as well as the final usage of energy.

Renewable energy sources constitute a key element for enabling global energy policy, with biomass in particular representing an important factor in meeting energy demand towards a more autonomous growth in global scale [86].

Biomass can be used to meet energy needs, either by direct combustion or by conversion to gaseous, liquid and/or solid fuels by thermochemical or biochemical processes. The exploitation of biomass usually encounters the disadvantages of large spreading, large volumes as well as the difficulties of collecting-processing-transporting-storing it is very important that the usage of biomass for energy production must be performed as close as possible to its place of production [37].

Therefore, it is common that the society's attitude toward such types of investments is sometimes negative, mainly due to the fact that the resulting energy production facility causes environmental disturbance to a certain level. Under this scope the Greek state has legislated an extensive framework which clearly defines the rules towards the installation locations of RES facilities in general and Biomass energy production plants in particular (law N2742/1999 regarding the National spatial planning and sustainable development).

The purpose of the paper was achieved by determining the locations where, after research, the optimal conditions exist, for biomass exploitation and therefore, managers can achieve the best possible results both in terms of energy production as well as in terms of minimization of environmental disturbance and economic impact. Additionally, the selected locations are compatible with the legislation regarding RES investments as it is described by N2742/1999.

The aforementioned were performed by providing a framework for a spatial Decision Support System which can be used in order to determine the optimal locations for biomass power plant installation. The research takes under consideration the current biomass production as well as environmental and techno economic constraints related to the final location selection.

The usage of AHP allows the determination of the weight coefficients of each constraint with great accuracy and thus makes the result of the analysis more objective. AHP can be used in two distinctive ways within a GIS environment. It can be either employed to derive the weights associated with suitability map layers or alternatively the AHP principle can be used to aggregate the priority for all level of hierarchy structure including the level representing alternatives. (Malczewski). Fuzzy datasets are used to further enhance the results. In a complex land-use suitability analysis, like the one performed in this paper, it is difficult, (or even impossible) to provide the precise numerical information required by the conventional methods based on the Boolean algebra. In conventional approaches a cut-off (a constraint or threshold) is defined as 'the acceptable site must be located within 1 km of a river,' for example. Such a cut-off however is not a natural one [85]. Fuzzy datasets enhance the results by allowing the determination of areas that are between these two cases. Thus, allowing mapping an area more accurately and combining results from different constraints with higher accuracy. This approach manages to deal with impression and ambiguity in the input data (attribute values and decision maker's preferences [85]. The end result of this approach is the creation of a detailed map which depicts the most suitable locations for the installation of a biomass energy power plant. After the proposal of the most suitable locations, managers should perform visits to the areas for selecting the final installation locations.

6. Conclusions

One of the advantages of the proposed framework is that it can be easily modified in order to study other types of RES by simply modifying the relating criteria. The results of the framework can be further enhanced by incorporating more criteria and alternatives, using high resolution satellite data etc.

The energy produced in the selected areas can be increased by using sludge and other urban waste water. Wastewater reuse in forest plantations has several advantages including rehabilitation of fragile ecological zones, reduced discharge of wastewater into natural water bodies and thus reduced pollution of aquatic ecosystems [87].

The usage of this DSS and other similar methodologies constitutes a valuable tool for energy planning in regional and national level, shaping the framework of investment activities, thus contributing to the implementation of the country's development policy.

It is important to form the appropriate policies which will promote and encourage at the same time the cultivation of energy crops for energy production. A key problem of biomass related projects is funding. Therefore, it is very important to investigate alternative means of funding. Additionally, it is also very important to simplify procedures and to minimize delays regarding the environmental licensing of these types of projects.

The results of this research are very useful, both to the scientific community for further research and review and to the managers which perform energy planning. Furthermore, the results can provide the Greek state with a tool for decision making both in energy planning as well as in environmental protection.

Additionally, it is very important for the Greek state to take under serious consideration the increasing biomass volumes and modify the legislation framework. Finally, it could be possible to modify the existing power transfer network for the exploitation of energy production from biomass and reduce the stress of Public Power Corporation, especially in high demand periods.

Further improvements of the proposed methodology can include the capability of scenarios with the aim of developing a sustainable methodology for alternate locations [88] or other mapping techniques [89], which can be used to further enhance the spatial multicriteria analysis.

Author Contributions: Konstantinos Ioannou and Georgios Tsantopoulos designed the model and the computational framework and analyzed the data. Konstantinos Ioannou carried out the implementation, performed the calculations and the computer programming. Konstantinos Ioannou and Georgios Tsantopoulos, Garyfallos Arabatzis and Zacharoula Andreopoulou, wrote the manuscript with input from all authors. Georgios Tsantopoulos and Zacharoula Andreopoulou, and Eleni Zafeiriou reviewed and discussed the results of the study.

Conflicts of Interest: The authors declare no conflicts of interest.

References

1. Pasten, C.; Santamarina, J.C. Energy and quality of life. *Energy Policy* **2012**, *49*, 468–476. [CrossRef]
2. Lambert, J.G.; Hall, C.A.S.; Balogh, S.; Gupta, A.; Arnold, M. Energy, EROI and quality of life. *Energy Policy* **2014**, *64*, 153–167. [CrossRef]
3. Mazur, A. Does increasing energy or electricity consumption improve quality of life in industrial nations? *Energy Policy* **2011**, *39*, 2568–2572. [CrossRef]
4. Arto, I.; Capellán-Pérez, I.; Lago, R.; Bueno, G.; Bermejo, R. The energy requirements of a developed world. *Energy Sustain. Dev.* **2016**, *33*, 1–13. [CrossRef]
5. Andreoni, V. Energy Metabolism of 28 World Countries: A Multi-scale Integrated Analysis. *Ecol. Econ.* **2017**, *142*, 56–69. [CrossRef]
6. International Energy Outlook, IEO 2017. U.S. Energy Information Administration, September 2017. Available online: https://www.eia.gov/outlooks/ieo/pdf/0484(2017).pdf (accessed on 10 January 2018).
7. Matsumoto, S. How do household characteristics affect appliance usage? Application of conditional demand analysis to Japanese household data. *Energy Policy* **2016**, *94*, 214–223. [CrossRef]
8. Webb, A.; Mayers, K.; France, C.; Koomey, J. Estimating the energy use of high definition games consoles. *Energy Policy* **2013**, *61*, 1412–1421. [CrossRef]
9. Richardson, I.; Thomson, M.; Infield, D. A high-resolution domestic building occupancy model for energy demand simulations. *Energy Build.* **2008**, *40*, 1560–1566. [CrossRef]
10. Viggers, H.; Keall, M.; Wickens, K.; Howden-Chapman, P. Increased house size can cancel out the effect of improved insulation on overall heating energy requirements. *Energy Policy* **2017**, *107*, 248–257. [CrossRef]
11. Stephan, A.; Crawford, R.H. The relationship between house size and life cycle energy demand: Implications for energy efficiency regulations for buildings. *Energy* **2016**, *116*, e221. [CrossRef]
12. Cartalis, C.; Synodinou, A.; Proedrou, M.; Tsangrassoulis, A.; Santamouris, M. Modifications in energy demand in urban areas as a result of climate changes: An assessment for the southeast Mediterranean region. *Energy Convers. Manag.* **2001**, *42*, 1647–1656. [CrossRef]
13. Isaac, M.; van Vuuren, D.P. Modeling global residential sector energy demand for heating and air conditioning in the context of climate change. *Energy Policy* **2009**, *37*, 507–521. [CrossRef]
14. Sailor, D.J.; Pavlova, A.A. Air conditioning market saturation and long-term response of residential cooling energy demand to climate change. *Energy* **2003**, *28*, 941–951. [CrossRef]
15. Alvarez-Herranz, A.; Balsalobre-Lorente, D.; Shahbaz, M.; Cantos, J.M. Energy innovation and renewable energy consumption in the correction of air pollution levels. *Energy Policy* **2017**, *105*, 386–397. [CrossRef]
16. Zecca, A.; Chiari, L. Fossil-fuel constraints on global warming. *Energy Policy* **2010**, *38*, 1–3. [CrossRef]
17. Hansen, J.; Ruedy, R.; Lacis, A.; Sato, M.; Nazarenko, L.; Tausnev, N.; Tegen, I.; Koch, D. Chapter 4 Climate modeling in the global warming debate. *Int. Geophys.* **2001**, *70*, 127–164.
18. Hoffert, M.I.; Covey, C. Deriving Global Climate Sensitivity from Paleoclimate Reconstructions. *Nature* **1992**, *360*, 573–576. [CrossRef]
19. Intergovernmental Panel on Climate Change. *Fourth Assessment Report (AR4), Working Group I, UNEP*; Cambridge University Press: New York, NY, USA, 2007.
20. Seinfeld, J.H.; Pandis, S.N. *Atmospheric Chemistry and Physics: From Air Pollution to Climate Change*; Wiley: New York, NY, USA, 1963.
21. Shafiee, S.; Topal, E. When will fossil fuel reserves be diminished? *Energy Policy* **2009**, *37*, 181–189. [CrossRef]
22. Mohr, S.H.; Wang, J.; Ellem, G.; Ward, J.; Giurco, D. Projection of world fossil fuels by country. *Fuel* **2015**, *141*, 120–135. [CrossRef]
23. Sorrell, S.; Speirs, J.; Bentley, R.; Brandt, A.; Miller, R. Global oil depletion: A review of the evidence. *Energy Policy* **2010**, *38*, 5290–5295. [CrossRef]

24. Hellenic Statistical Authority. 2015. Available online: http://www.statistics.gr/en/statistics/-/publication/ SPK33/ and http://www.statistics.gr/en/statistics/-/publication/SPG06/ (accessed on 5 December 2017).
25. Stram, B.N. Key challenges to expanding renewable energy. *Energy Policy* **2016**, *96*, 728–734. [CrossRef]
26. Malesios, C.; Arabatzis, G. Small hydropower stations in Greece: The local people's attitudes in a mountainous prefecture. *Renew. Sustain. Energy Rev.* **2010**, *14*, 2492–2510. [CrossRef]
27. World Energy Outlook 2012—Executive Summary—English version. IEA, 2012. pp. 6, 24, 28. Available online: https://www.iea.org/publications/freepublications/publication/weo-2012---executive-summary----english-version.html (accessed on 10 January 2018).
28. U.S. Energy Information Administration. Available online: https://www.eia.gov/beta/international/data/ browser/#/?pa=000000000000000000000000000001&c=4100000002000020000000000000g0002&ug=4&tl_id= 2-A&vs=INTL.29-12-AFRC-QBTU.A&vo=0&v=H&start=2004&end=2015 (accessed on 4 December 2017).
29. Eurostat. Share of energy from renewable sources. 2017. Available online: http://appsso.eurostat.ec. europa.eu/nui/show.do?query=BOOKMARK_DS-253950_QID_-7A8FB502_UID_-3F171EB0&layout= TIME,C,X,0;GEO,L,Y,0;UNIT,L,Z,0;INDIC_EN,L,Z,1;INDICATORS,C,Z,2;&zSelection=DS-253950UNIT, PC;DS-253950INDICATORS,OBS_FLAG;DS-253950INDIC_EN,119800;&rankName1=UNIT_1_2_-1_2& rankName2=INDICATORS_1_2_-1_2&rankName3=INDIC-EN_1_2_-1_2&rankName4=TIME_1_0_0_0& rankName5=GEO_1_2_0_1&sortC=ASC_-1_FIRST&rStp=&cStp=&rDCh=&cDCh=&rDM=true&cDM= true&footnes=false&empty=false&wai=false&time_mode=ROLLING&time_most_recent=true&lang= EN&cfo=%23%23%23%2C%23%23%23.%23%23%23 (accessed on 5 December 2017).
30. Koundouri, P.; Kountouris, Y.; Remoundou, K. Valuing a wind farm construction: A contingent valuation study in Greece. *Energy Policy* **2009**, *37*, 1939–1944. [CrossRef]
31. Arabatzis, G.; Kyriakopoulos, G.; Tsialis, P. Typology of regional units based on RES plants: The case of Greece. *Renew. Sustain. Energy Rev.* **2017**, *78*, 1424–1434. [CrossRef]
32. Coelho, S.T. Traditional biomass energy: Improving its use and moving to modern energy use. In *Renewable Energy: A Global Review of Technologies, Policies and Markets*; Earthscan: London, UK; Sterling, VA, USA, 2012; pp. 230–261. ISBN 9781849772341.
33. Demirbas, A.; Ozturk, T.; Demirbas, M.F. Recovery of energy and chemicals from carbonaceous materials. *Energy Sources Part A Recover. Util. Environ. Eff.* **2006**, *28*, 1473–1482. [CrossRef]
34. Demirbaş, A. Biomass resource facilities and biomass conversion processing for fuels and chemicals. *Energy Convers. Manag.* **2001**, *42*, 1357–1378. [CrossRef]
35. Demirbas, A. Conversion of corn stover to chemicals and fuels. *Energy Sources Part A Recover. Util. Environ. Eff.* **2008**, *30*, 788–796. [CrossRef]
36. Ruiz, J.A.; Juárez, M.C.; Morales, M.P.; Muñoz, P.; Mendívil, M.A. Biomass gasification for electricity generation: Review of current technology barriers. *Renew. Sustain. Energy Rev.* **2013**, *18*, 174–183. [CrossRef]
37. Center for Renewable Energy Sources (CRES). 2017. Available online: http://www.cres.gr/energy-saving/ images/pdf/biomass_guide.pdf (accessed on 3 January 2018).
38. Ellabban, O.; Abu-Rub, H.; Blaabjerg, F. Renewable energy resources: Current status, future prospects and their enabling technology. *Renew. Sustain. Energy Rev.* **2014**, *39*, 748–764. [CrossRef]
39. Kougkoulos, I.; Cook, S.J.; Jomelli, V.; Clarke, L.; Symeonakis, E.; Dortch, J.M.; Edwards, L.A.; Merad, M. Use of multi-criteria decision analysis to identify potentially dangerous glacial lakes. *Sci. Total Environ.* **2017**. [CrossRef] [PubMed]
40. Ioannou, K.; Lefakis, P.; Arabatzis, G. Development of a decision support system for the study of an area after the occurrence of forest fire International. *J. Sustain. Soc.* **2011**, *3*, 5–32.
41. Grigoroudis, E.; Petridis, K.; Arabatzis, G. RDEA: A recursive DEA based algorithm for the optimal design of biomass supply chain networks. *Renew. Energy* **2014**, *71*, 113–122. [CrossRef]
42. Arabatzis, G.; Petridis, K.; Galatsidas, S.; Ioannou, K. A demand scenario based fuelwood supply chain: A conceptual model. *Renew. Sustain. Energy Rev.* **2013**, *25*, 687–697. [CrossRef]
43. Kyriakopoulos, G.L.; Arabatzis, G.; Chalikias, M. Renewables exploitation for energy production and biomass use for electricity generation. A multi-parametric literature-based review. *AIMS Energy* **2016**, *4*, 762–803. [CrossRef]
44. Saaty, T.L. *Fundamentals of Decision Making and Priority Theory*; RWS Publications: Pittsburgh, PA, USA, 2001; ISBN 0-9620317-6-3.

45. Chaouachi, A.; Covrig, C.F.; Ardelean, M. Multi-criteria selection of offshore wind farms: Case study for the Baltic States. *Energy Policy* **2017**, *103*, 179–192. [CrossRef]
46. Aly, A.; Jensen, S.S.; Pedersen, A.B. Solar power potential of Tanzania: Identifying CSP and PV hot spots through a GIS multicriteria decision making analysis. *Renew. Energy* **2017**, *113*, 159–175. [CrossRef]
47. Kausika, B.B.; Dolla, O.; van Sark, W.G.J.H.M. Assessment of policy based residential solar PV potential using GIS-based multicriteria decision analysis: A case study of Apeldoorn, the Netherlands. *Energy Procedia* **2017**, *134*, 110–120. [CrossRef]
48. Al Garni, H.Z.; Awasthi, A. Solar PV power plant site selection using a GIS-AHP based approach with application in Saudi Arabia. *Appl. Energy* **2017**, *206*, 1225–1240. [CrossRef]
49. Tsoutsos, T.; Tsitoura, I.; Kokologos, D.; Kalaitzakis, K. Sustainable siting process in large wind farms case study in Crete. *Renew. Energy* **2015**, *75*, 474–480. [CrossRef]
50. Latinopoulos, D.; Kechagia, K. A GIS-based multi-criteria evaluation for wind farm site selection. A regional scale application in Greece. *Renew. Energy* **2015**, *78*, 550–560. [CrossRef]
51. Singh, R.P.; Nachtnebel, H.P. Analytical hierarchy process (AHP) application for reinforcement of hydropower strategy in Nepal. *Renew. Sustain. Energy Rev.* **2016**, *55*, 43–58. [CrossRef]
52. Yalcin, M.; Kilic Gul, F. A GIS-based multi criteria decision analysis approach for exploring geothermal resources: Akarcay basin (Afyonkarahisar). *Geothermics* **2017**, *67*, 18–28. [CrossRef]
53. Kiavarz, M.; Jelokhani-Niaraki, M. Geothermal prospectivity mapping using GIS-based Ordered Weighted Averaging approach: A case study in Japan's Akita and Iwate provinces. *Geothermics* **2017**, *70*, 295–304. [CrossRef]
54. Wang, Y.M.; Chin, K.S. Fuzzy analytic hierarchy process: A logarithmic fuzzy preference programming methodology. *Int. J. Approx. Reason.* **2011**, *52*, 541–553. [CrossRef]
55. Zyoud, S.H.; Kaufmann, L.G.; Shaheen, H.; Samhan, S.; Fuchs-Hanusch, D. A framework for water loss management in developing countries under fuzzy environment: Integration of Fuzzy AHP with Fuzzy TOPSIS. *Expert Syst. Appl.* **2016**, *61*, 86–105. [CrossRef]
56. Asakereh, A.; Soleymani, M.; Sheikhdavoodi, M.J. A GIS-based Fuzzy-AHP method for the evaluation of solar farms locations: Case study in Khuzestan province, Iran. *Sol. Energy* **2017**, *155*, 342–353. [CrossRef]
57. Kaya, T.; Kahraman, C. Multicriteria renewable energy planning using an integrated fuzzy VIKOR & AHP methodology: The case of Istanbul. *Energy* **2010**, *35*, 2517–2527. [CrossRef]
58. Sadeghi, A.; Larimian, T.; Molabashi, A. Evaluation of Renewable Energy Sources for Generating Electricity in Province of Yazd: A Fuzzy Mcdm Approach. *Procedia Soc. Behav. Sci.* **2012**, *62*, 1095–1099. [CrossRef]
59. Şengül, Ü.; Eren, M.; Eslamian Shiraz, S.; Gezder, V.; Sengül, A.B. Fuzzy TOPSIS method for ranking renewable energy supply systems in Turkey. *Renew. Energy* **2015**, *75*, 617–625. [CrossRef]
60. Haddad, B.; Liazid, A.; Ferreira, P. A multi-criteria approach to rank renewables for the Algerian electricity system. *Renew. Energy* **2017**, *107*, 462–472. [CrossRef]
61. Sindhu, S.; Nehra, V.; Luthra, S. Investigation of feasibility study of solar farms deployment using hybrid AHP-TOPSIS analysis: Case study of India. *Renew. Sustain. Energy Rev.* **2017**, *73*, 496–511. [CrossRef]
62. Aragonés-Beltrán, P.; Chaparro-González, F.; Pastor-Ferrando, J.P.; Pla-Rubio, A. An AHP (Analytic Hierarchy Process)/ANP (Analytic Network Process)-based multi-criteria decision approach for the selection of solar-thermal power plant investment projects. *Energy* **2014**, *66*, 222–238. [CrossRef]
63. Tavana, M.; Santos Arteaga, F.J.; Mohammadi, S.; Alimohammadi, M. A fuzzy multi-criteria spatial decision support system for solar farm location planning. *Energy Strateg. Rev.* **2017**, *18*, 93–105. [CrossRef]
64. Tasri, A.; Susilawati, A. Selection among renewable energy alternatives based on a fuzzy analytic hierarchy process in Indonesia. *Sustain. Energy Technol. Assess.* **2014**, *7*, 34–44. [CrossRef]
65. Heo, E.; Kim, J.; Boo, K.-J. Analysis of the assessment factors for renewable energy dissemination program evaluation using fuzzy AHP. *Renew. Sustain. Energy Rev.* **2010**, *14*, 2214–2220. [CrossRef]
66. Toklu, E. Biomass energy potential and utilization in Turkey. *Renew. Energy* **2017**, *107*, 235–244. [CrossRef]
67. Dominguez Bravo, J.; Garcia Casals, X.; Pinedo Pascua, I. GIS approach to the definition of capacity and generation ceilings of renewable energy technologies. *Energy Policy* **2007**, *35*, 4879–4892. [CrossRef]
68. Arnette, A.N.; Zobel, C.W. Spatial analysis of renewable energy potential in the greater southern Appalachian mountains. *Renew. Energy* **2011**, *36*, 2785–2798. [CrossRef]
69. Baban, S.M.J.; Parry, T. Developing and applying a GIS-assisted approach to locating wind farms in the UK. *Renew. Energy* **2001**, *24*, 59–71. [CrossRef]

70. Alabi, O. An investigation on using GIS to prospect for renewable energy in Nigeria. Ph.D. Thesis, University of Missouri-Kansas City, Kansas City, MO, USA, 2010.

71. Aydin, N.Y.; Kentel, E.; SebnemDuzgun, H. GIS-based site selection methodology for hybrid renewable energy systems: A case study from western Turkey. *Energy Convers. Manag.* **2013**, *70*, 90–106. [CrossRef]

72. Al-Yahyai, S.; Charabi, Y.; Gastli, A.; Al-Badi, A. Wind farm land suitability indexing using multi-criteria analysis. *Renew. Energy* **2012**, *44*, 80–87. [CrossRef]

73. Dunsford, H.; Macfarlane, R.; Turner, K. *The Development of a Regional Geographical Information System for the North East Renewable Energy Strategy*; Centre for Environmental and Spatial Analysis, University of Northumbria: England, UK, 2003.

74. Sánchez-Lozano, J.M.; Teruel-Solano, J.; Soto-Elvira, P.L.; Socorro García-Cascales, M. Geographical Information Systems (GIS) and Multi-Criteria Decision Making (MCDM) methods for the evaluation of solar farms locations: Case study in south-eastern Spain. *Renew. Sustain. Energy Rev.* **2013**, *24*, 544–556. [CrossRef]

75. Tsoutsos, T.; Frantzeskaki, N.; Gekas, V. Environmental impacts from the solar energy technologies. *Energy Policy* **2005**, *33*, 289–296. [CrossRef]

76. Brewer, J.; Ames, D.P.; Solan, D.; Lee, R.; Carlisle, J. Using GIS analytics and social preference data to evaluate utility-scale solar power site suitability. *Renew. Energy* **2015**, *81*, 825–836. [CrossRef]

77. Charabi, Y.; Gastli, A. PV site suitability analysis using GIS-based spatial fuzzy multi-criteria evaluation. *Renew. Energy* **2011**, *36*, 2554–2561. [CrossRef]

78. Watson, J.J.W.; Hudson, M.D. Regional Scale wind farm and solar farm suitability assessment using GIS-assisted multi-criteria evaluation. *Landsc. Urban Plan.* **2015**, *138*, 20–31. [CrossRef]

79. Ciolkosz, D.; Jacobson, M.; Heil, N.; Brandau, W. An assessment of farm scale biomass pelleting in the Northeast. *Renew. Energy* **2017**, *108*, 85–91. [CrossRef]

80. Purdy, A.; Pathare, P.B.; Wang, Y.; Roskilly, A.P.; Huang, Y. Towards sustainable farming: Feasibility study into energy recovery from bio-waste on a small-scale dairy farm. *J. Clean. Prod.* **2018**, *174*, 899–904. [CrossRef]

81. Ferreira, S.; Monteiro, E.; Brito, P.; Vilarinho, C. Biomass resources in Portugal: Current status and prospects. *Renew. Sustain. Energy Rev.* **2017**, *78*, 1221–1235. [CrossRef]

82. Villarini, M.; Bocci, E.; Di Carlo, A.; Savuto, E.; Pallozzi, V. The case study of an innovative small scale biomass waste gasification heat and power plant contextualized in a farm. *Energy Procedia* **2015**, *82*, 335–342. [CrossRef]

83. Arabatzis, G.; Myronidis, D. Contribution of SHP Stations to the development of an area and their social acceptance. *Renew. Sustain. Energy Rev.* **2011**, *15*, 3909–3917. [CrossRef]

84. Tampakis, S.; Tsantopoulos, G.; Arabatzis, G.; Rerras, I. Citizens' views on various forms of energy and their contribution to the environment. *Renew. Sustain. Energy Rev.* **2013**, *20*, 473–482. [CrossRef]

85. Skoulou, V. Design and development of a gasification reactor for energy production from biomass. Ph.D. Thesis, Department of Chemical Engineering, Aristotle University of Thessaloniki, Serres, Greece, 2009.

86. Malcezewski, J. GIS based lan-use suitability analysis: A critical overview. *Prog. Plan.* **2004**, *62*, 3–65. [CrossRef]

87. Kalavrouziotis, I.; Hortis, T.; Drakatos, P. The reuse of wastewater and sludge for cultivation of forestry trees in desert areas in Greece. *Int. J. Environ. Pollut.* **2004**, *21*, 425. [CrossRef]

88. Torrieri, F.; Bata, A. Spatial Multi-criteria Decision Support System and Strategic Impact Assessment: A case study. *Buildings* **2017**, *7*, 96. [CrossRef]

89. Chakhar, S.; Mousseau, V. Multicriteria Decision Making, Spatial. In *Encyclopedia of GIS*; Shekhar, S., Xiong, H., Eds.; Springer: Boston, MA, USA, 2008.

MDPI

St. Alban-Anlage 66

4052 Basel

Switzerland

Tel. +41 61 683 77 34

Fax +41 61 302 89 18

www.mdpi.com

Sustainability Editorial Office

E-mail: sustainability@mdpi.com

www.mdpi.com/journal/sustainability